Advances in Soil Science

Advances in Soil Science

B.A. Stewart, Editor

Advances in Soil Science

Volume 15

Edited by B.A. Stewart

With Contributions by
J.M. Barea, J.A.A. Berkhout, A. Bronger,
R. Lal, J.F. Power, R. Prasad, C.A. van Diepen,
H. van Keulen, J. Wolf

With 43 Illustrations

Springer-Verlag
New York Berlin Heidelberg London
Paris Tokyo Hong Kong Barcelona

B.A. Stewart
USDA Conservation & Production Research Laboratory
Bushland, TX 79012, U.S.A.

ISSN: 0176-9340

Printed on acid-free paper.

Typeset by David Seham Associates, Metuchen, New Jersey.

9 8 7 6 5 4 3 2 1

ISBN-13: 978-1-4612-7768-2 e-ISBN-13: 978-1-4612-3030-4
DOI: 10.107/978-1-4612-3030-4

Preface

The study of soils has taken on increased importance because a rapidly expanding population is placing demands on the soil never before experienced. This has led to an increase in land degradation. Land degradation is one of the most severe problems facing mankind. Volume 11 of *Advances in Soil Science* was devoted entirely to this critical area of soil science. The editors of that volume, R. Lal and B.A. Stewart, defined soil degradation as the decline in soil quality caused by its misuse by humans. They further stated that soil degradation is a major concern for at least two reasons. First, it undermines the productive capacity of an ecosystem. Second, it affects global climate through alterations in water and energy balances and disruptions in cycles of carbon, nitrogen, sulfur, and other elements. Through its impact on agricultural productivity and environment, soil degradation leads to political and social instability, enhanced rate of deforestation, intensive use of marginal and fragile lands, accelerated runoff and soil erosion, pollution of natural waters, and emission of greenhouse gases into the atmosphere. In fact, soil degradation affects the very fabric of mankind.

From the beginning of agriculture until about 1950, increased food production came almost entirely from expanding the cropland base. Since 1950, however, the yield per unit of land area for major crops has increased dramatically. Much of the increase in yields was because of increased inputs of energy. Between 1950 and 1985, the farm tractor fleet quadrupled, world irrigated area tripled, and fertilizer use increased ninefold. Between 1950 and 1985, total energy used in world agriculture increased 6.9 times.

Until recently, sustainability was seldom, if ever, mentioned in agricultural literature. Now, it is one of the most widely used terms. The high costs of irrigation development, escalating energy costs during the 1970s, public concern over potential negative impacts of fertilizer and pesticides on water supplies, soil erosion, soil compaction, and salinity problems, and other concerns have caused many people to question whether many of the present agriculture systems can be sustained. As a result, soil science is beginning to focus more on sustaining the resource base. The productivity level of an agricultural soil at any time is the resultant of the interaction of degradative processes and conservation/reclamation practices that are occurring simultaneously. A sustainable system is any system where the benefits from the soil conservation practices are equal to or greater than the negative effects of the soil degradative processes.

This series, *Advances in Soil Science*, was established to provide a forum for leading scientists to analyze and summarize the available scientific information on a subject, assessing its importance and identifying additional research needs. This goal seems even more appropriate today than in 1982, when the idea of the series was formulated. Much is known about our soil resources. The principles learned and the technology developed need to be used to increase food production and sustain the productivity of the resource base. *Advances in Soil Science* fills a gap between the scientific journals and the comprehensive reference books. Scientists can delve in depth on a particular subject relating to soil science. Contributors are asked in particular to develop and identify principles that have practical applications to both developing and developed agricultures.

Advances in Soil Science is international in scope and covers all subjects relating to soil science. This volume continues that format in that it contains reviews of diverse interests written by scientists from Spain, The Netherlands, India, Federal Republic of Germany, and the United States. Although we consider our audience to be primarily scientists and students of soil science, the series provides technical information to anyone interested in our natural resources and man's influence on these resources. Research in the future must focus on systems that are resource efficient and environmentally sound. The need to optimize crop production while conserving the resource base has never been greater.

The quick acceptance of *Advances in Soil Science* by both authors and readers has been very gratifying and confirms our perception that a need did exist for a medium to publish soil science reviews. I want to thank the authors for their excellent contributions and cooperation. I also want to thank members of the Editorial Board for their help in selecting such competent authors and the Springer-Verlag staff for their kind assistance and counsel. Lastly, and most importantly, I want to thank the readers for their acceptance and use of *Advances in Soil Science*.

B.A. Stewart

Contents

Contributors

J.M. Barea, Estacion Experimental del Zaidin, Departamento de Microbiologia, 18008 Granada 1, Spain.

J.A.A. Berkhout, Landscape and Environmental Research Group, University of Amsterdam, Amsterdam, The Netherlands.

A. Bronger, Department of Geography, University of Kiel, D-2300 Kiel, Federal Republic of Germany.

R. Lal, Department of Agronomy, The Ohio State University, Columbus, Ohio 43210, U.S.A.

J.F. Power, U.S. Department of Agriculture, Agricultural Research Service, Department of Agronomy, University of Nebraska, Lincoln, Nebraska 68583, U.S.A.

R. Prasad, Indian Agricultural Research Institute, New Delhi 110012, India.

C.A. van Diepen, The Winand Staring Centre for Integrated Land, Soil and Water Research, 6700 AC Wageningen, The Netherlands.

H. van Keulen, International Institute for Aerospace Survey and Earth Sciences, Enschede, and Centre for Agrobiological Research, Wageningen, The Netherlands.

J. Wolf, Agricultural University, 6700 AA Wageningen, The Netherlands.

Vesicular–Arbuscular Mycorrhizae as Modifiers of Soil Fertility

J.M. Barea

Contents

Advances in Soil Science, Volume 15

I. Introduction

It has become clear that microbial activity must be considered a key component among those conferring "soil fertility," i.e., the ability of a given soil to support plant development and nutrition (Pauli, 1967). The major components interacting to determine "soil fertility" are depicted in Figure 1. Accordingly, "fertility" can be considered an inherent property of a given soil. However, the plant itself is able to modify soil fertility in two different ways. One is based on the "rhizosphere effect" exerted by the plant, which can alter the fluxes of energy and the supply of substrates for soil microorganisms. The other way is based on the inherently different growth rates and metabolism of the different plant species that are known to "change" the capacity of the soil to provide each particular plant with nutrients (Hayman, 1975). In other words, the ability of a plant to exploit the nutritional supply of a given soil depends on the type of plant, mainly on the characteristics of its root system, and on the rates

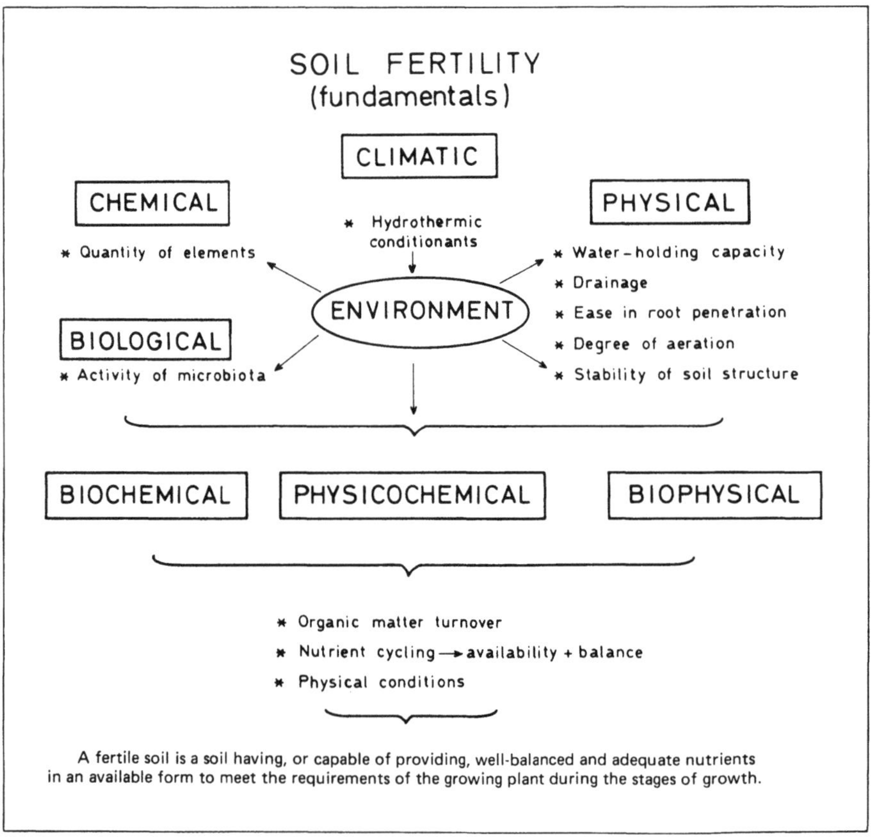

Figure 1. Summary of factors that affect soil fertility.

and patterns of exudation (Chapin, 1980; Clarkson, 1985). Thus, plants are able to "modify" the soil fertility.

It is widely known that a key determinant of the ability of a root system to acquire nutrients from the soil is the extent to which it is symbiotically colonized by appropriate mycorrhizal fungi. The mycorrhizal condition is actually the normal status of most terrestrial plants, and it greatly enhances the possibilities of the plant for nutrient uptake from soil (Mosse, 1973; Harley and Smith, 1983). The formation and activity of this symbiosis are in turn affected by the soil fertility level (Hayman, 1982; Gerschefske et al., 1988).

At this point it would be opportune to remember that soil microorganisms play vital roles in the root region, the rhizosphere, where they are invariably present and are stimulated by organic substrates supplied by the plant. The main beneficial activities of rhizosphere bacteria (actinomycetes are considered bacteria) and fungi include the increased availability of plant nutrients, the improvement of nutrient uptake, the production of plant growth regulators, plant protection against root pathogens, and so on. According to their relationships with the plant, the microorganisms can be divided into three groups: saprophites, parasitic symbionts ("pathogen"), and mutualistic symbionts ("symbionts").

Particularly, the mutualistic symbiosis between photosynthetic plants and specific microorganimsms plays a key role in both natural ecosystems and man-modified systems, mainly because the symbiotic microorganisms carry out functions for the plants that they are unable to perform for themselves (Barea and Azcón-Aguilar, 1983). The best-known examples of mutualistic symbiosis involving plant and microorganisms are (1) that established between bacteria and the roots of certain plant species to form N_2-fixing nodules, the *Rhizobium*–legume association being the model case, and (2) the mycorrhizae, the study of which is the central subject of this review.

The aim of the present chapter is, in fact, to carry out a comprehensive study of how mycorrhizae change the supply of mineral nutrients that a given soil can supply, thereby modifying soil fertility. As an introduction, a brief review of some general and universally accepted principles, the mycorrhizal types, their morphology, and the processes of mycorrhizae formation is presented to achieve a conceptual background for nonspecialist readers. To facilitate a better understanding of the mycorrhizal effects, some ideas are then sumarized concerning the processes of nutrient absorption by roots and their role at modifying the nutrient availability in a given soil. Current information on mycorrhizae functioning and its ecological conditionants is analyzed and summarized to constitute the core part of the review. A number of review articles are cited in this part of the review not only because of the conclusions they supply but also because they are a source of references, which have been reduced to a minimum in this part of the present chapter. Finally, the possibilities of application of mycorrhizae in agriculture, horticulture, fruit culture, and forestry are discussed. The research requirement to demonstrate rationally stated hypotheses is suggested, and current trends on the subject are also outlined.

This study is therefore devoted to describing processes taking place in the root–soil interfaces; hence, according to Clarkson (1985), it can be considered an attempt to gain information in a critical research area: that linking plant physiology and soil science.

II. Mycorrhizae

A. General Concepts

1. Universality of the Symbiosis

It is widely known today that the roots of most flowering plants live associated, in a form of mutualistic symbiosis, with certain soil fungi by establishing the so-called mycorrhizae. The fungus biotrophically colonizes the root cortex, becoming an integral part of these organs, where an extramatricial mycelium develops that helps the plant acquire mineral nutrients from the soil. The symbiosis is considered the most metabolically active part of the absorbing organs of the autotrophic host plant, which in turn furnish the heterotrophic fungal associate with organic nutrients and with an ecologically protected habitat. Mycorrhizae are recognized as playing a key role in plant survival and nutrient cycling in the ecosystem. They can be found in nearly all soils of the world. All but a few vascular plant species (these belonging mainly to Cruciferae, Chenopodiaceae, Cyperaceae, and Juncaceae) are able to form mycorrhizae. The physiology of the plant is greatly affected by the presence of the fungal symbionts (Harley and Smith, 1983; Smith and Gianinazzi-Pearson, 1988).

2. Mycorrhizal Types

It is obvious that the universality of the symbiosis implies a great diversity in the taxonomic features of the fungi and plants involved. There are, in fact, great differences in the morphology of mycorrhizal group, and this is reflected in the resulting physiological relationships.

Five types of mycorrhizae can be recognized; their structural and nutritional features have been detailed before (Smith, 1980; Harley and Smith, 1983). Only a brief consideration to differentiate these groups is given here.

About 3% of the higher plants, mainly forest trees in the *Fagaceae, Betulaceae, Pinaceae, Rosaceae, Eucalyptus,* and some woody legumes, form ectomycorrhizae. The fungi involved are mostly higher basidiomycetes and ascomycetes, which colonize the cortical cells of the root, the lack of intracellular penetration being characteristic. In general, the fungus develops a sheath or mantle around the feeder roots.

In addition, three types of mycorrhizae can be grouped as endomycorrhizae, where the fungus can colonize the root cortex intracellularly. One of these is restricted to some species in the *Ericaceae* ("ericoid" mycorrhizae), the second to the *Orchidaceae* ("orchid" mycorrhizae), and the third, the vesicular–arbuscular (VAM), which is by far the most widespread type. There is a fifth group, the ectendomycorrhizae, formed by plant species in families other than *Ericaceae* but in the Ericales. They form a sheath and produce intracellular penetrations ("arbutoid" mycorrhizae). The ecological and economic interest of VAM can be simply deduced from the fact that about four-fifths of all land plants, including agronomically important crops, form this type of mycorrhizae. The nomenclature refers to typical structures, the "vesicles" and the "arbuscules," of the fungus in symbiosis.

B. Vesicular–Arbuscular Mycorrhizae

1. Occurrence and Distribution

Both the fungus and the VAM are distributed worldwide. The fungi belong to the class Zigomicotina, order Endogonales, family Endogonaceae. Morton (1988) systematizes about 150 species in the only six genera able to form VAM *(Acaulospora, Entrophospora, Gigaspora, Glomus, Sclerocystis,* and *Scutellospora);* none of these fungi has yet been successfully cultured axenically (Siqueira, 1987).

Typical VAM-forming plants are legumes, cereals, temperate fruit trees, tropical timber trees, plantation crops; tropical, mediterranean, and "industrial" crops; and horticultural and ornamental crops (Barea and Azcón-Aguilar, 1983).

2. Characteristics of the Host–Fungus Relationships

Some properties of phenomena inherent to any symbiotic system are particularly relevant in VAM because the nutrient uptake by the plant can be strongly affected if associated with its suitable microsymbiont. Dependency, recognition phenomena, compatibility, and specificity are concepts that merit brief reference.

It seems (Pirozynski and Malloch, 1975; Malloch et al., 1980) that certain fungi played a critical role in the evolution of "plants" to colonize the land (Silurian and Devonian periods, 400 million years ago), because these fungi associated with such "plants" and helped them in the nutrient uptake processes. This may be plausible, since the Rhynie fossil (the earliest preserved plants—370 million years old) showed a fungal formation quite similar to modern VAM, suggesting a plant–VAM coevolution (Nicolson, 1975). These facts can explain the world wide spread of the VAM and several of their so-called "symbiotic properties," which are determined by (1) the ability of a plant to acquire nutrients through a fungus ("mycotrophy"), (2) the difficulties the fungus has in completing its life cycle independently of the host, being a physiologicaly obligate symbiont ("fungal dependency"), and (3) the characteristics of the plant as expressed by its need to be mycorrhizal to develop ("mycorrhizal dependency of a plant"). Such dependency on VAM varies with the plant species, since some of these need VAM to survive, others to improve their growth, and others to reach their maximum yield (Hayman, 1983).

There is a lack of "specificity" *(sensu strictu)* in VAM. Roughly, any VAM fungi (VAMF) can colonize any suitable plant species, a single root system can support different VAMF species, and roots of different plant species can be linked by the mycelium of a single VAMF (Mosse et al., 1981; Harley and Smith, 1983; Gianinazzi-Pearson, 1984). Nevertheless, the different plant species, and even cultivars within the same species (Azcón and Ocampo, 1981), vary greatly in the level of "susceptibility" to VAMF. This indicates that the plant genotype controls the amount of root tissue that is colonized to become a VAM (Gianinazzi-Pearson, 1984; Buwalda et al., 1984). Since the different VAMF also differ in the level of colonization that they can reach in the root system of a given host plant, it follows that a certain type of "specificity" can be recognized in VAM. This raises the concept of "compatibility" in VAM, which must be associated with that of "symbiotic effectiveness" to establish that of "functional compatibility" (Gianinazzi-Pearson, 1984). The last refers to the phenotypic expression of a VAM as a result of the environmental influences on the expression of the genotypic equipment of both the plant and the fugus involved (see Smith and Gianinazzi-Pearson, 1988). According to Gianinazzi-Pearson (1984), there is evidence of fungus–plant "recognition," as evidenced at several stages. Such evidence includes (1) cell-to-cell contact to form appresoria; (2) certain morphological and structural changes of the fungus, mainly cell wall composition, within the root tissues; (3) the integration of the physiology of both symbionts, and (4) the redistribution of enzymatic activities, especially those involved in nutrient exchange, between the components of the symbiosis.

III. Development of Vesicular–Arbuscular Mycorrhizae

A. The Processess of Vesicular–Arbuscular Mycorrhizae Formation

The VAM colonization originates from hyphae arising from soil-borne propagules (large resting spores of the VAMF or mycorrhizal root fragments) or from a VAM plant growing nearby. It seems that the network of VAM fungi mycelia in soil is an important source of inoculum. This can be reinforced by the fact that the hyphae can retain their infectivity after separation from the roots, even in dry soil (Jasper et al., 1989a,b). Consequently, soil disturbance disrupts the mycelium network and reduces infectivity.

On arrival of the fungal hyphae at the root surface, an appresorium is usually formed on the cortical cells. The infection units ("internal mycelium associated with a single entry point"; Wilson, 1984) grow as the hyphae spread between and through cells of the cortical root layers. On reaching the inner cortex, the hyphae can grow into cells and, by repeated dichotomic branching, form some tree-like structures called "arbuscules." The life span of individual arbuscules is about 4–14 days. When the internal colonization is spreading, the extramatricial hyphae ramify. These may grow along the root surface, forming more penetration points, and also out into the surrounding soil, to form a extensive tridimensional network of mycelium.

When the colonization process is well established, the fungus may form "vesicles," oval-to-spherical structures with a storage (mainly lipids) function. Most VAMF form large resting spores on the external mycelium (see Harley and Smith, 1983; Bonfante-Fasolo, 1984; for a detailed description of the anatomy of VAM and the formation of the symbiosis).

Intracellular colonizations, as in the case of arbuscules, have a characteristic feature, which is that the fungus is always surrounded by the intact host-cell plasmalemma. Therefore, arbuscule formation represents a large surface of cellular contact between the two symbionts. This facilitates the interchange of metabolites between host and fungus. In fact, the arbuscule is considered the main site of transfer of mineral nutrients from the fungus (which had taken these up from soil) to the plant (see Smith and Gianinazzi-Pearson, 1988, for the development of host–fungus interfaces).

B. Quantitative Estimates

The spread of VAMF is usually measured as the proportion of the root length that is colonized by VAM hyphae. For comparison purposes, the estimates of the fungal development within and around roots must consider the rate of root growth and the size of the root system to get a realistic idea of VAM size. Time-course quantifications of the fraction of the root length that has been converted into VAM usually follow a characteristic sigmoidal curve (Tinker, 1975). The length of the lag phase depends on the density of viable propagules in the medium, the rate of germination, stimulation, and germ-tube growth (Sanders and Sheikh, 1983). The subsequent exponential phase of VAM development follows the spread of secondary infections, when the hyphae grow along and between roots. Then the VAM extent reachs a plateau, and the resulting percentage is usually less than 100%. The plant–fungus combination and the onset of environmental factors and/or conditions affect the extent and/or rate of the processes (Mosse et al., 1981; Sanders and Sheikh, 1983). The different phases of the VAM development can be modeled (see Smith and Walker, 1981; Buwalda et al., 1982; Sanders and Sheikh, 1983; Walker and Smith, 1984; Tinker, 1985; Sanders, 1986; as examples of modeling approaches).

From the point of view of the role of VAM as modifiers of soil fertility, the main quantitative estimate to be considered is that concerning the extent of the external hyphae growing in soil, associated with mycorrhizal roots (Abbott and Robson, 1985). Data recorded from different publications by Smith and Gianinazzi-Pearson (1988) indicate that most calculations reach values of 1 m cm^{-1} root on average, but values of 10–14 m cm^{-1} root have also been recorded.

IV. Root–Soil Interactions

A. Nutrient Uptake Processes

Plant growth is largely regulated by the supply of nutrients available to the root system and by the efficiency of the active absorption of the root cells on nutrient arriving at the surface of these cells. The magnitude of the supply of a given nutrient to the root surface depends on (1) the concentration of the corresponding ions in the soil solution, (2) the sorption–desorption capacity of the soil, which allows ions in the exchangeable pools to replenish the soil solution as these ions (nutrients) are being take up by the plant, and (3) the efficiency of nutrient transport through the soil solution to the absorbing sites, either by mass flow or by diffusion (Tinker, 1980; Chapin, 1980). This is summarized in Figures 2 and 3.

In regard to the rate of nutrient movement to the root, it is important to realize that the bulk of the available forms of the major plant nutrients

NUTRIENT ABSORPTION BY PLANT ROOTS

The rate depends upon:

1. Nutrient supply to the root surface
2. Active absorption by root cortical cells

1. The nutrient supply depends, in turn, on:
   a. Soil solution concentration
   b. Buffering power of the soil
   c. Rate of nutrient movement to the root
      by: (i) Diffusion
      (ii) Mass flow of soil water

Figure 2. Summary of factors that affect nutrient absorption by plant roots.

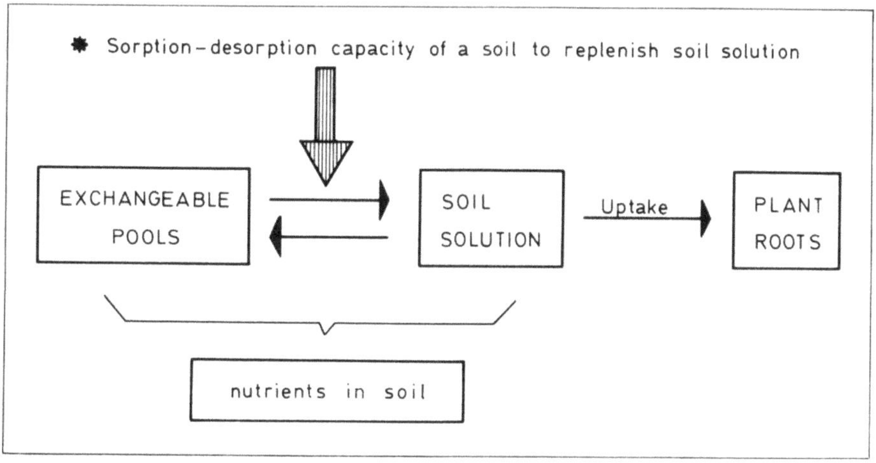

Figure 3. Buffering power of the soil.

(N, P, K) that are in the soil solution at low concentration (Nye and Tinker, 1977) move to the root surface by diffusion (Chapin, 1980). Therefore, the rate-limiting step in the absorption of low-mobility phosphate, ammonium, and posassium is actually the diffusion of these ions through the soil solution. Because the rate of diffusion of these ions is much lower than that of absorption once they arrive at the root surfaces, regions depleted of the nutrient frequently develop around roots (Chapin, 1980; Tinker, 1980; Hayman, 1983; Clarkson, 1985). This is illustrated in Figure 4.

Root properties greatly affect nutrient acquisition (Clarkson, 1985). First, plant metabolism exerts its influence through processes such as secretion of H^+ or HCO_3^- loss of substrates by root exudation, lysates, and sloughing of cells or tissue debris, and secretion of chelating substances, O_2, etc. Even plant shoots influence root activity by affecting the rate of photosynthesis by depleting mineral nutrients, by controlling (through feedback mechanisms) the source-to-sink equilibrium in the shoot-to-root relationship, and by synthesizing carriers and related substances (Chapin, 1980).

The characteristics of the root system as a whole, and of the root surfaces in particular, greatly affect the nutrient intake of the plant (mean uptake per unit root length). As reviewed by Clarkson (1985), two main properties of a root system modify nutrient inflow: (1) its size and distribution (i.e., morphological and geometric features) and (2) the capacity of the root surface for nutrient uptake (i.e., kinetic properties). It is known that the use of the nutrients present in the soil solution by the root follows a Michaelis–Menten pattern, expressed by the parameters K_m (indicating affinity) and I_{max} (related to V_{max}, which indicates capacity); this is summarized in Figure 5.

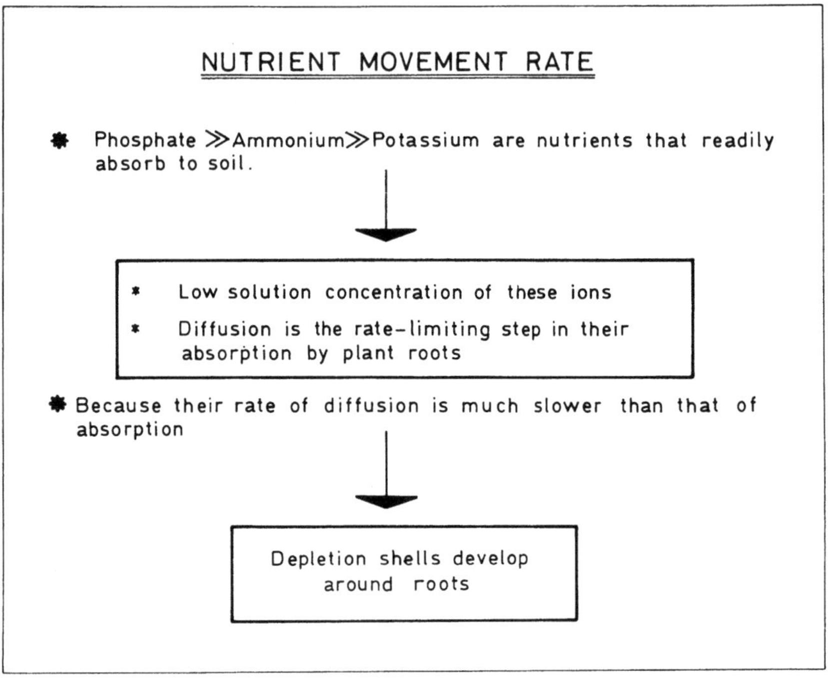

Figure 4. Motility of nutrients in soil.

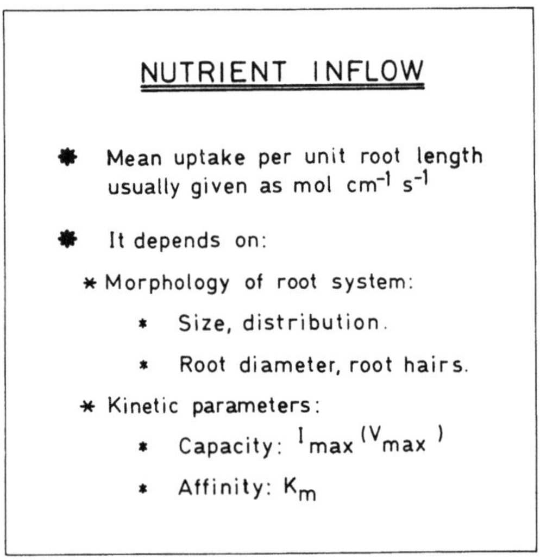

Figure 5. Summary of factors that affect nutrient inflow.

B. Vesicular–Arbuscular Mycorrhizae as Modified Root Systems

The development of an extensive network of extramatricial hyphae by the VAM in the soil surrounding the root, together with the capacity of these hyphae for nutrient absorption and transport to the cortical root cells, indicates that VAM modify the nutrient uptake properties of a root system. Actually, it is widely accepted that VAM play a recognized role in nutrient cycling in the ecosystem (Harley and Smith, 1983). Because the external mycelium extends several centimeters from the root surface, it can bypass the depletion zone surrounding the root and exploit soil microhabitats beyond the nutrient-depleted area where rootlets or root hairs cannot thrive (Rhodes and Gerdemann, 1980). This is discussed in detail later. It is evident that VAM have a greater exploring ability than the root and overcome limitations on acquisition of ions that diffuse slowly in the soil solution to the rhizosphere. The quantitative features of hyphal growth in soil have already been considered (Section III.B), and it is obvious that the size of the extramatricial mycelium is critical in defining its uptake characteristics. In addition, the turnover of hyphal development and the activity of these hyphae are also important (Smith and Gianinazzi-Pearson, 1988).

It is obvious that these facts allow one to envisage the VAM as a modified root system greatly improved for nutrient uptake. Besides, as discussed below, there is are some evidence that VAM can alter the kinetic properties of the root in regard to its absorption abilities (Harley and Smith, 1983; Bolan et al., 1987a).

It is commonly accepted that plants with profusely branched root systems having fine rootlets less than 0.1 mm in diameter and long root hairs (graminoids roots) are less dependent on VAM than those with coarse roots (magnoloid roots) having rootlets more than 0.5 mm in diameter (Baylis, 1975). This reinforces the idea that VAM represent a complement of the root system, being more critical when the latter is less developed or when the environment is stressed, nutrient-poor, or competitive (Mosse et al., 1981).

V. Vesicular–Arbuscular Mycorrhizae and Acquisition of Phosphate by Plants

A. Phosphate Transport by VAM, a Key Factor in Plant Nutrition

A great deal of work, recently reviewed (Barea and Azcón-Aguilar, 1983; Harley and Smith, 1983; Abbott and Robson, 1984; Hayman, 1986; Smith and Gianinazzi-Pearson, 1988) shows that VAM enhance plant growth as a result of improved mineral nutrition of the host plant, and this has been confirmed with the use of isotopic tracers. Mycorrhizal plants not only are large but also usually have an increased concentration and/or content of phosphorus compared to nonmycorrhizal controls. The VAM actually increase the rates of growth of plants and influence the partitioning of phytomass between shoot and root. This is related to the enhanced nutrient uptake by fungal activity; this is followed by nutrient translocation to the shoots, which increases the utilization of photosynthate in the aerial part of the plant. Therefore, relatively less of the products of photosynthesis are allocated to the root. Hence, the root/shoot ratio is usually lower in VAM plants than in their nonmycorrhizal counterparts (Smith, 1980).

Transitory growth depressions can be evident when the VAMF is colonizing, because there is drainage of carbon compounds without compensation for the mineral supply, as the VAM are not yet operative (Cooper, 1984). Persistent depressions could also take place when the VAMF behave as a parasite by consuming without returning, which can occur in certain situations, as discussed below. In regard to phosphate acquisition, the operation of VAM, considered as a whole, is usually named "phosphate transport." The general process consists of three subprocesses: (1) the absorption of phosphate from soil by VAMF hyphae; (2) the translocation of phosphate along the hyphae from the external to the internal (root cortex) mycelia; and (3) the transfer of phosphate to the cortical root cells, ready to be used by the plant. These subprocesses are analyzed separately after a discussion of the nature of the phosphorus fractions in soil used by VAM.

It is important to note that the processes related to the VAMF pathway for transport of phosphorus from the soil to the root cortical cells are not completely understood. It is well known that VAMF are able to take up, accumulate, and transfer large amounts of phosphate to the plant by releasing the nutrient in root cells containing arbuscules (Smith and Gianinazzi-Pearson, 1988). Nevertheless, the information is fragmentary with regard to the mechanisms involved and scarce concerning the molecular basis triggering or regulating phosphate absorption, translocation, and transfer.

The analysis by Abbott and Robson (1984) of the large number of studies of the effect of VAM on plant growth allows these authors to suggest a series of experiments on the topic and discuss those already done. These include (1) the careful selection of appropriate control treatments, (2) the need for plant growth response curves to added phosphate or other nutrients, and (3) sequential harvestings. Point 1 remains a matter of discussion (Fitter and Nichols, 1988; Baas et al., 1989; Koide and Li, 1989). Recommendations under point 2 are important, since response curves make possible horizontal rather than vertical comparisons. That is, as explained by Abbott and Robson (1984) the, comparison of growth of VAM plants and their controls at a particular level of applied nutrient (vertical comparison) versus the measurement of the amount of nutrient required for the same yield of VAM plants and controls (horizontal comparison). Secondly, such curves make it possible to ascertain effects of VAM other than those derived from increased phosphate uptake or to do physiological studies on VAM plants. The sequential harvestings permit the relating of VAM formation to growth effect or phosphate inflow.

1. Sources of Phosphate for VAMF in Soil

Phosphorus has a vital function in all biological systems (Westheimer, 1987) because it is a major plant nutrient required in relatively large amounts (Hayman, 1975; Tinker, 1980). However, some interrelated facts are known about plant phosphate nutrition. Under normal conditions the concentration of available phosphate in the soil solution is very low, usually around 10^{-6} M. This plant-available phosphate (labile or exchangeable phosphate, equilibrated with the phosphate in solution) accounts for only about 1–5% of the total phosphorus content. The remaining phosphorus is in forms (organics or inorganics) that are accepted as not being directly available for absorption by plant roots (Bieleski, 1973; Mosse, 1973; Tinker, 1975; Hayman, 1975).

A number of studies on this topic led to the conclusion that plants use the same available phosphate pool whether or not they are mycorrhizal. Such a conclusion was reached in experiments using isotopic dilution or fertilizer-labeling techniques (Sanders and Tinker, 1971; Hayman and Mosse, 1972; Mosse et al., 1973; Powell, 1975; Owusu-Bennoah and Wild, 1979). In general, these authors labeled the labile phosphate pool in the soil with ^{32}P to compare the specific activity ($^{32}P/^{31}P$) in plants growing in the test soil. The specific activity of phosphorus in plant tissues was similar for mycorrhizal and nonmycorrhizal plants, although the former took up more phosphate.

It is obvious that, if VAM plants obtain phosphorus from nonlabile (unlabeled) sources, the specific activity in these VAM plants would be expected to be lower than that in the nonmycorrhizal counterparts. However, the study by Bolan et al. (1984) suggests that there are forms of phosphorus in soil that can be labeled by ^{32}P and that are accessible to VAM, but not to nonmycorrhizal roots. These authors argued that the possibility exists for differences in availability of phosphorus to mycorrhizal and nonmycorrhizal plants that might not be reflected by differences in specific activity between VAM plant and their controls. Only if VAM plants utilize phosphorus in the organic fraction will they induce a lowering in the specific activity. Bolan et al. (1984) did not find evidence for such a possibility, as further confirmed by Martin (1985). The fraction of phosphorus that Bolan et al. (1984) suggested was better used by VAM is in the phosphate adsorbed to iron hydroxides, and the authors claim that the phosphate desorbed would be uniformaly labeled. Therefore, no changes in the specific activity in the plant would be shown. In any case, the use by VAM of phosphorus sources different from those used by nonmycorrhizal roots would seem to have a qualitative interest, and it can be accepted that VAM plants draw most of their phosphate from the soluble pool, although more efficiently than nonmycorrhizal plants.

In addition to these studies based on the use of radioactive phosphorus (^{32}P), another type of assay has been carried out to test the effectiveness of sparingly soluble added phosphorus sources for the growth and phosphate acquisition by VAM in comparison with nonmycorrhizal plants. In this context, early reports, as reviewed by Tinker ((1980) and Barea et al. (1983), seemed to indicate the possibility that VAM fungi might solubilize sources of phosphorus that would not otherwise be used by non-VAM roots. Obviously, isotopic studies, as discussed above, disregard such a possibility. Nevertheless, the fact that VAM plants can respond readily to additions of sparingly soluble phosphorus sources such as rock phosphate has been repeatedly shown (Tinker, 1980; Barea et al., 1983; Manjunath et al., 1989).

The explanation, as concluded in these reviews, is that VAM can improve the utilization of rock phosphate by the plant when, even slowly, some phosphate ions were physicochemically or biochemically dissociated from the rock phosphate into the soil solution. Then, and because the network of extramatricial hyphae make closer contact than roots with the surface of rock phosphate particles, they can benefit the plant by using these "naturally" dissolved phosphate ions. A low concentration of calcium in soil solution and an acid pH of the soil help rock phosphate dissociation (Khasawneh and Doll, 1978). In any case, although VAMF seem able to tap sparingly available sources of phosphate in soil and to absorb it more readily than roots (Gianinazzi-Pearson et al., 1981; Young et al., 1986), it is difficult to distinguish a direct fungal effect from an indirect one as induced by the mycorrhization of the plant or the rhizosphere.

Recently Bolan et al. (1987b) tested the effect of VAM on the availability of added particulate iron phosphate. They found that the uptake by the VAM plants was greater. To explain this fact, they accept the closer contact of VAM hyphae with localized sources near the particle surface but advance the further possibility that the fungal hyphae, by producing organic acids such as citrate, can affect phosphate solubilization. It is well documented that some chemical compounds produced either by the fungus or by the plant under the influence of VAM are involve l in the formation and maintenance of a "mycorrhizosphere" (Linderman, 1988). However, evidence that these substances can alter phosphate solubility is lacking in the case of VAMF.

Some time ago it was hypothesized (see review by Barea et al., 1983) that "phosphate-solubilizing" bacteria could cooperate with VAM to help plants use rock phosphate in nonacidic soils. Synergistic interactions between these bacteria and VAM have been found, and the positive responses were associated with low concentration of active calcium in the soils. Nevertheless, Azcón-Aguilar et al. (1986b) did not find that phosphate-solubilizing bacteria improved the utilization by mycorrhizal plants of a labeled source of insoluble phosphate added to the soil (pH = 7.4). Perhaps the high concentration of exchangeable calcium in the test soil precluded any potential bacterial "solubilization" of the labeled phosphate source.

2. Phosphate Absorption from Soil

It is known that VAM can absorb several times more phosphate and have greater phosphate inflow rates than roots. In addition, the hyphae use the same phosphate sources as the roots. Therefore, the problem is to define the mechanisms for VAM to improve the absorption of available phosphate. Several mechanisms have been proposed.

It has already been stated that plants take up phosphate much faster than these ions can diffuse to the absorption surfaces of the root system. This causes phosphate-depleted zones to develop around roots. These zones are 1–2 mm wide as documented by autoradiography (Owusu-Bennoah and Wild, 1979). The VAM hyphae growing through soil pore spaces are able to affect phosphate absorption beyond the depletion zone. Thus, this mechanism of fungal action is merely physical and based on the increased number of sites for phosphate absorption, which allows the exploration of a greater soil volume. The fungal hyphae actually transport phosphate over large distances (several centimeters) into the root cortical cells (Pearson and Tinker, 1975; Rhodes and Gerdemann, 1980), as stated earlier (Section III.B). The considerable extent of the extramatricial network of mycelium fits the ability of VAM for phosphate uptake. Obviously, once inside the hyphae, phosphate is protected from refixation by soil components.

Other, more physiological mechanisms have also been suggested to account for the increased phosphate uptake by VAM. As discussed by Harley and Smith (1983), Tinker and Gildon (1983), and Bolan et al. (1984) (as some representative examples), there are indications that VAM are able to take up phosphate from soil solutions with low phosphate concentration more efficiently than simple roots. This could be explained if (1) VAM were able to absorb phosphate from a lower threshold concentration from soil (Mosse et al., 1973) and/or (2) VAM take up phosphate from solution faster than roots do. In fact, Bolan et al. (1983) documented the existence of a threshold concentration for effective phosphate uptake by nonmycorrhizal, coarse-rooted clover but not for VAM clover or for finely rooted ryegrass, whether or not mycorrhizal.

On the other hand, the kinetic analyses carried out by Cress et al.

(1979) suggest that VAM apparently have a pathaway of phosphate uptake with a much higher affinity for the ion. This was indicated by a lower Michaelis constant (K_m) in VAM than in roots in spite of the fact that maximum rates of phosphate uptake (V_{max}) were quite similar for VAM and control plants. These conclusions were reached from studies at low phosphate concentration in the solution. As already stated, a lower K_m would indicate a higher affinity of uptake sites in VAM to acquire phosphate from transitory and diluted sources. However, the discussion by Tinker and Gildon (1983) points out that the results of Cress et al. (1979) do not necessarily imply a physiological ability (smaller K_m) of fungal hyphae but merely a consequence of the increased uptake rate in VAM, which caused an apparent decrease in K_m. Tinker and Gildon (1983) also support the presence of a threshold for phosphate uptake that is most likely related to the rate of diffusion through soil solution rather than to the ability of VAM relative to roots to absorb the nutrient from solutions at low concentrations.

The studies by Karumaratne et al. (1986) introduced some additional conflicting results concerning the K_m values for phosphate uptake by VAM versus simple roots, since they found a higher K_m value for phosphate uptake by VAM. The differences in the experimental treatments and methodologies could account for the different results obtained. The topic deserves further research, however. In any case, VAM hyphae can take advantage of their geometry and better distribution than roots to acquire phosphate from transitory, localized, and diluted sources of the element (Harley and Smith, 1983). This undoubtedly can account for the efficiency of VAM whether or not they have a lower K_m for phosphate uptake than that of simple roots.

The qualitative and quantitative changes in the root exudation patterns (Harley and Smith, 1983) and the differences between VAM and nonmycorrhizal plants in the absorption of anions and cations (Buwalda et al., 1983), which can change the pH of the rhizosphere, are indirect mechanisms that Bolan et al. (1984) argued would explain the effect of VAM to increase phosphate availability to the plant.

All in all, it must be stated that the uptake mechanisms are poorly understood and that there have been few studies to gain information on the topic. For example, the process of phosphate uptake by hyphae arising from germinating spores in temperature sensitive (Bowen et al., 1975). The concentration of orthophosphate (Pi) in VAM fungi mycelium in comparison with that in the soil solution indicates a cell-to-soil concentration gradient of 1000 : 1 (Gianinazzi-Pearson and Gianinazzi, 1986). It follows that an active mechanism for phosphate uptake will be necessary. By comparison with other organisms, such a mechanism is probably via a proton symport located on the plasma membrane of the fungus, coupled to a proton ATPase system (Beever and Burns, 1980; Clarkson, 1985; Smith and Smith, 1986).

3. Phosphate Translocation along Hyphae

The phosphate absorbed by VAMF from soil solution is accumulated in the vacuoles of the fungus as polyphosphate (poly-P) granules (Callow et al., 1978; Cox et al., 1975, 1980). Poly-P is an important phosphate reserve in fungi (Beever and Burns, 1980) and can represent 16–40% of total phosphate in VAMF (Callow et al., 1978; Capaccio & Callow, 1982). These granules, being a form of sequestered soluble phosphate, are osmotically inactive. Therefore, they avoid interfering with cell metabolism (Gianinazzi-Pearson and Gianinazzi, 1986). This mechanism to control phosphate concentrations in fungal cytoplasm operates by means of enzymatic systems responsible for the synthesis and breakdown of poly-P, the presence of which has been demonstrated in VAM roots (Capaccio and Callow, 1982).

Poly-P forms not only are storage forms but are also implicated in the translocation of the nutrient. The flux rates of translocation have been calculated to be in the range of $0.1–2.0 \times 10^{-9}$ mol cm^{-2} s^{-1} (see Tinker, 1980; Cooper, 1984; Smith and Gianinazzi-Pearson, 1988). The poly-P granules are propelled through the hyphae by cytoplasmic streaming to the arbuscules. The process is metabolically dependent, being slowed by low temperatures and stopped by cytochalasin B, which is an inhibitor of cytoplasmic streaming. The specific mechanisms for phosphate loading, translocation, and unloading are not only active but also very efficient. According to Bieleski (1973), the calculated inflow of phosphate through external hyphae is approximately 1000-fold faster than the phosphate diffusion rate through soil pores. The translocation of poly-P appears to occur down a concentration gradient between the phosphate source, as affected by the phosphate uptake by the extramatricial hyphae, and the sink, as affected by the Pi removal rates from the fungus to the root cells. It is also important that when the phosphate supply is in excess, poly-P accumulates the nutrient as a storage pool that can be used at times of increasing demand to support later stages of plant growth (Smith and Gianinazzi-Pearson, 1988).

4. Phosphate Transfer to Root Cells

The poly-P granules in the fine branches of the arbuscules are broken down by specific enzymatic activities (see Cooper, 1984), releasing Pi into the cytoplasm. However, the Pi concentration in fungal cytosol did not increase because the nutrient was transferred to the host plant. In fact, it has been demonstrated that the arbuscules are the main sites of phosphate transfer to the host. Actually, the arbuscules are well adapted, both metabolically and structurally, for nutrient exchange in VAM. The plasma membranes of both symbionts, representing an intracellular contact, are separated by the apoplastic spaces. Such an interface is favorable for metabolite interchange. The contact between the symbionts to form the ar-

- UPTAKE, from soil solution by the external mycelium.
- TRANSLOCATION, as polyphosphate granules, to the internal mycelium.
- TRANSFER, from fungus to the host cells (arbuscules).

Figure 6. Summary of the main phosphate transport processes in VA mycorrhizae.

buscule induces a redistribution of ATPase activities on the plasma membranes of both root and fungus. These enzymes are energy-generating systems, both for the host actively to take up the phosphate that the fungus releases into the interfacial apoplast and for the fungus to acquire hexose as supplied by the host to the interface (see Smith and Gianinazzi-Pearson, 1988, for a review of related literature).

In summary, the overall significance of VAM to phosphorus nutrition is that the symbiosis represents a change in the phosphate uptake properties of the root systems. The key aspects of the VAM role are summarized in Figures 6 and 7.

B. Factors Affecting the Processes of Phosphate Acquisition

Some well-established principles account for the significance of VAM under natural conditions (Mosse et al., 1981). First, the establishment, development, and function of VAM are all dependent on interactions among the prevailing fungal, plant, and environmental factors. Second, the VAM fungi are ubiquitous, but several factors, especially some agricultural

*More than 1 m of hyphae per 1 cm of root can be formed. This is critical for P uptake from soil solution.
*Translocation across several cm (7–8) has been demonstrated (^{32}P).
*Inflow (mol cm^{-1} s^{-1}) into VAM is, on average, 3 to 4 times greater than that into uninfected roots (^{32}P).
*Rate of translocation can be 1000-fold faster than that of diffusion in soil solution.
*VAM-specific enzymatic activities involved in the formation and degradation of polyphosphate granules.
*Active transfer of Pi in both fungal and postplasma membranes (arbuscules).
*Redistribution of root cells' membrane-bound ATPases as arbuscules develop.

Figure 7. Transport of phosphate in VA mycorrhizae.

practices, affect the indigenous mycorrhizal populations both quantitatively and qualitatively. Third, in spite of the lack of specificity, there is great variation in the symbiotic response. Therefore, if the mycorrhizal efficiency is the result of interactions among fungus–plant–soil–environmental conditions (Hayman, 1983), the factors affecting phosphate acquisition by VAM must be considered in light of these interactions.

1. Characteristics of Symbiotic Partners

As stated earlier (Section II.B.2), the ability of a given VAMF to supply phosphate to the host plant depends on a series of interrelationships developed between the symbiotic partners. Among those, the degree of mycorrhizal dependency of the plant and the specificity or compatibility of the association are properties that significantly affect mycorrhizae formation and/or functioning and need to be considered in this concern.

Smith and Gianinazzi-Pearson (1988) analyzed the determinants of symbiotic efficiency to satify plant phosphate demand and gave a comprehensive view of the fungal, plant, and interactional factors involved. (1) *Fungal factors* include the growth rate of the fungus to colonize the root cortex (mainly arbuscule production) and the root-surrounding soil, the actual extent of development of the extramatricial and intrarradical mycelia, the capacity of the fungus for phosphate uptake and translocation, the poly-P turnover, and so on. Assays of competitiveness between different VAMF, especially between introduced and native strains for root colonization, are very important. Difficulties arise in distinguishing the endophytes involved. Anatomic characteristics are useful in some cases (Lopez-Aguillon and Mosse, 1987), and a diagnostic isoenzymatic analysis has also been applied (Hepper et al., 1988). (2) *Plant factors* mainly include the size and growth rate of the root system, taking into account its morphology, geometry, and distribution in the soil profile and the phosphate requirement of the plant as a whole, which conditions the nutrient demand. (3) *Factors regarding the symbiosis itself* are those affecting the formation and functioning of the interfaces and the nutrient exchange. The results of the interactions of these factors and/or conditions in a given situation will determine the role of VAM to supply phosphate to the plant.

2. Fertilizer Applications

One of the topics receiving the most attention in VAM research has been the effect of the application of soluble phosphate on the formation and functioning of the symbiosis. In fact, a large number of papers have been published on the topic from that by Daft and Nicolson (1966) to the recent ones by Rajapakse et al. (1989) and Amijee et al. (1989). In summary, it can be concluded that increasing soluble phosphate levels in the soil reduces the overall percentage of VAM colonization. Of special importance, the observations indicate that soluble phosphate decreases both the extent of the extramatricial mycelium (Abbott et al., 1984) and the number of arbuscules formed (Smith and Gianinazzi-Pearson, 1988).

The experimental design followed in most of the studies did not permit determination of whether the decrease in the percentage of VAM colonization was the result of a reduction in fungal growth or of an increase in root growth (Amijee et al., 1989). They conclude that the formation of entry points of the fungus on the root is the rate-limiting step and that this rate is reduced by soluble phosphate to interfere with colonization.

It appears that the effects of phosphate are exerted via the plant (Azcón et al., 1978, for details and references). It has been argued that the phosphate effect on root colonization is mediated by changes that the limiting nutrient (phosphate) exerts on membrane permeability, which affect the root exudation rates to the rhizosphere soil or to the cellular spaces in the root cortex (Cooper, 1984; Barea, 1986). The direct effect of suluble phosphate on fungal metabolism, mainly by regulating enzymatic activities related to phosphate transfer to the host, has been recently discussed (Smith and Gianinazzi-Pearson, 1988).

3. Other Components of the Ecosystem

Both the formation and the activity of VAM can be affected by other components of the soil–plants–atmosphere system. Some influences are exerted from the soil, and others are plant-mediated.

Among the former, factors of either physicochemical or biological nature are involved. Soil conditions are obviously important, and in this context it is known that VAM formation is favored by low to moderate soil fertility levels. However, there are fungal adaptations to higher levels of soil fertility (Mosse et al., 1981; Hayman, 1982).

In spite of fact that no general correlation has been found between VAM and soil pH, it is well known that particular soil pHs favor particular VAMF species. This is important because it significantly affects the effectivenees of mycorrhizal fungi (Mosse et al., 1981; Hayman, 1982; Arines, 1990). These papers also showed that soil pH, by changing the solubility status of plant nutrients, can indirectly influence VAM formation and/or activity.

Organic matter content, mainly because it alters other physicochemical soil properties, has a striking influence on VAM functioning (Arines, 1990). In general, however, it is difficult to establish clear-cut conclusions on correlations between the organic matter content and quantitative or qualitative parameters of VAM development (Mosse et al., 1981).

As obligate aerobes, VAMF are affected by O_2 concentration, and flooding tends to reduce VAM formation. But in some cases there are adaptations, and VAM can occur in watterlogged conditions. On the other hand, VAM are also formed in very dry situations (Mosse et al., 1981).

The dependence of VAM functioning on plant photosynthesis and the interactions of light intensity or photon irradiance with VAM development and activity need to be considered, and there are several recent studies on this topic (Son & Smith, 1988; Son et al., 1988).

Soil temperature affects several processes at different stages of VAM development (Smith and Gianinazzi-Pearson, 1988).

With regard to the effects of biological soil factors, soil microorganisms are known to influence greatly the establishment and activity of VAM. The recent reviews by Oliveira and Garbaye (1989) and Azcón-Aguilar and Barea (1990) record the most significant information on the topic. This can be summarized as follows. (1) The soil microbiota exert an effect on the formation of VAM. The mycorrhizal fungi are immersed in the framework of interactions taking place in soil microhabitats, and therefore, soil microorganisms affect, either positively or negatively, the epidemiology of root colonization by VAMF (Barea and Azcón-Aguilar, 1982; Bagyaraj, 1984). (2) The soil microorganisms influence the VAM effects on plant nutrient uptake. Nitrogen-fixing bacteria such as *Rhizobium,* root-nodulating actinomycetes, cyanobacteria, in symbiosis with the plant, and the free-living *Azotobacter* or *Azospirillum* interact with VAMF, developing activities of relevance to plant growth (Barea and Azcón-Aguilar, 1982, 1983; Bagyaraj, 1984), as discussed in Section VI.A. The so-called "phosphate-solubilizing microorganisms" can release some phosphate ions from otherwise sparingly soluble phosphate sources (Kucey et al., 1989), and it was postulated that VAMF hyphae can tap these ions and translocate then to the plant (Barea et al., 1983).

Other microorganisms such as plant hormone or siderophore producers and, in general, the "plant growth-promoting rhizobacteria" (PGPR) can affect the formation of the symbiosis or the nutrient uptake activity. These influences are usually exerted through plant-mediated mechanisms (Azcón-Aguilar and Barea, 1990). Plant pathogens are also known to interact with VAM (Perrin, 1985) by competing for colonization sites. Other ecological factors also influence VAM. The most important is the rate of photosynthesis and related factors. The fungus, as a heterotrophic organism, is supplied with carbohydrates from the photosynthate of the host plant. It is obvious that factors such as light intensity, temperature, CO_2 concentration, etc. that affect the photosyntesis rate or the carbon allocation in the plant would affect VAM development and functioning (Harley and Smith, 1983). Feedback controls exerted by nutrients, as supplied by VAM, can in turn regulate VAM activity (Smith, 1980).

Agricultural practices, especially biocide application and drastic changes in the surface soil horizons such as excavation and mining, can affect VAM (Mosse et al., 1981; Mosse, 1986). Crop rotation involving fallow periods and nonhost plants is an influencing factor, since it alters both the size and the species composition of VAMF population (Ocampo and Hayman, 1981; Harinikumar and Bagyaraj, 1988). The use of biocides can affect VAMF populations. Thus, the selection of an appropriate biocide and its application rate are important for VAM functioning (Mosse, 1986).

VI. Vesicular–Arbuscular Mycorrhizae and Nitrogen Nutrition

A. Processes Involved

A number of reports show that VAM increase nitrogen concentration and/or content in plant shoots (see Barea et al., 1988). To explain such an effect a number of mechanisms were suggested and investigated by using ^{15}N-based methodologies. These mechanisms include (1) the improvement of symbiotic biological N_2 fixation (an indirect VAM activity based on the supply of phosphate for N_2 fixation functioning), (2) direct uptake of combined nitrogen by VAMF, (3) facilitated "N transfer," a process by which part of the nitrogen, as biologically fixed by nodulated plants, benefits the nonfixing plants growing nearby, and (4) enzymatic activities involved in N metabolism. This last mechanism has been discussed elsewhere (Harley and Smith, 1983; Smith and Gianinazzi-Pearson, 1988).

1. Improvement of N_2 Fixation

There is much information from greenhouse and field studies to show that VAM improve growth, nodulation, and N_2 fixation in legume–*Rhizobium* symbiosis and in actinorrhizae (Barea and Azcón-Aguilar, 1983; Hayman, 1986; Barea et al., 1988). This effect probably arises from the fact that N_2 fixation depends on steady adequate supply of phosphate to the root and nodules. Methodologies using ^{15}N-labeled fertilizers are now applied to evaluate N_2 fixation in the field (Danso, 1988), and it is widely accepted that these are the only direct methods to distinguish the relative contributions of the nitrogen sources to "fixing" plants, i.e., soil, fertilizer, and atmosphere. These methodologies determine if the effect of a given treatment is exerted directly on N_2 fixation or on another of the nitrogen sources. Pot and field studies carried out in this laboratory (Barea et al., 1987, 1989a,b) demonstrated, by using ^{15}N-labeled inorganic material, that VAM inoculation enhanced Biological Nitrogen Fixation (BNF) similarly to a phosphate fertilizer (75–100 kg P ha^{-1}).

2. Nitrogen Uptake from Soil

The effect of VAM fungi in the uptake of nitrogen compounds from soil is a topic of recent interest. In fact, Ames et al. (1983) demonstrated that VAM hyphae were able to take and translocate $^{15}NH_4^+$, a nitrogen form that can be assimilated by VAMF because they have the appropriate enzymes (Smith et al., 1985). Later, Smith et al. (1986) showed that VAM increased the nitrogen inflow to the plant. The field experiment by Barea et al. (1987) and Kucey and Bonetti (1988) confirmed, by using ^{15}N-labeled fertilizer, that VAM hyphae took up nitrogen from soil, thereby increasing the nitrogen content in the VAM plant in comparison with controls receiving phosphate. A further experiment under controlled conditions (Barea et al., 1989a) confirmed this fact and demonstrated that the apparent soil nitrogen pool size (An value) was significantly higher in mycorrhizal pots, supporting a role of VAM in nitrogen uptake.

The important question is the form of nitrogen used by VAM in this particular case and in general, according to established principles concerning nitrogen uptake by plants. It is known that VAM can use both NH_4^+ and NO_3^- (Bowen and Smith, 1981). However, the soil used by Barea et al. (1987, 1989a) had a pH of 7.5. Therefore, NO_3^- was probably the predominant form of assimilable nitrogen. Since nitrate is much more mobile in soil than ammonium (Chapin, 1980; Harley and Smith, 1983), it seems unlikely that VAM would exert any special effect on nitrate uptake. Nevertheless, several points must be considered. First, because of the great demand for nitrogen by plants, the soil surrounding the root can also be deficient in nitrate (Harley and Smith, 1983). The fungus, in fact, can absorb nitrate ions from beyond the more deficient shells around roots. It is also known, at least indirectly, that VAM increases nitrate reductase in the plant (Smith and Gianinazzi-Pearson, 1988), which is necessary for NO_3^- assimilation.

Second, ammonium, the slowly diffusing nitrogen form, though existing in low concentration in soils on the alkaline side of neutrality, can be involved in nitrogen uptake by VAM under such conditions. This is because fungi are known to have the ability to accumulate ammonium from low external concentrations (Smith et al., 1985). The soil used by Barea et al. (1987, 1989a) also had a high clay content (about 45%), and it is known that such soils can retain ammonium ions, preventing their volatilization (Stevenson, 1986). Work by Ames et al. (1984) indicated that VAM plants can derive nitrogen from sources that are less available to nonmycorrhizal plants. Thus, the possibility exists that VAM fungi can take up ammonium ions retained by clay particles in soils. This possibility is presently being studied in this laboratory using ^{15}N.

In neutral and slightly acid soils, ammonium is the predominant nitrogen form (Harley and Smith, 1983). It is likely that VAM play a key role in nitrogen uptake by plants because of the low mobility of NH_4^+ in soils.

3. Nitrogen "Transfer"

There is great ecological and economic interest in biological N_2 fixation as the main route by which nitrogen enters the biosphere. This interest also applies to the case of intercropping, the age-old technique of growing two or more crop species simultaneously in the same plot (Ofori and Stern, 1987). Nitrogen-fixing plants, usually legume–*Rhizobium* associations, are key components of intercropping systems because they enrich soil with nitrogen to benefit succeeding phases of the cropping sequence or to provide nitrogen (N transfer) directly to companion plants by sharing some of the fixed N_2 with them (Haynes, 1980; Heichel, 1987). Pastures and agroforestry, as examples of intercropping systems, have received some attention from the point of view of the role of VAM (Barea, 1988).

Apart from the VAM role improving biological N_2 fixation, it has been hypothesized that VAM hyphae can improve N transfer, since the network of VAM mycelia can link different plant species growing nearby and help overlap the pool of available nutrients for these plants (see Newman, 1985; Barea, 1988). Therefore, the nitrogen released into the overlapping mycorrhizospheres by legume root exudation or by nodule decay can be used by nonfixing plants. Some greenhouse and field experiments using ^{15}N have been carried out to ascertain the role of VAM on N transfer (Kessel et al., 1985; Haystead et al., 1988; Barea et al., 1988, 1989a,b). In some cases, but not in others, VAM appear to improve N transfer. This difference is currently a matter of discussion. Isotopic evidence to demonstrate a VAM effect in promoting N transfer conclusively is difficult to obtain because interactions between nitrogen nutrition and VAM possibly mask such effects (Barea et al., 1989a). In fact, assessment of N transfer is deduced from mathematical calculations based on principles that do not take into account the VAM as a particular mechanism of the plant to scavenge nitrogen in soil and that plants differ in their degree of mycotrophy. Therefore, calculations based on nitrogen uptake profiles exhibited by different plant species could be affected by the degree of VAM that these species can develop.

B. Factors Affecting VAM Processes of Nitrogen Nutrition

The processes whereby VAM can improve plant nitrogen nutrition are influenced by factors and/or conditions of the environment. The role of VAM in N_2 fixation, is mainly mediated by phosphate. Therefore, arguments given in Section V.B are valid for this process. Since biological N_2 fixation depends on the availability of some micronutrients, the role of VAM in supplying these elements is relevant (Munns and Mosse, 1980; Hayman, 1986).

The different degrees of mycotrophy of "fixing" and "nonfixing" plants, mostly legumes and graminaceous crops, affect VAM status in mixed cropping in comparison with pure stand and, consequently, influence the role of VAM in N transfer (Barea et al., 1989b).

The effect of VAM hyphae on nitrogen uptake from soil is affected by a number of factors that in turn influence the predominant available form of nitrogen, i.e., NH_4^+ or NO_3^-. Thus, factors such as the organic matter content, pH, soil texture (mainly clay content), microbial mineralization and nitrification have a great influence on nitrogen uptake via extramatricial mycelia of VAM (Mosse et al., 1990; Arines, 1990). The application of NH_4^+ versus NO_3^- as a fertilizer has an additional implication, since the different ways of assimilation in the plant alter the pH of exudates and therefore change differentially the pH of the rhizosphere of VAM plants (Smith, 1980).

VII. Vesicular–Arbuscular Mycorrhizae and the Acquisition of Other Nutrients

There have been relatively few studies to ascertain directly the role of VAM in plant uptake of nutrients other than phosphate and nitrogen. However, many papers give concentration and/or content data on diverse major and trace elements in plant tissues, usually the shoots. In most instances these experiments lacked appropriate controls and thus do not allow one to distinguish whether a VAM effect is the result of an improvement in nutrient uptake via the extramatricial mycelium or is an indirect consequence of a VAM effect balancing the phosphate status of the plants. Another import topic is the interaction among nutrients at uptake levels, since this can alter the acquisition patterns of the absorbing system, including VAM. Reviews by Rhodes and Gerdemann (1980), Tinker and Gildon (1983), and Harley and Smith (1983) discuss these topics and conclude that VAM are involved in the uptake of Zn and Cu, trace elements having low mobility in soil. An increase of Fe uptake has also been shown in some cases (Rai, 1988).

It is also clear (Rhodes and Gerdemann, 1980) that VAM colonization affects sulfate (^{35}S) uptake by plants, although hyphal translocation does not seem critical in sulfate nutrition. Since sulfate is rather mobile in soil solution, the increase in sulfate concentration can be the result of an improved phosphate nutrition mediated by VAM (Harley and Smith, 1983). These authors also indicate that there is no conclusive support for a role of VAM in K^+ uptake in spite of the fact that the diffusion rate of these ions is rather low in soil solution (Chapin, 1980).

Comments on related papers by Smith and Gianinazzi-Pearson (1988) suggest an association of Ca^{2+} distribution in plants with the synthesis and breakdown of poly-P granules, since the cation is a secondary constituent of these granules.

It has also been suggested that Br^-, Cl^-, and anions in general are increased in plants as a VAM response but not to a phosphate addition (Buwalda et al., 1982). It seems unlikely that these increases are specifically related to VAM in view of their mobility in soil solution. A more probable explanation is that they play a role in the regulation of cellular pH, which is different in VAM plants than in nonmycorrhizal ones (Smith, 1980). It has also been shown that VAM influence plant uptake of Cs and Co (Rogers and Williams, 1986).

VIII. Vesicular–Arbuscular Mycorrhizae and Plant Growth under Stress Situations

A. Vesicular–Arbuscular Mycorrhizae in Nutrient Deficient Ecosystem

In many situations a plant must cope with stress situations caused by the infertility of the soil. Plants have developed a number of strategies to adapt to these situations (Chapin, 1980). Among the different strategies are three that relate to this review: (1) changes in the root absorption capacity, (2) modification of the root-to-shoot ratio, and (3) rhizosphere interactions. From the well-known consequences of mycorrhiza formation, the involvement of VAM in these three types of adaptations can be deduced. It appears that plant species evolving from low-fertility habitats have a low root absorption capacity (Chapin, 1980). This is because at the low nutrient availability and slow nutrient diffusion characteristic of the ecosystem, the plants do not need a higher absorption rate. However, these plants appear to exibit efficient absorption at low nutrient concentrations, and this suggests that they have a lower apparent K_m. Following Chapin's (1980) observations, plants from infertile habitats possess a high root-to-shoot ratio and mycorrhizae. The symbiosis appears to confer a greater root longevity and to help plants absorb nutrients.

A special case of infertile habitat is that of eroded or degraded soils where the soil profiles are disrupted, thereby affecting the arable top layer where most of the VAMF propagules are contained. Work by Tisdall and Oades (1979) reported a direct effect of VAM on soil aggregation, and a more recent paper confirmed such an effect (Thomas et al., 1986). The effective role of VAM in land rehabilitation has been well documented in the literature (Gardner, 1986; Skujins and Allen, 1986; Allen et al., 1987; Allen and Allen, 1988; Sylvia and Will, 1988; Habte et al., 1988; White et al., 1989).

B. Vesicular–Arbuscular Mycorrhizae and Water Stress

Several experiments, as reviewed by Cooper (1984), indicate that VAM improve water relations in many situations. However, these studies and more recent ones (Augé et al., 1987; Azcón et al., 1988; Bethlenfalvay et al., 1988; Peña et al., 1988) do not distinguish whether the mechanisms by which VAM act are concerned with an increase of drought resistance or an improvement of water flow through the plant. Even more, it is not clear whether VAM improve water relations in plants or merely facilitate phosphate uptake when the diffusion rate of phosphate in the soil is lowered under drought conditions.

In phosphate-deficient situations VAM plants can exhibit a lower resistance to water flow. However, phosphate additions, in general, produce a similar effect. Nevertheless, there are indications of a positive effect of VAM on water relations in plants that were not equalized by addition of phosphate (Hardie and Leyton, 1981). To explain the cases where VAM seem to improve water flow through hyphae, Cooper (1984) argued that the external VAM hyphae bypass the dry zones surrounding the root during the drought period. Thus, VAM can maintain a water continuity across the soil–root interface. This is also supported by Hardie (1985). However, inflow calculations by Allen (1982), taking into account those by Sanders and Tinker (1973) together with data obtained by using tritiated water (Safir and Nelsen, 1981), do not support an improvement of water uptake as a result of increased hyphal translocation (see Cooper, 1984). The topic continues to be a matter of discussion, but it is likely that VAM, by maintaining the uptake of slowly diffusing nutrients under water stress situations (Azcón et al., 1988), do help plants cope with drought stress.

C. Vesicular–Arbuscular Mycorrhizae and Soil Salinity

Excess of soluble salt in agricultural soils is a special problem in arid and semiarid conditions, as is well known. The consequences of salinity, which causes a nutritional imbalance for the plants, are also well documented. For example, an excess in Cl^- can interfere with NO_3^- and PO_4^- uptake, and a high Na^+ concentration can affect Ca^{2+} and Mg^{2+} acquisition (Plaut and Grieve, 1988). In view of their significance in nutrient uptake and also in inducing certain physiological changes in the plant (Smith and Gianinazzi-Pearson, 1988), VAM could be expected to alleviate some of the negative effects of salinity. Several papers report effects of VAM in saline situations. For example, Hirrel and Gerdemann (1980) and Pond et al. (1984) found that VAM increase plant tolerance to salinity, thereby increasing the yield. Again, the mechanism could be merely an improvement in phosphate nutrition, although other reasons have been argued. For example, Allen and Cunningham (1983) and Poss et al. (1985) support a role of VAM based on an increase of K^+ concentration in plants. They found a higher K^+/Na^+ ratio in VAM plants than in control plants, which results in plant protection against a salinity effect as exerted by increased Na^+ concentration in plant tissues.

One fact concerning the effect of VAM against salinity is noteworthy. Since VAM inoculation can increase Cl^- concentration in the plant (Buwalda et al., 1983), it might be thought that VAM would enhance the toxic effect of salinity. The situation cannot be generalized, because the opposite effect, decreasing Cl^- concentration by VAM activity, has also been reported (Hartmond et al., 1987). Therefore, it can be suggested that VAM help plants grow at certain levels of salinity. The mechanisms remain unclear but are a matter for further research.

D. Vesicular–Arbuscular Mycorrhizae and Other Stresses

Heavy metal pollution, acid rain, ozone, biocides, etc. are known to introduce stress problems in the soil–plant ecosystem. Therefore, they must interact with VAM formation and function (Mosse, 1986). Reports by Tinker and Gildon (1983) and Dakessian et al. (1986) describe the effect of heavy metals on VAM establishment and activity. It is interesting that VAMF can develop strains able to adapt to polluted soil (Mosse et al., 1981). Even more, in spite of the fact that VAM can improve the uptake of an element such as Zn (Tinker and Gildon, 1983), it also appears that they can alleviate Zn toxicity. This situation merits further attention, since it also seems to take place with other ions. The increased absorption of them by VAM can theoretically be detrimental for the plant. However, the VAM buffer the toxicity, so the damage is often less than expected. In the case with Zn, the mechanisms have been discussed but the evidence is not substantial (Dueck et al., 1986).

The attack on a plant by a pathogenic microorganism can also be considered a stress situation for plant development. Several studies have shown that VAM can decrease the severity of diseases caused by root pathogenic fungi, bacteria, and nematodes. The VAM appear to decrease plant susceptibility or increase tolerance against the attack of root pathogens (see Bagyaraj, 1984; Perrin, 1985; Zambolin, 1987, for details and references). The proposed mechanisms of VAM prophylaxis include improved nutrient uptake, which enhances plant resistance by a more balanced nutritional status, and induced physiological changes in the plant that deter the pathogen or compete with it for colonization/infection sites (Dehn and Dehne, 1986).

IX. Managing Vesicular–Arbuscular Mycorrhizae under Natural Conditions

Once it was appreciated that VAM improve plant growth and nutrition and play a key role in helping the plant under several stress situations, such as nutrient deficiency, drought, and salinity or soil disturbance, a logical "next step" was to ascertain the possibilities of harnessing the symbiosis to challenge its potential in agriculture and horticulture systems or in the establishment and maintenance of forest and grassland ecosystems. The first point to be considered is the size and effectiveness of naturally existing VAMF populations in relation to the test plant of interest.

Several circunstances influence this point. First, agricultural practices that cause stress, soil disturbance, erosion, etc. can diminish the population size and/or the species composition (Hayman, 1982; Mosse et al., 1981; Mosse, 1986; Skujins and Allen, 1986) Second, VAMF and, in general, all biological components of ecosystems tend to concentrate on survival rather than maintaining their effectiveness for productivity (Bowen, 1980). Therefore, the strains naturally present are not necessarily the most effective. Third, the VAMF present may not have the appropriate attributes to express a "functional compatibility" (Gianinazzi-Pearson, 1984). In the case of suitable VAMF being sparse or inefficient, or when the plant production system uses sterilized substrates, it is necessary to introduce appropriate VAMF to the plant rhizosphere by means of inoculation. The basic principles for successful mycorrhizal inoculation are outlined in Figure 8; production of a "high-quality" inoculum is the limiting step for large-scale inoculation with VAMF.

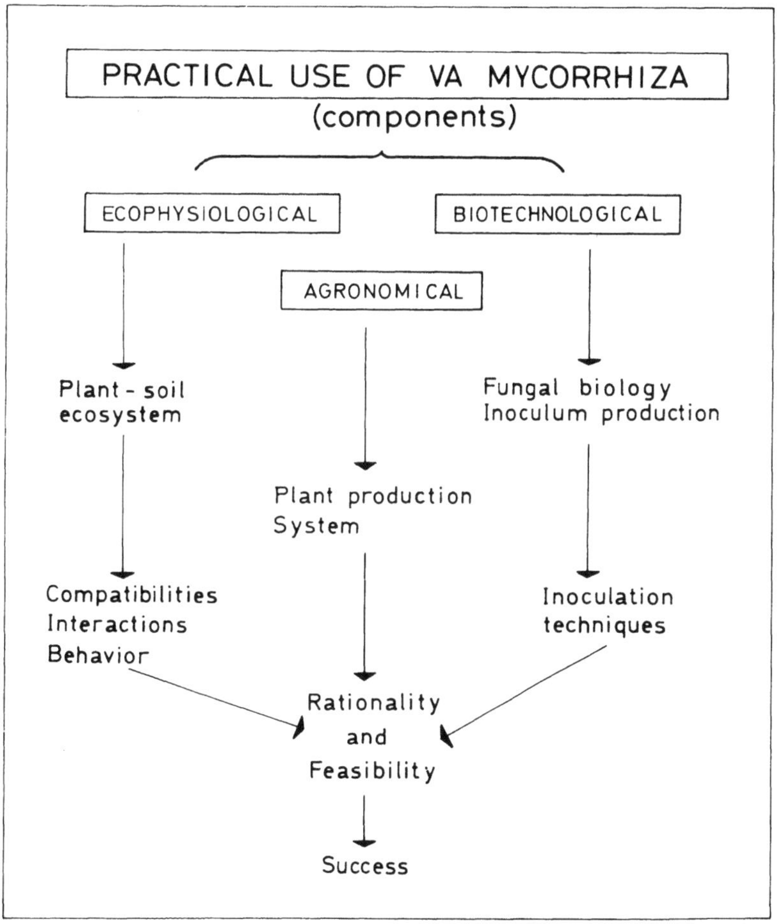

Figure 8. Practical uses of VA mycorrhizae.

It is important to assess situations where VAMF inoculation is needed, feasible, and rewarding (Azcón-Aguilar et al., 1986a). Criteria for selection of VAMF, technology for inoculum production, and inoculation techniques have been reviewed (Hayman, 1984; Menge, 1984). New methodologies have been introduced (Hayman et al., 1986; Dehne, 1987). Recent reviews by Howeler et al. (1987), Gianinazzi et al. (1988), and Sieverding and Barea (1990) analyze the problems on the topic. As a summary, the VAMF inocula can be available for crops using a transplant stage, as usually happens in fruit culture, horticulture, and forestry, where plants are produced in nursery beds, in containers, by tissue culture, etc. Methods of inoculum production using inorganic carriers such as "expanded clay" or inoculation techniques under "pelleting" forms appear suitable to introduce VAM as an integral component of production agriculture systems.

To finish these considerations, a sentence we wrote elsewhere (Barea and Azcón-Aguilar, 1983) can be paraphrased:

> Mycorrhizae therefore can be regarded as an alternative strategy for a more rational agricultural program. However, because the mycorrhizal condition is nearly universal, the natural mycorrhizal potential of a soil needs first to be preserved (avoiding detrimental practices), second to be optimized (manipulating soil conditions to be conducive to the symbiosis), and third, finally, to be considered when inoculation is required.

X. Conclusions and Perspectives

The roots of most of world's plants are colonized by specific soil fungi to form vesicular–arbuscular mycorrizae (VAM), a mutualistic symbiosis that can be considered an integral part of the plant. This can be explained on the basis that the symbiosis seems to have an ecological and evolutionary significance in the origin and development of plants on earth, and, in turn, the coevolution of plant and fungus makes the last an obligate plant symbiont. The VAM induce physiological changes that influence plant growth and survival. Both formation and function of VAM can be affected by the level of soil fertility, which in turn is "modified" by the VAM by changing the ability of a plant to use the nutritional potential of a given soil. The "VAM effect" is mainly accounted for by the changes they induce in the phosphate uptake properties of the root system. The amount and turnover rates of the external mycelium, which usually grows beyond the zone depleted of slowly diffusing nutrients, are critical in VAM activity. Therefore, VAM contribute to a better exploitation of soil phosphate and to more efficient use of added fertilizers. The VAM operativity is self-regulated by mechanisms that assure that the phosphate supply to the plant is optimized over a range of soil phosphate levels. They also improve the uptake of other low-mobility ions such as some nitrogen forms, Cu, and Zn. By a phosphate-mediated mechanism VAM also enhances N_2 fixation by legumes and actynorrhizal plants, as assessed using ^{15}N.

It is now recognized that VAM can be harnessed in order to improve productivity in agriculture, horticulture, fruit culture, and forestry by reducing the input of fertilizers and/or by enhancing plant survival, thus saving ecological and environmental costs. The VAM can help plants to become established in nutrient-deficient soils or degraded (eroded) habitats, to thrive under arid conditions, and to endure plant stresses (drought, salinity, pathogen attack).

Further research is needed in several areas concerning the aims and scope of this review. These include (1) the ecology and epidemiology of VAMF, (2) the establishment of the determinant of symbiotic efficiency for particular fungus–plant–environment combination, (3) the application of isotopic techniques to ascertain the VAM effect on nutrient cycling, (4) the role of mycorrhizal interconections in nutrient transfer in plant communities (isotope aided), and (5) improvement of the techniques to introduce selected VAMF in plant-production systems.

References

Abbott, L.K., and A.D. Robson. 1984. The effect of mycorrhizas on plant growth. pp. 113–130. In: C.L. Powell and D.J. Bagyaraj (eds.), *VA mycorrhiza*. CRC Press, Boca Raton, FL.

Abbott, L.K., and A.D. Robson. 1985. Formation of external hyphae in soil by four species of vesicular–arbuscular mycorrhizal fungi. *New Phytol.* 99:255–255.

Abbott, L.K., A.D. Robson, and G. De Boer. 1984. The effect of phosphorus on the formation of hyphae in soil by the vesicular-arbuscular mycorrhizal fungus, *Glomus fasciculatum*. *New Phytol.* 97:437–446.

Allen, M.F. 1982. Influence of vesicular–arbuscular mycorrhizae on water movement through *Bouteloua gracilis* (H.B.J.) Lag ex Steud. *New Phytol.* 91:191–196.

Allen, E.B., and M.F. Allen. 1988. Facilitation of succession by the nonmycotrophic colonizer *Salsola kali (Chenopodiaceae)* on a harsh site: Effects of mycorrhizal fungi. *Am. J. Bot.* 75:257–266.

Allen, E.B., J.C. Chambers, K.F. Connor, M.F. Allen, and R.W. Brown. 1987. Natural reestablishment of mycorrhizae in disturbed alpine ecosystems. *Arctic Alpine Res.* 19:11–20.

Allen, E.B., and G.L. Cunningham. 1983. Effects of vesicular–arbuscular mycorrhizae on *Distichlis spicata* under three salinity levels. *New Phytol.* 93:227–236.

Ames, R.N., L. Porter, T.V. St. John, and C.P.P. Reid. 1984. Nitrogen sources and 'A' values for vesicular–arbuscular and non-mycorrhizal sorghum grown at three rates of ^{15}N ammonium sulphate. *New Phytol.* 97:269–276.

Ames, R.N., C.P.P. Reid, L.K. Porter, and C. Cambardella. 1983. Hyphal uptake and transport of nitrogen from two ^{15}N-labelled sources by *Glomus mosseae*, a vesicular–arbuscular mycorrhizal fungus. *New Phytol.* 95:381–395.

Amijee, F., P.B. Tinker, and D.P. Stribley. 1989. The development of endomycorrhizal root systems. VII. A detailed study of effects of soil phosphorus on colonization. *New Phytol.* 111:435–446.

Arines, J. 1990. Aspectos físico químicos de la fijación y movilización biológica de nutrientes en el suelo y su incidencia en la formación y efectos de las micorrizas VA. In: J.L. Gorgé, J.M. Barea, and J. Olivares (eds.), *Fijación y movilización biológica de nutrientes. Nuevas tendencias.* C.S.I.C., Madrid (in press).

Augé, R.M., K.A. Schekel, and R.L. Wample, 1987. Leaf water and carbohydrate status of VA mycorrhizal rose exposed to drought stress. *Plant Soil* 99:291–302.

Azcón, R., F. El-Atrach, and J.M. Barea. 1988. Influence of mycorrhiza vs. soluble phosphate on growth nodulation and N_2 fixation (^{15}N) in alfalfa under different levels of water potential. *Biol. Fertil. Soils* 7:28–31.

Azcón, R., A. Marin, and J.M. Barea. 1978. Comparative role of phosphate in soil or inside the host on the formation and effects of endomycorrhiza. *Plant Soil* 49:561–567.

Azcón, R., and J.A. Ocampo. 1981. Factors affecting the vesicular–arbuscular infection and mycorrhizal dependency of thirteen wheat cultivars. *New Phytol.* 87:677–685.

Azcón-Aguilar, C., and J.M. Barea. 1990. Interactions between mycorrhizal fungi and other rhizosphere microorganisms. In: F. Allen (ed.), *Mycorrhizal functioning* Routledge, Chapman and Hall. Inc. New York (in press).

Azcón-Aguilar, C., J.M. Barea, R. Azcón, and R.M. Díaz Rodriguez. 1986a. Assessment of field situations for the feasibility of vesicular–arbuscular mycorrhizal inoculation using a forage legume as test plant. *Agric. Ecosyst. Environ.* 15:241–252.

Azcón-Aguilar, C., V. Gianinazzi-Pearson, J.C. Fardeau, and S. Gianinazzi. 1986b. Effect of vesicular–arbuscular mycorrhizal fungi and phosphate-solubilizing bacteria on growth and nutrition of soybean in a neutral–calcareous soil given $^{32}P^{45}Ca$ tricalcium phosphate. *Plant Soil* 96:3–15.

Baas, R., C. Van Dijk, and S.R. Troelstra. 1989. Effects of rhizosphere soil, vesicular–arbuscular mycorrhizal fungi and phosphate on *Plantago major* L. ssp. *pleiosperma Pilger. Plant Soil.* 113:59–67.

Bagyaraj, D.J. 1984. Biological interactions with VA mycorrhizal fungi. pp. 131–153. In: C.L. Powell, and D.J. Bagyaraj (eds.), *VA mycorrhiza.* CRC Press, Boca Raton, FL.

Barea, J.M. 1986. Importance of hormones and root exudates in mycorrhizal phenomena. pp. 177–187. In: V. Gianinazzi-Pearson and C. Gianinazzi (eds.), *Physiological and genetical aspects of mycorrhizae.* INRA, Paris.

Barea, J.M. 1988. Endomycorrhiza biotechnology to improve plant productivity in intercropping systems. pp. 9–10. In: *2nd European Symposium on Mycorrhizae, Prague.*

Barea, J.M., and C. Azcón-Aguilar. 1982. Interactions between mycorrhizal fungi and soil microorganisms. pp. 181–193. In: S. Gianinazzi, V. Gianinazzi-Pearson, and A. Trouvelot (eds.), *Les mycorhizes: Biologie et utilization.* INRA, Paris.

Barea, J.M., and C. Azcón-Aguilar. 1983. Mycorrhizas and their significance in nodulating nitrogen-fixing plants. pp. 1–54. In: N.C. Brady (ed.), *Advances in agronomy*. Academic Press, New York.

Barea, J.M., R. Azcón, and C. Azcón-Aguilar. 1983. Interactions between phosphate solubilizing bacteria and VA mycorrhiza to improve the utilization of rock phosphate by plants in non acidic soils. In: *3rd International Congress on Phosphorus Compounds, Brussels*.

Barea, J.M., C. Azcón-Aguilar, and R. Azcón. 1987. Vesicular–arbuscular mycorrhiza improve both symbiotic N_2 fixation and N uptake from soil as assessed with a ^{15}N technique under field conditions. *New Phytol.* 106:717–725.

Barea, J.M., C. Azcón-Aguilar, and R. Azcon. 1988. The role of mycorrhiza in improving the establishment and function of the *Rhizobium*–legume system under field conditions. pp. 153–162. In: D.P. Beck and L.A. Materon (eds.), *Nitrogen fixation by legumes in Mediterranean agriculture*. ICARDA, and Martinus Nijhoff. Dordrecht

Barea, J.M., R. Azcón, and C. Azcón-Aguilar. 1989a. Time-course of N_2-fixation (^{15}N) in the field by clover growing alone or in mixture with ryegrass to improve pasture productivity, and inoculated with vesicular–arbuscular mycorrhizal fungi. *New Phytol.* 112:399–404.

Barea, J.M., F. El-Atrach, and R. Azcón. 1989b. Mycorrhiza and phosphate interactions as affecting plant development, N_2-fixation, N-transfer and N-uptake from soil in legumes–grass mixtures by using a ^{15}N dilution technique. *Soil Biol. Biochem.* 21:581–589.

Baylis, G.T.S. 1975. The magnoloid mycorrhiza and mycotrophy in root systems derived from it. pp. 373–389. In: F.E. Sanders, B. Mosse, and P.B. Tinker (eds.), *Endomycorrhizas*. Academic Press, London.

Beever, R.E., and D.J.W. Burns. 1980. Phosphorus uptake storage and utilisation by fungi. *Adv. Bot. Res.* 8:128–219.

Bethlenfalvay, G.J., M.S. Brown, R.N. Ames, and R.S. Thomas. 1988. Effects of drought on host and endophyte development in mycorrhizal soybeans in relation to water use and phosphate uptake. *Physiol. Plant.* 72:565–571.

Bieleski, R.L. 1973. Phosphate pools, phosphate transport, and phosphate availability. *Annu. Rev. Plant Physiol.* 24:225–252.

Bolan, N.S., A.D. Robson, and N.J. Barrow. 1983. Plant and soil factors including mycorrhizal infection causing sigmoidal response of plants to applied phosphorus. *Plant Soil* 73:187–201.

Bolan, N.S., A.D. Robson, N.J. Barrow, and L.A.G. Aylmore. 1984. Specific activity of phosphorus in mycorrhizal and non-mycorrhizal plants in relation to the availability of phosphorus to plants. *Soil Biol. Biochem.* 16:299–304.

Bolan, N.S., A.D. Robson, and N.J. Barrow. 1987a. Effects of phosphorus application and mycorrhizal inoculation on root characteristics of subterranean clover and ryegrass in relation to phosphorus uptake. *Plant Soil* 104:294–298.

Bolan, N.S., A.D. Robson, N.J. Barrow. 1987b. Effects of vesicular–arbuscular mycorrhiza on the availability of iron phosphates to plants. *Plant Soil* 99:401–410

Bonfante-Fasolo, P. 1984. Anatomy and morphology of VA mycorrhizae. pp. 5–34. In: C.L. Powell and D.J. Bagyaraj (eds.), *VA mycorrhiza*. CRC Press, Boca Raton, FL.

Bowen, G.D. 1980. Misconceptions, concepts and approaches in rhizosphere biology. pp. 283–304. In: D.C. Ellwood, J.N. Hedger, M.J. Latham, J.M. Lynch, and J.H. Slater (eds.), *Contemporary microbial ecology*. Academic Press, London.

Bowen, G.D., D.I. Bevege, and B. Mosse. 1975. Phosphate physiology of vesicular–arbuscular mycorrhizas. pp. 241–260. In: F.E. Sanders, B. Mosse, and P.B. Tinker (eds.), *Endomycorrhizas*. Academic Press, London.

Bowen, G.D., and S.E. Smith. 1981. The effects of mycorrhizas on nitrogen uptake by plants. pp. 237–247. In: F.W. Clark and T. Rosswall (eds.), *Terrestrial nitrogen cycles: Processes, ecosystem strategies and management impacts*. Ecological Bulletin No 33, Swedish Natural Science Research Council, Stockholm.

Buwalda, J.G., G.J.S. Ross, D.P. Stribley, and P.B. Tinker. 1982. The development of endomycorrhizal root systems. III. The mathematical representation of the spread of vesicular–arbuscular mycorrhizal infection in root systems. *New Phytol*. 91:669–682.

Buwalda, J.G., D.P. Stribley, and P.B. Tinker. 1983. Increased uptake of anions by plants with vesicular–arbuscular mycorrhiza. *Plant Soil* 71:463–467.

Buwalda, J.G., D.P. Stribley, and P.B. Tinker. 1984. The development of endomycorrhizal root systems V. The detailed pattern of infection and the control of development of infection level by host in young leek plants. *New Phytol*. 96:411–427.

Callow, J.A., L.C.M. Capaccio, G. Parish, and P.B Tinker. 1978. Detection and estimation of polyphosphate in vesicular–arbuscular mycorrhizas. *New Phytol*. 80:125–134.

Capaccio, L.C.M., and J.A. Callow. 1982. The enzymes of polyphosphate metabolism in vesicular–arbuscular mycorrhizas. *New Phytol*. 91:81–91.

Chapin, F.S. III. 1980. The mineral nutrition of wild plants. *Annu. Rev. Ecol. Syst*. 11:233–260.

Clarkson, D.T. 1985. Factors affecting mineral nutrient acquisition by plants. *Annu. Rev. Plant. Physiol*. 36:77–115.

Cooper, K.M. 1984. Physiology of VA mycorrhizal associations. pp. 155–186. In: C.L. Powell and D.J. Bagyaraj (eds.), *VA mycorrhiza*. CRC Press, Boca Raton, FL.

Cox, G., F.J. Moran, F.E. Sanders, C. Nockolds, and P.B. Tinker. 1980. Translocation and transfer of nutrients in vesicular–arbuscular mycorrhizas. III. Polyphosphate granules and phosphorus translocation. *New Phytol*. 84:649–659.

Cox, G., F.E. Sanders, P.B. Tinker, and J.A. Wild. 1975. Ultrastructural evidence relating to host–endophyte transfer in a vesicular–arbuscular mycorrhiza. pp. 297–312 In: F.E. Sanders, B. Mosse, and P.B. Tinker (eds.), *Endomycorrhizas*. Academic Press, London.

Cress, W.A., G.O. Throneberry, and D.L. Lindsey. 1979. Kinetics of phosphorus absorption by mycorrhizal and non-mycorrhizal tomato roots. *Plant Physiol*. 64:484–487.

Daft, M.J., and T.H. Nicolson. 1966. Effect of *Endogone* mycorrhiza on plant growth. *New Phytol*. 65:343–350.

Dakessian, M.S., M.S. Brown, and G.J. Bethlenfalvay. 1986. Relationship of mycorrhizal growth enhancement and plant growth with soil water and texture. *Plant Soil*. 94:439–443.

Danso, S.K.A. 1988. The use of ^{15}N enriched fertilizers for estimating nitrogen fixation in grain and pasture legumes. pp. 245–358. In: D.P. Beck and L.A.

Materon (eds.), *Nitrogen fixation by legumes in Mediterranean agriculture*. ICARDA, and Martinus Nijhoff. Dordrecht.

Dehn, B., and H.W. Dehne. 1986. Development of VA mycorrhizal fungi and interactions with *Cochliobolus sativus* in roots of gramineae. pp. 773–778. In: V. Gianinazzi-Pearson and S. Gianinazzi (eds.), *Physiological and genetical aspects of mycorrhizae*. INRA, Paris.

Dehne, H.W. 1987. Improvement of the VA mycorrhiza status in agriculture and horticulture. pp. 817–830. In: *Transaction of the XIII. Congress of International Society of Soil Science, Hamburg*.

Dueck, T.A., P. Visser, W.H.O. Ernst, and H. Schat. 1986. Vesicular–arbuscular mycorrhizae decrease zinc-toxicity to grasses growing in zinc-polluted soil. *Soil Biol. Biochem*. 18:331–333.

Fitter, A.H., and R. Nichols. 1988. The use of benomyl to control infection by vesicular–arbuscular mycorrhizal fungi. *New Phytol*. 110:201–206.

Gardner I.C. 1986. Mycorrhizae of actinorrhizal plants. *MIRCEN J*. 2:147–160.

Gerschefske, D.K., B.A. Daniels-Hetrick, and G.W.T. Wilson. 1988. Relationship of soil fertility to suppression of the growth response of mycorrhizal big bluestem in non-sterile soil. *New Phytol*. 109:473–481.

Gianinazzi S., V. Gianinazzi-Pearson, and A. Trouvelot. 1988. Conceptual approaches for the rational use of VA endomycorrhizae in agriculture: Possibilities and limitations. pp. 39–40. In: *2nd European Symposium on Mycorrhizae, Prague*.

Gianinazzi-Pearson, V. 1984. Host–fungus specificity in mycorrhizae. pp. 225–253. In: D.P.S. Verma, and T.H. Hohn (eds.), *Genes involved in plant–microbe interactions*. Springer, Vienna.

Gianinazzi-Pearson, V., J.C. Fardeau, S. Asimi, and S. Gianinazzi. 1981. Source of additional phosphorus absorbed from soil by vesicular–arbuscular mycorrhizal soybean. *Physiol. Veg*. 19:33–43.

Gianinazzi-Pearson, V., and S. Gianinazzi. 1986. The physiology of improved phosphate nutrition in mycorrhizal plants. pp. 101–109. In: V. Gianinazzi-Pearson and S. Gianinazzi (eds.), *Physiological and genetical aspects of mycorrhizae*. INRA, Paris.

Habte, M., R.L. Fox, R. Aziz, and S.A. El-Swaify. 1988. Interaction of vesicular–arbuscular mycorrhizal fungi with erosion in an oxisol. *Appl. Environ. Microbiol*. 45:945–950.

Hardie, K. 1985. The effect of removal of extraradical hyphae on water uptake by vesicular–arbuscular mycorrhizal plants. *New Phytol*. 101:677–684.

Hardie, K., and L. Leyton. 1981. The influence of vesicular–arbuscular mycorrhiza on growth and water relations of red clover. I. In phosphate deficient soil. *New Phytol*. 89:599–608.

Harinikumar, K.M., and D.J. Bagyaraj. 1988. Effect of crop rotation on native vesicular arbuscular mycorrhizal propagules in soil. *Plant Soil*. 110:77–80.

Harley, J.L., and S.E. Smith. 1983. *Mycorrhizal symbiosis*. Academic Press. London.

Hartmond, U., N.V. Schaesberg, J.H. Graham, and JP. Syvertsen. 1987. Salinity and flooding stress effects on mycorrhizal and non-mycorrhizal citrus rootstock seedlings. *Plant Soil*. 104:37–43.

Hayman, D.S., 1975. Phosphorus cycling by soil micro-organisms and plant roots. pp. 67–91. In: N.D. Walker (ed.), *Soil microbiology*. Butterworths, London.

Hayman, D.S. 1982. Influence of soils and fertility on activity and survival of vesicular–arbuscular mycorrhizal fungi. *Phytopathology* 72:1119–1125.
Hayman, D.S. 1983. The physiology of vesicular–arbuscular endomycorrhizal symbiosis. *Can. J. Bot.* 61:944–963.
Hayman, D.S. 1984. The physiology of vesicular–arbuscular mycorrhiza. pp. 44–47. In: D.J. Thomson (ed.), *Forage legumes, British Grassland Society Occasional Symposium 16,* Hurley.
Hayman, D.S. 1986. Mycorrhizae of nitrogen-fixing legumes. *MIRCEN J.* 2:121–145.
Hayman, D.S., J.M. Day, and M. Dye. 1986. Preliminary observations on dual inoculation of white clover. pp. 467–473. In: V. Gianinazzi-Pearson and S. Gianinazzi (eds.), *Physiological aspects of mycorrhizae.* INRA Press, Paris.
Hayman, D.S., and B. Mosse. 1972. Plant growth responses to vesicular–arbuscular mycorrhiza. III. Increased uptake of labile P from soil. *New Phytol.* 71:41–47.
Haynes, R.J. 1980. Competitive aspects of the grass–legume association. pp. 227–261. In: N.C. Brady (ed.), *Advances in agronomy.* Academic Press, New York.
Haystead, A., N. Malajczuk, and T.S. Crove. 1988. Underground transfer of nitrogen between pasture plants infected with vesicular–arbuscular mycorrhizal fungi. *New Phytol.* 108:417–423.
Heichel, G.H. 1987. Legume nitrogen: Symbiotic fixation and recovery by subsequent crops. pp. 63–80. In: A.R. Helsel (ed.), *Energy in plant nutrition and pest control.* Elsevier, Amsterdam.
Hepper, C.M., C. Azcón-Aguilar, S. Rosendahl, and R. Sen. 1988. Competition between three species of *Glomus* used as spatially separated introduced and indigenous mycorrhizal inocula for leek *(Allium porrum L.). New Phytol.* 110:207–215.
Hirrel, M.C., and J.W. Gerdemann. 1980. Improved growth of onion and bell pepper in saline soils by two vesicular–arbuscular mycorrhizal fungi. *Proc. Soil Sci. Soc. Am.* 44:654–655.
Howeler, R.H., E. Sieverding, and S. Saif. 1987. Practical aspects of mycorrhizal technology in some tropical crops and pastures. *Plant Soil.* 100:249–283.
Jasper, D.A., L.K. Abbott, and A.D. Robson. 1989a. Soil disturbance reduces the infectivity of external hyphae of vesicular–arbuscular mycorrhizal fungi. *New Phytol.* 112:93–99.
Jasper, D.A., L.K. Abbott, and A.D. Robson. 1989b. Hyphae of a vesicular–arbuscular mycorrhizal fungus maintain infectivity in dry soil except when the soil is disturbed. *New Phytol.* 112:101–107.
Karunaratne, S., J.H. Baker, and A.V. Barker. 1986. Phosphorus uptake by mycorrhizal and nonmycorrhizal roots of soybean. *J. Plant Nutr.* 9:1303–1313.
Kessel, C.V., P.W. Singleton, and H.J. Hoben. 1985. Enhanced N-transfer from a soybean to maize by vesicular–arbuscular mycorrhizal (VAM) fungi. *Plant Physiol.* 79:562–563.
Khasawneh, F.E., and E.D. Doll. 1978. The use of phosphate rock for direct application to soils. pp. 159–206. In: N.C. Brady (ed.), *Advances in agronomy.* Academic Press, New York.
Koide, R.T., and M. Li. 1989. Appropriate controls for vesicular–arbuscular mycorrhiza research. *New Phytol.* 111:35–44.
Kucey, R.M.N., and R. Bonetti. 1988. Effect of vesicular–arbuscular mycorrhizal

fungi and captan on growth and N_2 fixation by *Rhizobium*-inoculated field beans. *Can. J. Soil Sci.* 68:143–149.

Kucey, R.M.N., H.H.K, Janzen, and M.E. Leggett. 1989. Microbially mediated increases in plant-available phosphorus. pp. 199–228. In: N.C. Brady (ed.), *Advances in agronomy*. Academic Press, New York.

Linderman, R.G. 1988. Mycorrhizal interactions with the rhizosphere microflora: The mycorrhizosphere effect. *Phytopathology* 78:366–371.

Lopez-Aguillon, R., and B. Mosse. 1987. Experiments on competitiveness of three endomycorrhizal fungi. *Plant Soil*. 97:155–170.

Malloch, D.W., K.A. Pirozynski, and P.H. Raven. 1980. Ecological and evolutionary significance of mycorrhizal symbioses in vascular plants. *Proc. Natl. Acad. Sci. U.S.A.* 77:2113–2118.

Manjunath, A., N.V. Hue, and M. Habte. 1989. Response of *Leucaena leucocephala* to vesicular–arbuscular mycorrhizal colonization and rock phosphate fertilization in an oxisol. *Plant Soil* 114:127–133.

Martin, J.K. 1985. A single plant technique for field studies of distribution of ^{32}P-labelled phosphate between plant and soil pools. *Plant Soil* 86:415–432.

Menge, J.A. 1984. Inoculum production. pp. 187–203. In: C.L. Powell and D.J. Bagyaraj (eds.), *VA mycorrhiza*. CRC Press, Boca Raton, FL.

Morton, J.B. 1988. Taxonomy of VA mycorrhizal fungi: Classification, nomenclature, and identification. *Mycotaxon* 32:267–324.

Mosse, B. 1973. Advances in the study of vesicular–arbuscular mycorrhiza. *Annu. Rev. Phytopathol.* 11:171–196.

Mosse, B. 1986. Mycorrhiza in a sustainable agriculture. *Biol. Agri. Hortic.* 3:191–209.

Mosse, B., D.S. Hayman, and D.J. Arnold. 1973. Plant growth responses to vesicular–arbuscular mycorrhiza. V. Phosphate uptake by three plant species from P-deficient soils labelled with ^{32}P. *New Phytol.* 72:809–815.

Mosse, B., D.P. Stribley, and F. Le Tacon. 1981. Ecology of mycorrhizae and mycorrhizal fungi. *Adv. Microb. Ecol.* 5:137–210.

Munns, D.N., and B. Mosse. 1980. Mineral nutrition of legume crops. pp. 115–125. In: R.J. Summerfield and A.H. Bunting (eds.), *Advances in legume science*. Her Majesty's Stationery Office, London.

Newman, E.I. 1985. Some factors affecting the abundance of mycorrhizas in grassland. pp. 63–68. In: S.K. Danso (ed.), *Nuclear techniques to study the role of mycorrhiza in increasing food crop production*. IAEA-TECDOC-338. International Atomic Energy Agency, Vienna.

Nicolson, T.H. 1975. Evolution of vesicular–arbuscular mycorrhizas. pp. 25–34. In: F.E. Sanders, B. Mosse, and P.B. Tinker (eds.), *Endomycorrhizas*. Academic Press, London.

Nye, P.H., and P.B. Tinker. 1977. *Solute movement in the soil–root system*. Blackwell, Oxford.

Ocampo, J.A., and D.S. Hayman. 1981. Influence of plant interactions of vesicular-arbuscular mycorrhizal infections. II. Crop rotations and residual effects of non-host plants. *New Phytol.* 87:333–343.

Ofori, F., and W.R. Stern. 1987. Cereal–legume intercropping systems. pp. 41–48. In: N.C. Brady (ed.), *Advances in agronomy*. Academic Press, New York.

Oliveira, V.L., and J. Garbaye. 1989. Les microorganismes auxiliares de l'etablissement des symbioses mycorhiziennes. *Eur. J. For. Pathol.* 19:54–64

Owusu-Bennoah, E., and A. Wild. 1979. Autoradiography of the depletion zone of phosphate around onion roots in the presence of vesicular–arbuscular mycorrhiza. *New Phytol.* 82:133–140.
Pauli, F.W. 1967. *Soil fertility. A biodynamical approach.* Adam Hilger, London.
Pearson, V., and P.B. Tinker. 1975. Measurement of phosphorus fluxes in the external hyphae of endomycorrhizas. pp. 277–287. In: F.E. Sanders, B. Mosse, and P.B. Tinker (eds.), *Endomycorrhizas.* Academic Press, London.
Peña, J.I., M. Sanchez-Díaz, J. Aguirreolea, and M. Becana. 1988. Increased stress tolerance of nodule activity in the *Medicago–Rhizobium–Glomus* symbiosis under drought. *J. Plant. Physiol.* 133:79–83.
Perrin, R. 1985. Peut-on computer sur les mycorrhizes pour lutter contre les maladies des plantes ligneuses? *Eur. J. For. Pathol.* 15:372–379.
Pirozynski, K.A., and D.W. Malloch. 1975. The origin of land plants: A matter of mycotrophism. *Biosystems* 6:153–164.
Plaut, Z., and C.M. Grieve. 1988. Photosynthesis of salt-stressed maize as influenced by Ca:Na ratios in the nutrient solution. *Plant Soil* 105:283–286.
Pond, E.C., J.A. Menge, and W.M. Jarrell. 1984. Improved growth of tomato in salinized soil by vesicular–arbuscular mycorrhizal fungi collected from saline soils. *Mycologia* 79:74–84.
Poss, J.A., E. Pond, J.A. Menge, and W.M. Jarrell. 1985. Effect of salinity on mycorrhizal onion and tomato in soil with and without additional phosphate. *Plant Soil* 88:307–319.
Powell, C.L. 1975. Plant growth responses to vesicular-arbuscular mycorrhiza. VIII. Uptake of P by onion and clover infected with different *Endogone* spore types in ^{32}P-labelled soil. *New Phytol.* 75:563–566.
Rai, R. 1988. Interaction response of *Glomus albidus* and *Cicer–Rhizobium* strains on iron uptake and symbiotic N_2 fixation in calcareous soil. *J. Plant. Nutr.* 11:863–869.
Rajapakse, S., D.A. Zubeber, and J.C. Miller, Jr. 1989. Influence of phosphorus level on VA mycorrhizal colonization and growth of cowpea cultivars. *Plant Soil* 114:45–52.
Rhodes, L.H., and Gerdemann, J.W. 1980. Nutrient translocation in vesicular–arbuscular mycorrhizae. pp. 173–195. In: C.B. Cooks, P.W. Pappas, and E.D. Rudolp (eds.), *Cellular interactions in symbiosis and parasitism.* Ohio State University Press, Columbus, OH.
Rogers, R.D., and S.E. Williams. 1986. Vesicular–arbuscular mycorrhiza: Influence on plant uptake of cesium and cobalt. *Soil Biol. Biochem.* 18:371–376.
Safir, G.R., and C.E. Nelsen. 1981. Water and nutrient uptake by vesicular–arbuscular mycorrhizal plants. pp. 25–81. In: R. Myers (ed.), *Mycorrhizal associations and crop production.* Rutgers University Press, New Brunswick.
Sanders, F.E. 1986. Quantitative approaches to the analysis of the development of mycorrhizal roots systems. pp. 209–216. In: V. Gianinazzi-Pearson and S. Gianinazzi (eds.), *Physiological and genetical aspects of mycorrhizae.* INRA, Paris.
Sanders, F.E., and N.A. Sheikh. 1983. The development of vesicular–arbuscular mycorrhizal infection in plant root systems. *Plant Soil* 71:223–246.
Sanders, F.E., and P.B. Tinker. 1971. Mechanism of absorption of phosphate from soil by *Endogone* mycorrhizas. *Nature* 233:278–279.
Sanders, F.E., and P.B. Tinker. 1973. Phosphate flow into mycorrhizal roots. *Pestic. Sci.* 4:385–395.

Sieverding, E., and J.M. Barea. 1990. Perspectivas de la inoculatión de sistemas de producción vegetal con hongos formadores de micorrizas VA. In: J.L. Gorgé, J.M. Barea, and J. Olivares (eds.), *Fijación y movilización biológica de nutrientes. Nuevas tendencias*. C.S.I.C., Madrid (in press).

Siqueira, J.O. 1987. Cultura axénica e monoxénica dos fungos micorrízicos vesículo–arbusculares. pp. 44–70. In: *II Reuniao Brasileira sobre Micorrizas, São Paulo*.

Skujins, J., and M.F. Allen. 1986. Use of mycorrhizae for land rehabilitation. *MIRCEN J.* 2:161–176.

Smith, F.A., and S.E. Smith. 1986. Movement across membrane: Physiology and biochemistry. pp. 75–84. In: V. Gianinazzi-Pearson and G. Gianinazzi (eds.), *Physiological and genetical aspects of mycorrhizae*. INRA, Paris.

Smith, F.A., S.E. Smith, B.J. St. John, and D.J.D. Nicholas. 1986. Inflow of N and P into roots of mycorrhizal and non-mycorrhizal onions. pp. 371–375. In: V. Gianinazzi-Pearson and S. Gianinazzi (eds.), *Physiological and genetical aspects of mycorrhizae*. INRA, Paris.

Smith, S.E. 1980. Mycorrhizas of autotrophic higher plants. *Biol. Rev.* 55:475–510.

Smith, S.E., and Gianinazzi-Pearson, V. 1988. Physiological interactions between symbionts in vesicular–arbuscular mycorrhizal plants. *Annu. Rev. Plant Physiol. Plant Mol. Biol.* 39:221–244.

Smith, S.E., B.J. St. John, F.A. Smith, and J.D. Nicholas. 1985. Activity of glutamine synthetase and glutamate dehydrogenase in *Trifolium subterraneum* L. and *Allium cepa* L: Effects of mycorrhizal infection and phosphate nutrition. *New Phytol.* 99:211–227.

Smith, S.E., and N.A. Walker. 1981. A quantitative study of mycorrhizal infection in *Trifolium:* Separate determination of the rates of infection and of mycelial growth. *New Phytol.* 89:225–240.

Son, C.L., and S.E. Smith. 1988. Mycorrhizal growth responses: interactions between photon irradiance and phosphorus nutrition. *New Phytol.* 108:305–314.

Son, C.L., F.A. Smith, and S.E. Smith. 1988. Effect of light intensity on root growth, mycorrhizal infection and phosphate uptake in onion (*Allium cepa* L.). *Plant Soil.* 111:183–186.

Stevenson, F.J. 1986. *Cycles of soil.* John Wiley & Sons, New York.

Sylvia, D.M., and M.E. Will. 1988. Establishment of vesicular–arbuscular mycorrhizal fungi and other microorganisms on a beach replenishment site in Florida. *Appl. Environ. Microbiol.* 54:348–352.

Thomas, R.S., S. Dakessian, R.N. Ames, M.S. Brown, and G.J. Bethlenfalvay. 1986. Aggregation of a silty clay loam soil by micorrhizal onion roots. *Soil Sci. Soc. Am. J.* 50:1494–1499.

Tinker, P.B. 1975. Soil chemistry of phosphorus and mycorrhizal effects on plant growth. In: pp. 353–371. F.E. Sanders, B. Mosse, and P.B. Tinker (eds.), *Endomycorrhizas*. Academic Press, London.

Tinker, P.B. 1980. The role of rhizosphere microorganisms in phosphorus uptake by plants. pp. 617–654. In: F. Kwasaneh and E. Sample (eds.), *The role of phosphorus in agriculture*. American Society of Agronomy, Madison.

Tinker, P.B. 1985. Modelling mycorrhizal development. pp. 140–141. In: R. Molina (ed.), *Proceedings of the 6th North American conference on mycorrhizae*. Fores Research Laboratory, Corvallis, OR.

Tinker, P.B., and A. Gildon. 1983. Mycorrhizal fungi and ion uptake. pp. 21–32.

In: D.A. Robb and W.S. Pierpoint (eds.), *Metals and micronutrients. Uptake and utilization by plants*. Academic Press, London.
Tisdall, J.M., and J.M. Oades. 1979. Stabilisation of soil aggregates by the root systems of ryegrass. *Aust. J. Soil Res.* 17:449–460.
Walker, N.A., and S.E. Smith. 1984. The quantitative study of mycorrhizal infection. II. The relation of rate of infection and speed of fungal growth to propagule density, the mean length of infection unit and the limiting value of the fraction of the root infected. *New Phytol.* 96:55–69.
Westheimer, F.H. 1987. Why nature chose phosphates. *Science*. 235:1173–1178.
White, J.A., L.C. Munn, and S.E. Williams. 1989. Edaphic and reclamation aspects of vesicular–arbuscular mycorrhizae in Wyoming red desert soils. *Soil. Sci. Soc. Am. J.* 53:86–90.
Wilson, J.M. 1984. Comparative development and interaction between vesicular–arbuscular mycorrhizal fungi. *New Phytol.* 97:413–426.
Young, C.C., T.C. Juang, and H.Y. Guo. 1986. The effect of inoculation with vesicular–arbuscular mycorrhizal fungi on soybean yield and mineral phosphorus utilisation in subtropical–tropical soils. *Plant Soil* 95:245–253.
Zambolim, L. 1987. Tolerancia de plantas micorrizadas a fitonematóides. pp. 103–125. In: *II Reuniao brasileira sobre micorrizas, Sao Paulo*.

Argillic Horizons in Modern Loess Soils in an Ustic Soil Moisture Regime: Comparative Studies in Forest–Steppe and Steppe Areas from Eastern Europe and the United States

A. Bronger

Contents

I. Introduction

The study of the general or at least regular relationships between the distribution of soils (or, more exactly, the different categories distinguished by soil classification systems) and the soil-forming factors of climate, vegetation and fauna, parent material, relief (with additional surface runoff), man, and time (Jenny, 1941, 1980) is the subject of *"general soil geography."* *"Regional soil geography"* is based on this. We understand regional soil geography to mean the investigation of the causal relationships between the soil-forming factors and the regular spatial distribution pattern of soils. The expression "structure of the soil cover" is commonly

Advances in Soil Science, Volume 15

used for this regular spatial distribution pattern (Fridland, 1967, 1974; Haase and Schmidt, 1970). Thus, in their subject matter these two fields far exceed a mere description of the spatial distribution of the different soils on the earth.

More than 100 years ago Docucaëv (1883) recognized the *zonal* principle in the structure of the soil cover of the eastern European plains (cf. Ehwald, 1984). By this we mean a regular sequence of soils with the same or very similar properties in several successive vegetation zones following each other from north-northwest to south-southeast. This zonal arrangement is the result of a rapid decrease in precipitation, at least in the deciduous forest, forest–steppe, and steppe zones, and a simultaneous increase in temperature, accompanied by an increase in potential evapotranspiration (Figures 3 and 4; for locations see Figure 1). From this it is evident that climate and the closely related potential natural vegetation dominate the process of soil development here. In addition, the parent material, in the form of fine-grained loose calcareous sediments, does not vary greatly, on the whole. The other soil-forming factors cause only specific, albeit in individual cases often not inconsiderable, deviations from this zonal sequence.

In the western Central Lowland* and the Great Plains* of the United States as well, there is a pronounced zonal sequence of climate and with it vegetation, from deciduous forest to forest–steppe or forest–prairie (Knapp, 1965), tall-grass steppe (prairie), long-grass steppe (prairie), and desert–steppe or semidesert. In the United States, however, the zones follow each other from east to west as a result of a very pronounced decrease in precipitation in the Great Plains. When making a comparison one must additionally take into consideration that the northern boundary of the American Great Plains lies at the same geographic latitude as the central part of the Ukraine, although in the western and even in the central part of the European USSR the additional warming effect of the Gulf Stream can still be felt, though in a very weakened form. These differences in the zonal sequence are presumably an important reason for the fact that the steppe regions of the European USSR are on the whole comparable to those of the northern and central Great Plains (up to and including Nebraska), whereas it is very difficult to compare the Udolls of the United States with the forest–steppe soils of eastern Europe (cf. Sections II.B–C).

Since large parts of both regions are covered by loess, the soil-forming factor parent material is not so important when we are comparing the soil sequences. In both regions the distribution of soils is quite well known (cf. Aandahl, 1972, 1982; USDA, 1960; Ivanova et al., 1962; *Fiziko-Geograficheskiy Atlas Mira,* 1964), but few attempts have been made to com-

---

*Physical divisions according to the U.S. Geological Survey (Fenneman, 1931, 1938; Hunt, 1974).

pare the soils of these two steppe regions in order to interpret the differences in structure and genesis of the soil cover (Florov, 1929; Stremme, 1930; Fridland and Erokhina, 1963).

Out of this entire complex we single out one special aspect: In which soils of the climatic sequence from forest to dry steppe does the process of clay illuviation take place? In the southeastern part of Central Europe, recent clay illuviation can be observed micromorphologically only in loess soils under forest but no longer in areas in which forest–steppe is the potential natural vegetation (Bronger, 1976; cf. section III.A). Does the process of clay illuviation in the loess soils of eastern Europe also take place only under forest, or does it also take place in the forest–steppe or even the steppe? In the Central Lowland of the United States not only forest soils (Udalfs) and soils of the forest–prairie, i.e., some of the Udolls, have a ≥20% higher clay content in the subsoil, which is then referred to as the argillic horizon. Especially in the Great Plains, including the short-grass prairie, Ustolls (formerly Chernozems and Chestnut soils) and even Aridisols with an argillic horizon are widespread (Aandahl, 1972, 1982; cf. Figure 2). How are we to explain these differences in the structure and genesis of the soil cover? Can the higher clay content in the subsoil compared with the A horizon even in the Ustolls and Aridisols, which leads to the designation of the argillic horizon as a diagnostic horizon, be explained by clay illuviation alone? Might there be other possible explanations?

As we show in section II, the argillic horizon, and with it the process of clay illuviation, is of the utmost importance for a comparison of soil classifications and thus for a comparison of the soil genesis and soil geography of the forest–steppe and steppe regions of the USSR and the United States (Section III).

II. Geographic Comparison of the Forest–Steppe and Steppe Regions of Eastern Europe and the United States

A. The Argillic Horizon as a Diagnostic and Genetic Soil Horizon

The argillic horizon is defined in *Soil Taxonomy* (Soil Survey Staff, 1975) as the key *diagnostic* horizon in the orders of Alfisols and Ultisols as well as in the Argids, one of the two suborders of the Aridisols. It is also the key diagnostic horizon for several great groups in the order of Mollisols, not only in the udic but especially in the ustic soil moisture regime. The latter are the main subject of this chapter.

The argillic horizon is considered *genetically* to be a subsurface horizon "in which layer-lattice silicate clays have accumulated by illuviation to a significant extent" (Soil Survey Staff, 1975; Holzhey et al., 1973). The most important criterion, which evidently is generally considered sufficient, is an increase in clay content of ≥20% compared with the overlying horizon, if this has 15% to 40% total clay. Also, in many cases macroscopically visible clay cutans and a higher <0.2 μm : <2 μm ratio than in the overlying or underlying horizon are considered to be additional criteria (e.g., Anderson, 1987, Fig. 10). If a soil has a lithologic discontinuity between the eluvial and the argillic horizon, thin sections should show that some parts of the horizon have about 1% or more of oriented clay bodies (Soil Survey Staff, 1975, 1987).

Because of the strong *genetic* bias of the argillic horizon (Cline and Johnson, 1963; Smith, 1983), and also because a lithologic discontinuity is often difficult to exclude, *micromorphology* is essential for identifying clay illuviation (Bronger, 1966, 1976, 1978; Bullock and Thompson, 1985). In particular, microlaminated fine clay plasma or illuviation argillans (Brewer, 1964; Bullock et al., 1985) are unequivocal signs of clay illuviation. This is particularly conspicuous in Holocene but also in Pleistocene soils derived from loess, where the fine clay plasma, or the illuviation argillans, contrasts sharply with the rest of the fabric. This was already pointed out by Kubiena (1956: his "Braunlehm–Teil plasma"; also cf. Smith, 1986, p. 87). For this reason our thoughts and investigations concentrate primarily on soils derived from loess or loess-like parent material in both eastern Europe and the United States.

Even when the illuviation argillans are transformed by changing environmental conditions (e.g., Smolikova, 1967), they still remain visible as a very stable element of the fabric. They are even found in the excrement of earthworms (e.g., Zachariae, 1964; Bronger, 1976, photo 3) or in "Fließerde" material from a buried mid-Pleistocene Udalf (Bronger, 1969/70, Figure 2A,B). Other features of illuvial clay may be less clear; interpretations are subjective (McKeague, 1983; Bullock and Thompson, 1985). Different micromorphologists estimate different amounts of illuvial clay when analyzing the same thin section (McKeague et al., 1978, 1980, 1981), which is a problem for the abovementioned 1% criterion. Therefore, Murphy and Kemp (1987) suggest an increase to 2% illuvial clay in a thin section to separate argillic from nonargillic horizons. Nettleton et al. (1969) assumed, however, that in some argillic horizons where no illuvial clay coatings could be observed in thin sections, but where clay illuviation was nevertheless postulated, that the missing illuviation argillans ("clay skins") were never formed because of significant shrink–swell potentials; this assumption appears questionable (cf. section IV). The clay material in these cases shows only oriented birefringence caused by argillipedoturbation (cf. Hole, 1961). Bullock and Thompson (1985) emphasize the overall identification of stress-oriented versus illuvial clay as a remaining problem. In our opinion it is necessary to separate the basically climatogenic (see below) process of clay illuviation, in the form of illuviation argillans, from the process of argillipedoturbation, which is predominantly a lithogenic process determined by the soil material (for more about this problem, cf. Bronger, 1977).

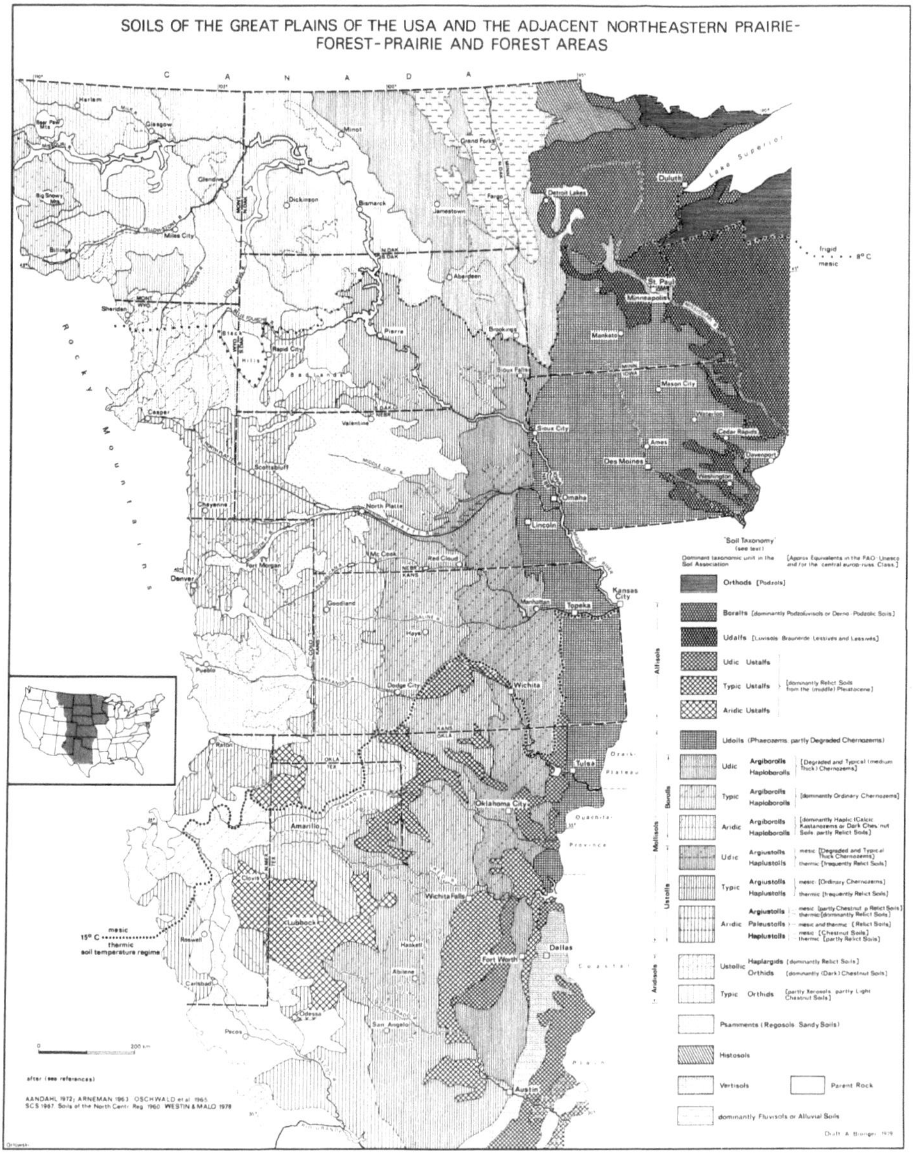
SOILS OF THE GREAT PLAINS OF THE USA AND THE ADJACENT NORTHEASTERN PRAIRIE-FOREST-PRAIRIE AND FOREST AREAS
C A N A D A
Rocky Mountains
Lake Superior
Minot
Grand Forks
Fargo
Jamestown
Bismarck
Dickinson
Glendive
Miles City
Billings
Glasgow
Harlem
Duluth
Detroit Lakes
St. Paul
Minneapolis
Mankato
Aberdeen
Brookings
Pierre
Rapid City
Sheridan
Sioux Falls
Mason City
Sioux City
Ames
Des Moines
Cedar Rapids
Davenport
Washington
Omaha
Lincoln
Casper
Valentine
Scottsbluff
Cheyenne
North Platte
Mc Cook
Red Cloud
Denver
Fort Morgan
Goodland
Hays
Manhattan
Topeka
Kansas City
Pueblo
Dodge City
Wichita
Raton
Tulsa
Oklahoma City
Amarillo
Clovis
Lubbock
Wichita Falls
Roswell
Haskell
Abilene
Fort Worth
Dallas
Carlsbad
Odessa
San Angelo
Pecos
Austin
frigid
8° C
mesic
mesic
15° C
thermic
soil temperature regime
0
200 km
after (see references)
AANDAHL 1972; ARNEMAN 1963; OSCHWALD et al. 1965
SCS 1967; Soils of the North Centr. Reg. 1960; WESTIN & MALO 1978
'Soil Taxonomy'
(see text)
Dominant taxonomic unit in the Soil Association
[Approx. Equivalents in the FAO-Unesco and for the central europ.-russ. Class.]
Orthods [Podzols]
Boralfs [dominantly Podzoluvisols or Derno-Podzolic Soils]
Udalfs [Luvisols, Braunerde, Lessivés and Lessivés]
Udic Ustalfs
Typic Ustalfs [dominantly Relict Soils from the (middle) Pleistocene]
Aridic Ustalfs
Alfisols
Udolls (Phaeozems, partly Degraded Chernozems)
Udic Argiborolls Haploborolls [Degraded and Typical (medium Thick) Chernozems]
Typic Argiborolls Haploborolls [dominantly Ordinary Chernozems]
Aridic Argiborolls Haploborolls [dominantly Haplic (Calcic) Kastanozems or Dark Ches nut Soils, partly Relict Soils]
Borolls
Udic Argiustolls Haplustolls mesic [Degraded and Typical Thick Chernozems] thermic [frequently Relict Soils]
Typic Argiustolls Haplustolls mesic [Ordinary Chernozems] thermic [frequently Relict Soils]
Aridic Argiustolls Paleustolls Haplustolls mesic [partly Chestnut, p. Relict Soils] thermic [dominantly Relict Soils] mesic and thermic [Relict Soils] mesic [Chestnut Soils] thermic [partly Relict Soils]
Ustolls
Mollisols
Ustollic Haplargids [dominantly Relict Soils] Orthids [dominantly (Dark) Chestnut Soils]
Typic Orthids [partly Xerosols, partly Light Chestnut Soils]
Aridisols
Psamments (Regosols, Sandy Soils)
Histosols
Vertisols
Parent Rock
dominantly Fluvisols or Alluvial Soils
Draft A. Bronger 1979

Figure 2

B. Problems in Comparing the Russian Soil Classifications with That of FAO–Unesco and Soil Taxonomy

When we attempt to correlate the soils in the Russian classifications with those of the FAO–Unesco and *Soil Taxonomy,* several difficulties arise. Until recently several Russian soil classifications have evidently existed simultaneously (among others, Rozov and Ivanova, 1967a,b; Kovda et al., 1969; Kovda, 1975; Gavriluk, 1979; Liverovskiy, 1969; Glazovskaya, 1983/1984; Fridland, 1981), but these do not differ greatly from one another according to Volobuyev (1980). Above all, they hardly differ from the classification of soils in Ivanova et al. (1962) and in the *Fiziko-Geograficheskiv Atlas Mira* (1964), so that this classification was used in Figure 1 (cf. below), though in a partially simplified form. No attempt has been made on the part of the Russians to correlate these systems with *Soil Taxonomy,* or rather such a possibility was already assessed quite critically at a very early time (Gerasimov, 1969). Occasional attempts have been made at a correlation with parts of the FAO–Unesco classification (Volobuyev, 1976; Gerasimov, 1980; Rozanov, 1982). This is sometimes quite helpful for a comparison with *Soil Taxonomy* (e.g., in the case of the Greyzems; cf. below), since both classifications use the same diagnostic horizons, although they are occasionally more precisely defined in *Soil Taxonomy.*

In *Soil Taxonomy,* however, an attempt was made (Soil Survey Staff, 1975, pp. 443–455) at a correlation with the "Soviet taxonomy" of Rozov and Ivanova (1967a; also cf. Ivanova and Rozov, 1967). In this, the comparison of "Soviet temperature units" ($\Sigma T° > 10°C$) with the soil temperature regimes in *Soil Taxonomy* was not completely successful; therefore, the correlations between, e.g., the Derno- (Sod-)Podzolic soils and the Gray Forest soils and the categories in *Soil Taxonomy* are unsatisfying. For this reason the boundary between the frigid and the mesic soil temperature regime is shown in Figure 1. In contrast to the Ukraine, in central Kazakhstan it already lies in the region of the Aridisols (also cf. Figures 3 and 4; see below and Section II.C). On the other hand, only "groups" of soils were compared and not "subgroups," such as, for instance, those of the Chernozems (Rozov and Ivanova, 1967b). The further subdivision of the Chernozems (see below; cf. Figure. 1), which is tantamount to the subdivision of the Borolls or Ustolls, is, however, particularly important for the question of clay illuviation in the ustic soil moisture regime.

In a comparison of the Russian classifications, especially with *Soil Taxonomy,* many more problems arise from the fact that in Russian soil science even today, the process of podzolization is interpreted much more broadly than in Central or Western European and American concepts of soil genesis and often even includes the process of clay illuviation. This is true not only of the *Derno-Podzolic* soils (also Sod-Podzolic or Soddy-Podzolic soils in the map in Kovda et al., 1975), the very widespread climaphytomorphic soils (cf. Schroeder, 1984) of the European–southwestern Siberian mixed forest region (Figure 1). Thus, according to Gerasimov (1975), in these soils the process of podzolization takes place along with other processes but not that of "illimerization," which has been synonymous with the process of clay illuviation since Fridland (cf. Table 3 in Gerasimov, 1975).

In contrast, Kundler (1959) had already shown that at least in the southern part of the eastern European mixed forest zone—the profile he investigated lies in the environs of Moscow under a *Picea–Quercus robur* forest—it is above all the transport of trimorphic silicate clay minerals that contributes to the differentiation of texture. For this reason, Kundler sees a close relationship between the Derno-Podzolic soils and the *Sols lessivés* in western Europe despite the differences in climate. Breburda et al. (1972) came to very similar conclusions (also cf. Breburda, 1966). Yarilova and Rublina (1975) arrived independently of the abovementioned authors at the conclusion that clay illuviation is the most important process in Derno-Podzolic soils; Tonkogonov and Rubilina (1980) even decided "that podzolization does not take place in the profile of Sod-Podzolic soils at the present time." Although these results were attained more in the southern part of the range of these soils, it appears justifiable to correlate the Derno-Podzolic soils with the Cryoboralfs (in part Glossoboralfs) in the northern and eastern parts of their range and the Eutroboralfs in the southwestern part. Only small areas in the northern part of the Ukraine can be classified as Udalfs (cf. Figure 1). They correspond to the Podzoluvisols in the FAO–Unesco classification.

The genesis of the *Gray Forest soils*, the climaphytomorphic soils (cf. above) of the deciduous forest zone and under forest in the forest–steppe (cf. Sections II.D and III.A), has been discussed extensively since Docucaev (1883). This discussion was recently summarized by Ivanova et al. (1969, 1970) and especially by Akhtyrtsev (1979). Until recently several authors assumed that the process of podzolization takes place in addition to clay illuviation (e.g., Lebedeva, 1969; Gerasimov, 1975; Shubina et al., 1979). On the other side, the Gray Forest soils are correlated with the Greyzems of the FAO–Unesco classification, which are, however, defined by a mollic A horizon, among other things. Thus, they should be classified especially among the Albic, Udic, and Boralfic Udic subgroups of the Argiborolls. Small areas, particularly of the Dark Gray Forest soils of the forest–steppe in the Ukraine, already have a mesic soil temperature regime, however, so they should be classified as Argiudolls, which again most nearly correspond to the Luvic Phaeozems (cf. Figure 1 and Sections II.D and III.A).

But even up to the present, in all abovementioned Russian classifications, the process of podzolization is extended to some of the Chernozems; i.e., "Podzolized" Chernozems are distinguished from "Leached," "Typical Thick," "Ordinary," and "Southern" Chernozems as climaphytomorphic soils (cf. above) along with "Meadow" Cher-

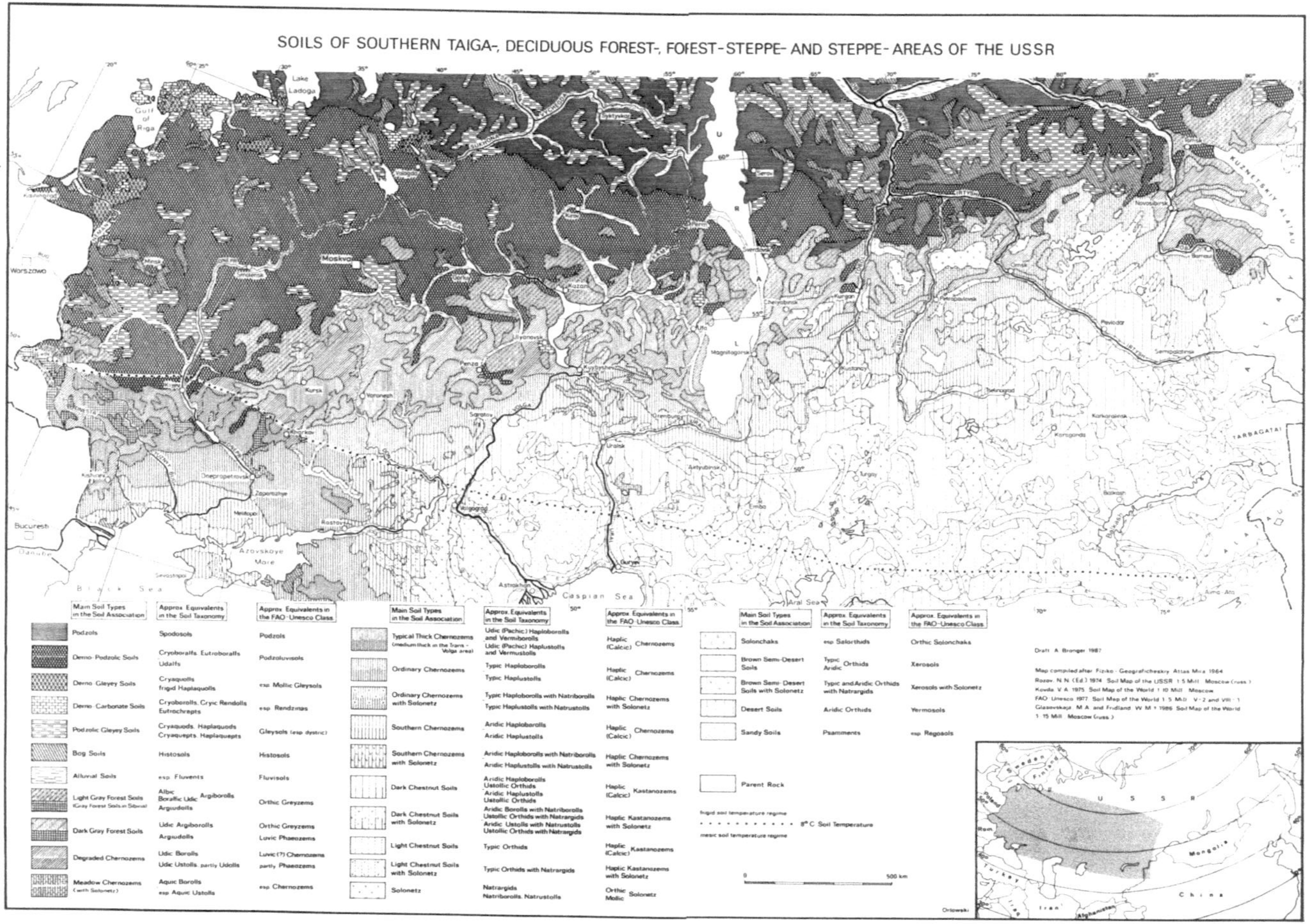
SOILS OF SOUTHERN TAIGA-, DECIDUOUS FOREST-, FOREST-STEPPE- AND STEPPE-AREAS OF THE USSR
Main Soil Types in the Soil Association
Approx. Equivalents in the Soil Taxonomy
Approx. Equivalents in the FAO-Unesco Class.
Podzols
Spodosols
Podzols
Demo-Podzolic Soils
Cryoboralfs, Eutroboralfs Udalfs
Podzoluvisols
Demo-Gleyey Soils
Cryaquolls frigid Haplaquolls
esp. Mollic Gleysols
Demo-Carbonate Soils
Cryoborolls, Cryic Rendolls Eutrochrepts
esp. Rendzinas
Podzolic Gleyey Soils
Cryaquods, Haplaquods Cryaquepts, Haplaquepts
Gleysols (esp. dystric)
Bog Soils
Histosols
Histosols
Alluvial Soils
esp. Fluvents
Fluvisols
Light Gray Forest Soils (Gray Forest Soils in Sibiria)
Albic Boralfic Udic Argiborolls Argiudolls
Orthic Greyzems
Dark Gray Forest Soils
Udic Argiborolls Argiudolls
Orthic Greyzems Luvic Phaeozems
Degraded Chernozems
Udic Borolls Udic Ustolls, partly Udolls
Luvic(?) Chernozems partly Phaeozems
Meadow Chernozems (with Solonetz)
Aquic Borolls esp. Aquic Ustolls
esp. Chernozems
Typical Thick Chernozems (medium thick in the Trans-Volga area)
Udic (Pachic) Haploborolls and Vermiborolls Udic (Pachic) Haplustolls and Vermustolls
Haplic (Calcic) Chernozems
Ordinary Chernozems
Typic Haploborolls Typic Haplustolls
Haplic (Calcic) Chernozems
Ordinary Chernozems with Solonetz
Typic Haploborolls with Natriborolls Typic Haplustolls with Natrustolls
Haplic Chernozems with Solonetz
Southern Chernozems
Aridic Haploborolls Aridic Haplustolls
Haplic (Calcic) Chernozems
Southern Chernozems with Solonetz
Aridic Haploborolls with Natriborolls Aridic Haplustolls with Natrustolls
Haplic Chernozems with Solonetz
Dark Chestnut Soils
Aridic Haploborolls Ustollic Orthids Aridic Haplustolls Ustollic Orthids
Haplic (Calcic) Kastanozems
Dark Chestnut Soils with Solonetz
Aridic Borolls with Natriborolls Ustollic Orthids with Natrargids Aridic Ustolls with Natrustolls Ustollic Orthids with Natrargids
Haplic Kastanozems with Solonetz
Light Chestnut Soils
Typic Orthids
Haplic (Calcic) Kastanozems
Light Chestnut Soils with Solonetz
Typic Orthids with Natrargids
Haplic Kastanozems with Solonetz
Solonetz
Natrargids Natriborolls, Natrustolls
Orthic Mollic Solonetz
Solonchaks
esp. Salorthids
Orthic Solonchaks
Brown Semi-Desert Soils
Typic Aridic Orthids
Xerosols
Brown Semi-Desert Soils with Solonetz
Typic and Aridic Orthids with Natrargids
Xerosols with Solonetz
Desert Soils
Aridic Orthids
Yermosols
Sandy Soils
Psamments
esp. Regosols
Parent Rock
frigid soil temperature regime
8° C Soil Temperature
mesic soil temperature regime
0
500 km
Draft: A. Bronger 1987
Map compiled after: Fiziko - Geograficheskiy Atlas Mira 1964
Rozov, N.N. (Ed.) 1974: Soil Map of the USSR 1:5 Mill. Moscow (russ.)
Kovda, V.A. 1975: Soil Map of the World 1:10 Mill. Moscow
FAO-Unesco 1977: Soil Map of the World 1:5 Mill. V-2 and VIII-1
Glasovskaja, M.A. and Fridland, W.M. 1986: Soil Map of the World 1:15 Mill. Moscow (russ.)
Orlowski
Moskva
Kazan
Lake Ladoga
Gulf of Riga
Black Sea
Azovskoye More
Caspian Sea
Aral Sea
Warszawa
Bucuresti
Danube
Dnepropetrovsk
Zaporozhye
Rostov
Voronezh
Kursk
Penza
Ulyanovsk
Saratov
Volgograd
Astrakhan
Kuybyshev
Orenburg
Uralsk
Guryev
Aktyubinsk
Magnitogorsk
Chelyabinsk
Kurgan
Petropavlovsk
Pavlodar
Semipalatinsk
Novosibirsk
Barnaul
Karaganda
Tselinograd
Alma-Ata
KUZNETSKIY ALATAU
TARBAGATAI
U S S R
Mongolia
China
Iran
Turkey
Sweden
Finland
Afghanistan

Figure 1

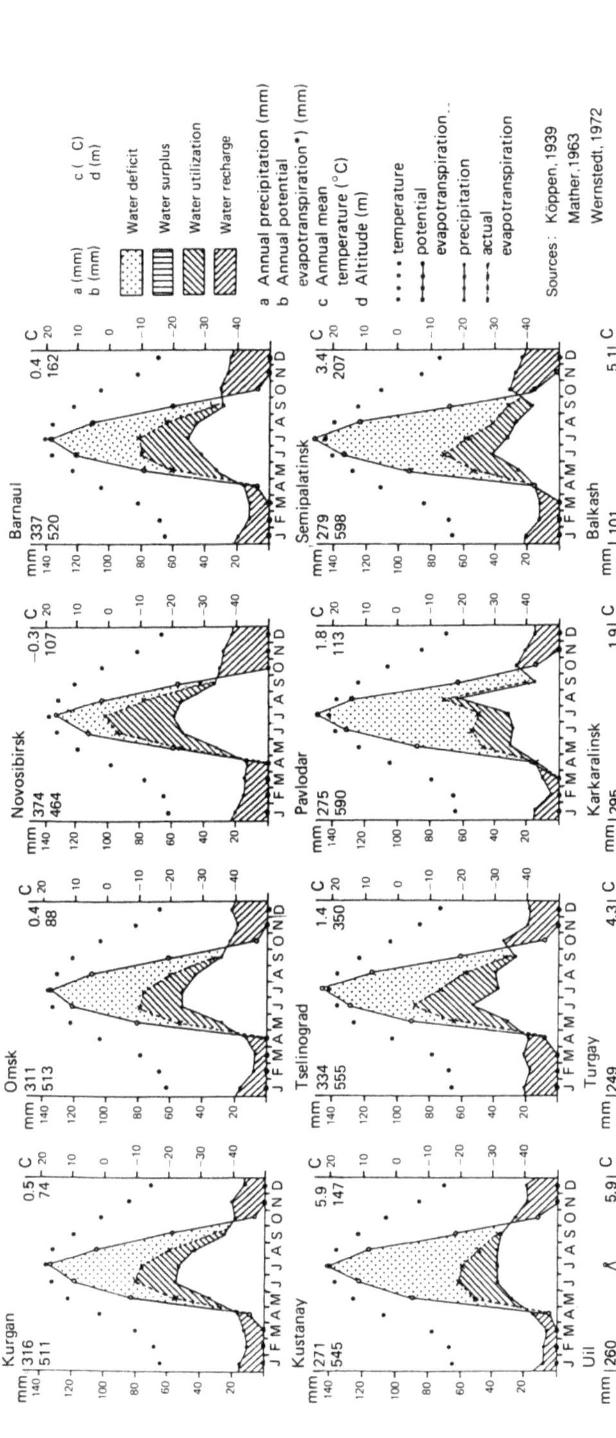

Figure 3. Climatic data and soil water balance of stations in the forest–steppe, tall-grass steppe, and short-grass steppe to semidesert in the west-Asian part of the USSR (cf. Figure 1).

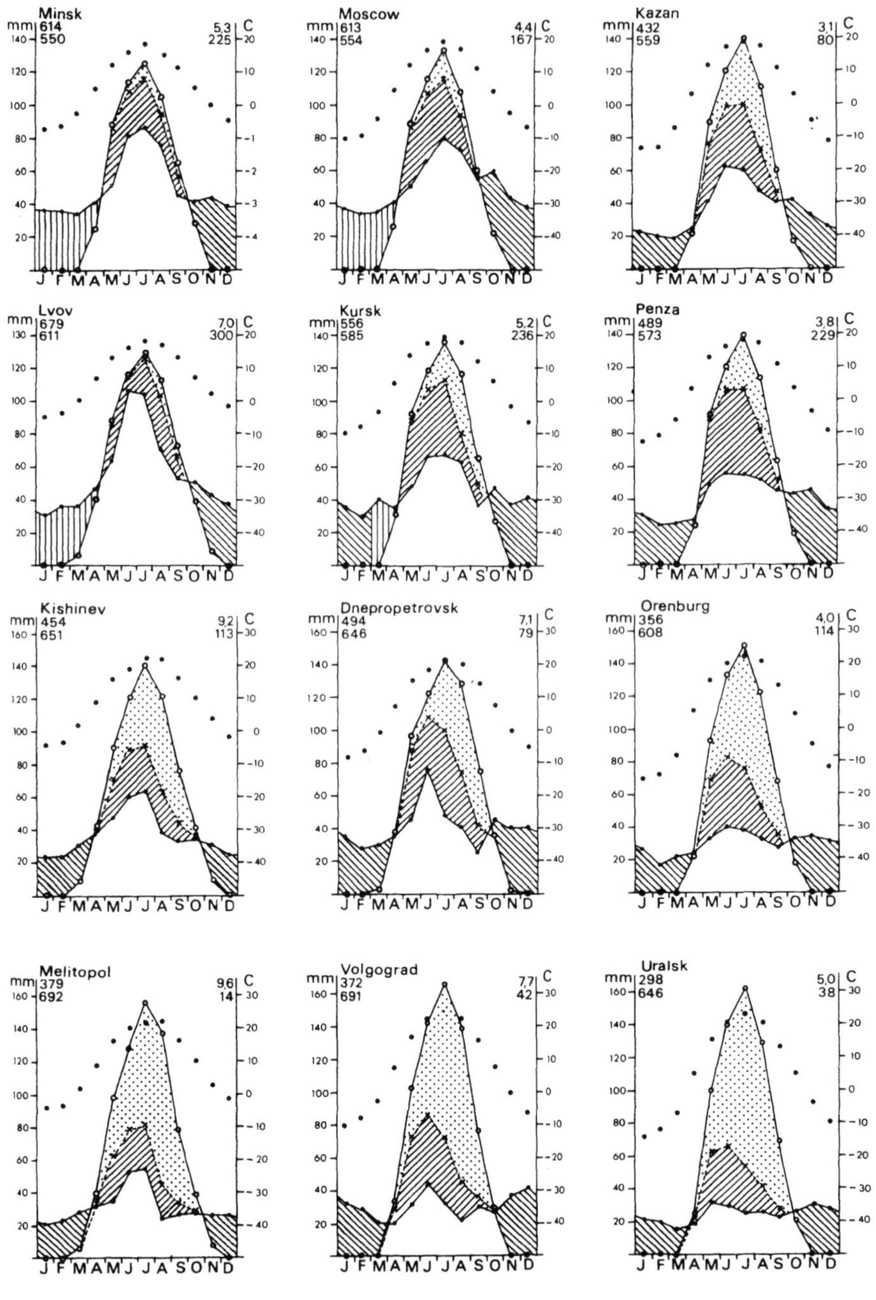

Figure 4. Climatic data and soil water balance of stations in the mixed forest, forest–steppe, tall-grass steppe, and short-grass steppe in the European part of the USSR (cf. Figures 1 and 5; for explanation see Figure 3).

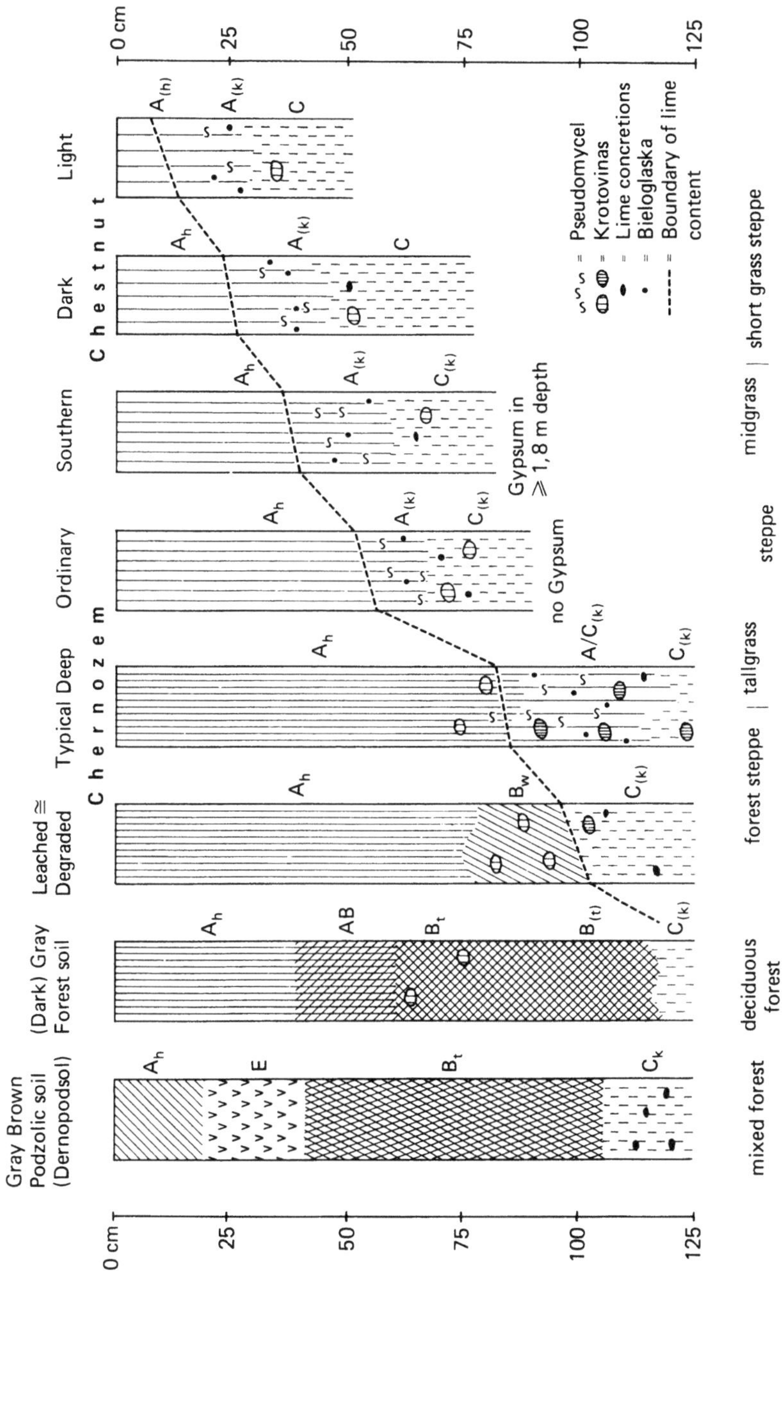

Figure 5. Climatic (bio-) sequences of loess soils in eastern Europe.

nozems. According to Rozov and Ivanova (1967b, Table 2), the pH of the water extract in the "Podzolized" Chernozems in the "east European facies" lies between 5.5 and 6.0, and in the "warm Podzolized (deep mycelial carbonate)" Chernozems of the "southern European facies" between 6.0 and 6.5. For our comparison of the soil classifications, the "Podzolized" and "Leached" Chernozems are combined with the Degraded Chernozems (cf. Figure 1), a term that is very often used synonymously with Leached Chernozems in the Russian literature.

Crucial for the question of clay illuviation in the ustic soil moisture regime is that clay illuviation has been demonstrated (Yarilova, 1972; cf. Section III.A) or assumed (e.g., Fridland, 1981) *only* in the soils that are combined here under the term Degraded Chernozem, which lie partly in the udic and partly in the ustic soil moisture regime (compare Figure 1 with Figures 2 and 4). According to Gerasimov (1975), clay illuviation no longer takes place in "Leached" Chernozems, in contrast to the "Podzolized" Chernozems. Later, however, he correlates (1980, Table 4) the "Leached" Chernozems with the Luvic Chernozems of the FAO–Unesco classification. On the other hand, to classify the Luvic Chernozems and the Luvic Phaeozems with the "Podzolized" Chernozems of the Russian classification (FAO–Unesco, 1974, p. 17) is problematic, to say the least.

In the Typical Thick, Ordinary, and Southern Chernozems and the Dark and Light Chestnut soils, (Figure 5) clay illuviation is not mentioned among the soil-forming processes by Andreyeva and Redkov (1983), Borovskij and Uspanov (1972), or Durasov and Almayev (1970) for Kasakhstan or by Valkov and Kryshchenko (1973), Ponomareva (1974), or Fridland (1981). On the basis of micromorphological investigations, other authors expressly rule out the possibility of clay illuviation in these Chernozems (Yarilova, 1974; Yarilova and Bystritskaya, 1976; Samoilova and Markeyeva, 1979; Gerasimov, 1980; Glasovskaya, 1983, Vol. I, p. 149). Volobuyev (1976) writes that only Haplic and Calcic Kastanozems can be correlated with the Chestnut soils of the USSR, but not the Luvic Kastanozems (also cf. FAO–Unesco, 1974, p. 17). For this reason (cf. Figure 1), the Typical Thick, Ordinary, and Southern Chernozems are correlated with the Udic (Pachic), Typic, and Aridic *Haplo*borolls or—in the mesic soil temperature regime—with corresponding *Hapl*ustolls. The Dark Chestnut soils are correlated with Aridic *Haplo*borolls, Aridic *Hapl*ustolls, or Ustollic *Orthids*, and the Light Chestnut soils with the Typic *Orthids*.

The age of the Chernozems and Chestnut soils in the European portions of the USSR is given as maximally 11,000 to 12,000 ^{14}C years (Rubilin et al., 1974; Rubilin and Kozyreva, 1980). According to Zolotun (1974), even in the short-grass steppe of the southern Ukraine in the region of the Southern Chernozems and Dark Chestnut soils, there was no loess sedimentation any more in the "Middle and Upper Holocene." This is very important for a comparison with the soils of the short-grass prairie in the central and northern Great Plains of the United States (cf. Sections III.B and IV).

C. Clay Illuviation in Soils of the Great Plains?

As we briefly mentioned in Section I, in the western Central Lowland, and particularly in the Great Plains of the United States, there is a rapid decrease in precipitation in an east–westerly direction (cf. Figure 6). This results in a pronounced sequence of vegetation zones from oak–hickory forests via a transitional region of forest and prairie (called "forest–prairie" by Knapp, 1965, on the basis of the term forest–steppe), tall-grass prairie, mixed-grass prairie, to short-grass prairie. This changes over into semidesert, semiscrub formations in some parts of the western Great Plains: in the south with mesquite (introduced by man) and in the central and northern part of the western Great Plains predominantly with *Artemisia tridentata*. The generally exact knowledge of the distribution of the original vegetation formations of this region, which probably cannot be reconstructed in such detail in any other agricultural area of this expanse, has resulted from the fortunate circumstance that thorough geodetic, cartographic, and vegetation surveys were done before the land was divided up among the farmers and ranchers coming from the East. These surveys are utilized in Shantz and Zon's (1924) map of natural vegetation and later in Küchler's (1964) much more precise map of potential natural vegetation, although particularly for the transitional region of the "mixed prairie," this map can only reproduce an average condition (cf. Küchler, 1972). For individual states there are even much more detailed maps of the natural vegetation, e.g., Minnesota (Marschner, 1930).

For a great majority of the counties in the states of the Great Plains, especially up to the western boundary of North and South Dakota, Nebraska, Kansas, Oklahoma, and Texas, there are soil maps, 1:20,000, covering the whole area; in the regions farther to the west up to the foot of the Rocky Mountains, the detailed surveys have not progressed as far. These are combined for the county in question in a map, generally on a scale of 1:200,000, at the taxonomic level of the soil series associations. On this basis, Aandahl (1972) drew up a very detailed map, 1:2,500,000, of the Great Plains on which soil associations (a total of almost 200!) are represented at the next higher level of abstraction, the soil families and, above all, the subgroups. All soil associations were later (Aandahl, 1982) described in more detail on the basis of the soil series (pp. 6–121); 70 particularly important soil series including colored soil profiles and landscape pictures with the range of each are presented in detail (pp. 133–272).

This map is the basis for Figure 2 for the region of the Great Plains. Very slight changes were made in the western part of South Dakota on the basis of a map in Westin and Malo (1978). Because of the degree of generalization necessary, associations of subgroups had to be combined.

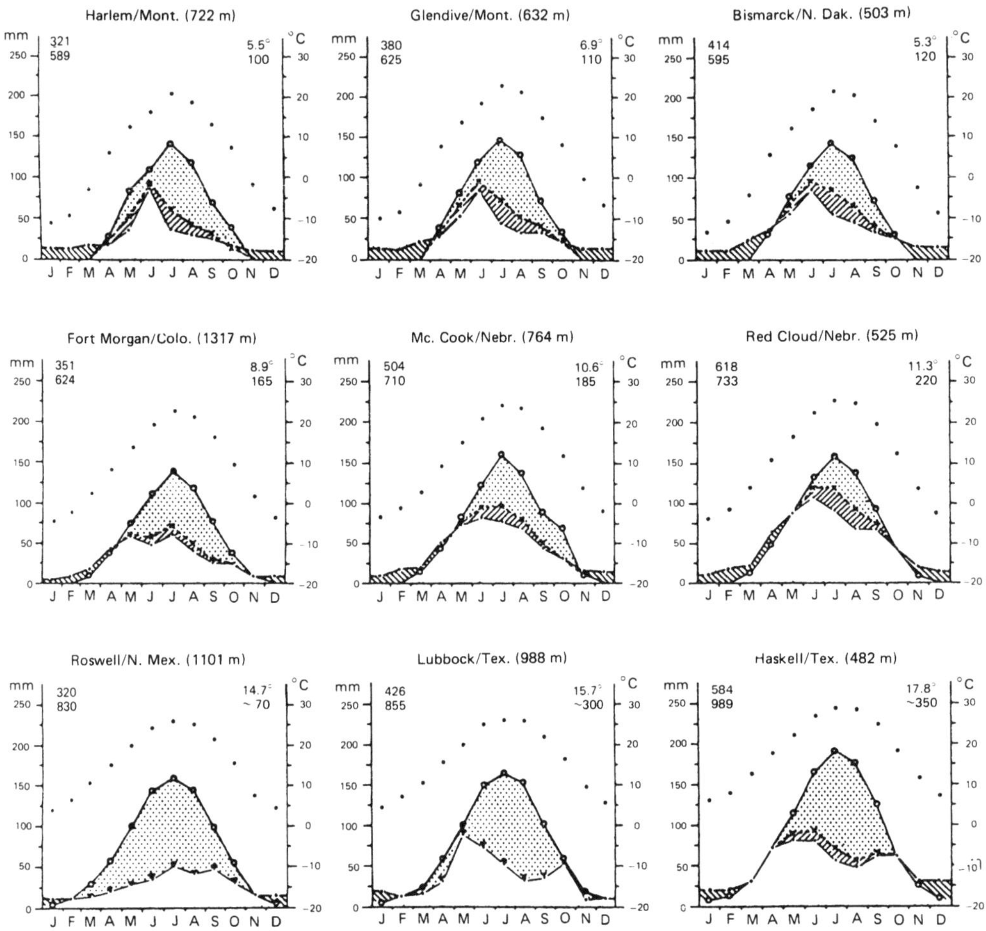

We attached great importance to the distinctions among the udic, typic, and aridic subgroups of Borolls, Ustolls, and Ustalfs to accentuate this climatic differentiation. At the taxonomic level of the great groups, an additional symbol (cf. legend) was used to set off the soil associations in which soils with an argillic horizon dominate, i.e., Argiborolls, Argiustolls, and Haplargids, from the other soil associations in which they are only secondary. To be able to represent the transition from the Borolls to the Boralfs of the forest regions in the northern part of the Great Plains, or from the Ustolls to the Udolls of the forest–prairie to the Udalfs of the original oak–hickory forests in the central part (Küchler, 1964), the soils (soil associations) of Minnesota and Iowa were additionally included in Figure 2; the sources are given in the map.

Soil Taxonomy (Soil Survey Staff, 1975) provides an almost complete system of six well-coordinated taxonomic levels. All or almost all catego-

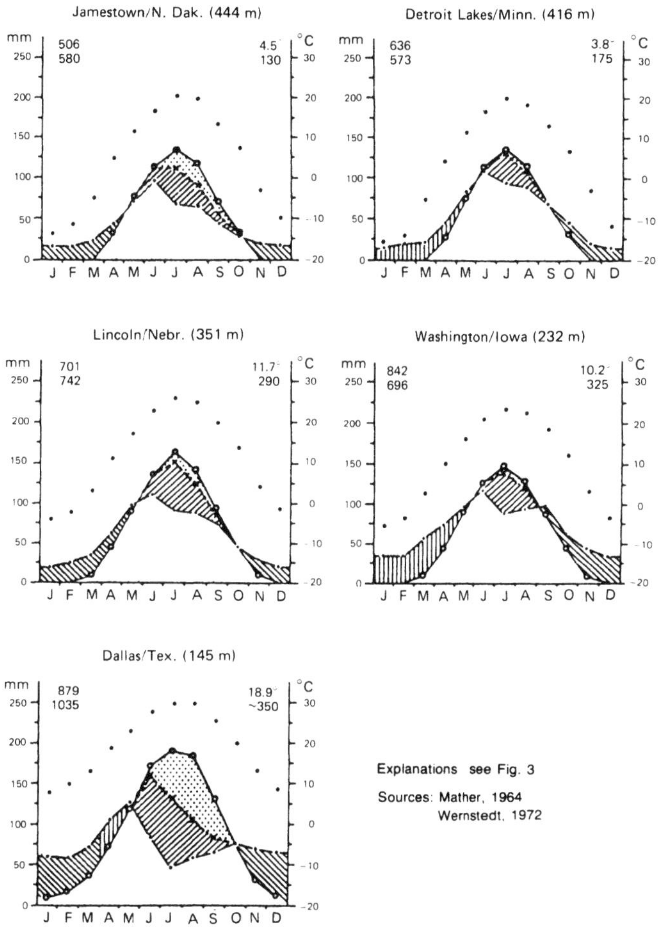

Figure 6. Climatic data and soil water balance of stations in the northern and central Great Plains of the United States with adjacent forest–prairie and forest areas and of the southern Great Plains (cf. Figure 2).

ries are defined strictly according to quantitative properties, an advantage that no other international soil classification offers (cf. Bronger, 1979, 1980). In the Great Plains with its accentuated decrease in precipitation (cf. Figure 6) from the forest–prairie to the short-grass prairie (see above), not only the soil temperature regime but, above all, the soil moisture regime is an important diagnostic criterion. In the uppermost category the Aridisols are distinguished from the other orders (with the exception of the Entiosols). At the level of suborders the Udolls are distinguished from the Ustolls, and these are again separated from the Borolls by the soil temperature regime. The Ustolls and Borolls, which are particularly widespread in the Great Plains, are then further subdivided at the level of subgroups according to the soil moisture regime. There is no further exact quantitative distinction among the Udic, Typic, and Aridic Ustolls, however. The Udic Borolls extend, in contrast to the Udic Ustolls, even into the udic soil moisture regime: the borderline between the udic and the ustic soil moisture regime, which separates the Udolls from the Udic Ustolls, ends with the 8°C isotherm in the southwestern corner of Minnesota (cf. Figure 2).

In a comparison with the soils of the forest–steppe and steppe regions of the USSR (cf. Figure 1), the Udic Borolls and even the Udic Ustolls in the mesic soil temperature regime can be compared approximately with the Degraded and Typical (medium) Thick Chernozems.* The Typic Borolls and Typic Ustolls (mesic) correspond approximately to the Ordinary and the Southern Chernozems, the Aridic Borolls or Ustolls (mesic) most nearly to the Dark Chestnut soils (compare Figures 1 and 2 with Figures 3, 4, and 6). This attempt at a correlation is not completely valid in the opposite direction; thus, the Typical Thick Chernozem, e.g., in the type locality of the forest–steppe in the vicinity of Kursk (cf. Figures 1, 4, and 5 and Section III.A), also does not occur among the Pachic Udic Borolls of the Great Plains. It would go beyond the scope of this chapter to discuss the reasons in greater detail at this point.

Let us point out again that the entire climatic sequence from the Degraded Chernozems to the (Light) Chestnut soils in Kasakhstan lies in the cryic or frigid soil temperature regime. Therefore, it is difficult to understand why in the FAO–Unesco (1974) *Soil Map of the World,* No II2, Kastanozems directly succeed the Phaeozems in the central part of the United States as well, for example, as in the tall-grass prairie in the eastern part of Nebraska. Kastanozems or Chestnut soils are, however, characteristic soils of the short-grass steppe or short-grass prairie (cf. Bronger, 1978, p. 45).

The abovementioned subtypes of the Chernozems can for the time being be correlated only with the abovementioned subgroups of the *Haplo*borolls or *Hapl*ustolls (cf. Section II.B). However, in the central and northern parts of the Great Plains, soils are widespread among the Ustolls and Borolls and even the Aridisols that supposedly have an *argillic horizon* resulting from clay illuviation (cf. Section II.A; see also Aandahl, 1972, 1982; Figure 2). The parent material was loose calcareous sediments, either (sandy) loess or loess-like sediments or, in the north, calcareous glacial till (cf. Thorp and Smith, 1952; see Section III.B). Particularly among the Ustolls and especially in the short-grass prairie, Aridic *Argiustolls* are more common than aridic *Hapl*ustolls; some have been categorized as Aridic Paleustolls, which, probably being older soils, have a particularly pronounced argillic horizon. Among the Aridisols the

---

*For the rather unfortunate term Typical in contrast to Ordinary Chernozem, cf. Gavriluk (1979).

Haplargids prevail in extent over the Orthids. Among the Borolls as well, the Typic *Argiborolls* and even the Aridic *Argiborolls* are more common that the corresponding subgroups without argillic horizon. These findings are surprising not only in comparison with those previously mentioned in the Chernozems and Chestnut soils—likewise on loess or loess-like sediments—of the forest–steppe and steppe regions of the USSR.

In the American literature it has been generally assumed that in calcareous soils the carbonate must be removed before clay can be mobilized. This was stressed, e.g., by Arnold (1965) on the basis of earlier investigations by Smith et al. (1950) and Arnold and Riecken (1964), particularly in Iowa. Allan and Hole (1968) reached very similar conclusions in Wisconsin, as did Bartelli and Odell (1960) in Illinois, since Ca^{2+} has a coagulating effect; i.e., it prevents the peptization of the clays. Likewise, Culver and Gray (1968) postulated from investigations in Oklahoma, and Dankert and Drew (1970) in Nebraska, that prior to clay illuviation carbonates must be removed. The question arises, however, how far below 7.0 must the pH drop, and above all what role does faunal pedoturbation (Hole, 1961) play (cf. Section III.A)?

It is in all probability a different question when suspended clay derived from calcareous dust is added (e.g., Gile, 1970; Holliday, 1985a,b). Goss et al. (1973a,b) have shown in experiments that clay can be moved through calcareous material if there are at least 20% large pores (0.1 to 1.0 mm in diameter) and if already suspended clay is added with considerable amounts of water. In the first experiment 2.15 L/cm^2 water was added containing 0.54 g sediment with ≥80% clay, ≥30% even <0.8μm. In the second experiment 1.34 L/cm^2 (this would still correspond to 13,400 mm of precipitation) was added with 0.18 g sediment/cm^2. In this experiment 91% of the suspended clay that was added remained in the upper 2.5 cm, the explanation given being that the openings of the larger pores had been destroyed. Only in the first experiment, in which the large pores remained open down into greater depths, was 50% of the clay transported into depths of >46 cm. From this they deduced that "large pores were necessary for clay movement in calcareous materials." To this we may remark that under natural conditions in steppe soils there is considerable faunal pedoturbation, so that large pores are not likely to be preserved into great depths.

The question of the genesis of the sometimes much higher clay content in the subsoils of steppe soils on loose calcareous sediments, which are explained as an argillic horizon (see above), are investigated in more detail in the following.

D. Material and Methods

The above questions (cf. also Section I, particularly the last two paragraphs) were investigated in soil sequences between Moscow and the Crimea (cf. Figure 5). During the 10th International Congress of Soil Science (cf. Afanasyeva et al., 1974) samples were taken for micromorphological investigations of the most important climaphytomorphic soils of the deciduous forest zone, the forest–steppe, and the steppe in the strict sense of the word (cf. Walter 1968, 1974). The samples included a (Light) Gray Forest soil from near Pushino/Oka (north of Tula, cf. Figure 1) on glacial till that is very low in sand (<1%) and rich in silt (60% to 80%). The parent material of all other soils was calcareous loess. In the central forest–steppe region in the vicinity of Kursk, three Typical Thick Chernozems under original steppe, under 200 to 300-year-old forest with a grass cover, and under cultivated land were investigated. In an original forest patch in the forest–steppe near Kharkov, a Dark Gray Forest soil was studied. Additionally, samples were taken of an Ordinary Chernozem from the Festuca–Stipa steppe between Dnepropetrovsk and Zaporozhye, a Southern Chernozem just north of Melitopol, and a Dark Chestnut soil associated with Solonetz soils near Askania Nova (approximately 150 km WSW of Melitopol) in an even drier Festuca–Stipa steppe (cf. Afanasyeva et al., 1974). Most of the soils mentioned here are shown in color in the *Atlas Poch CCCP* (Kauritschewa and Gromyko, 1974), at least at the same subgroup level. These studies were compared with results in loess soils from Central Europe (in particular, Bronger, 1976; Bronger et al., 1976; Schöhals et al., 1982; Rau, 1965).

The author spent a total of 5 months in the central and northern Great Plains of the United States and in the forest–prairie regions east of there in Iowa and Minnesota in 1976 and 1978.* During these two stays 60 soils were selected. They were restricted to soils from sites on the level, to exclude additional surface runoff. Another requirement was that they lie far above the water table (cf. Section III.A), so that Na^+ ions do not play any role in the adsorption complex. Results from catenas, which often contain sodium soils in the interior of depressions (e.g., potholes), were recently summarized by Anderson (1987). The parent material of all of these soils was either loess or loess-derived sediments, some of which had a relatively high proportion of sand. In the north the parent material was calcareous glacial till containing very little sand and with a particle size distribution similar to the loess-derived sediments (for examples, see Figures 7–15). Thus, it was possible to investigate the influence of the two soil-forming factors, vegetation and climate, on any potentially occurring clay illuviation.

For this purpose small-scale *biosequences* of soils were selected to minimize the influence of climate: in Iowa one biosequence each on calcareous Wisconsin-age glacial till and on loess (cf. Hallberg and Kemmis, 1986) ranging from a Udoll ("Prairie soil") via a Mollic Udalf ("Transitional soil") to a Typic Hapludalf (Forest soil). The choice was based primarily on studies by Smith et al. (1950), White (1953), White and

*Including a short stay in 1980.

Riecken (1955), and Oschwald et al. (1965; cf. also Fenton, 1983; Rust, 1983). For comparison, further profiles of the same widespread "Prairie soil" (Tama series) in eastern Iowa (Miller, 1974) were included in the investigations. Further biosequences were selected, one also a sequence Udoll–Mollic Udalf–Typic Hapludalf on loess in the forest–prairie (cf. Marschner, 1930) in the extreme southeastern part of Minnesota (Houston County). Another sequence was taken at the relatively sharp boundary between forest and prairie near Detroit Lakes (cf. Figures 2 and 6; cf. also Arneman, 1963; Severson and Arneman, 1973), running from a Typic Eutroboralf via a Mollic Eutroboralf to a Udic Haploboroll somewhat farther to the west but still in northwestern Minnesota. Just south of Devils Lake (North Dakota), approximately 140 km west of Grand Forks (cf. Figure 2) and already in an ustic soil moisture regime, a Boralfic Udic Argiboroll or Albic Argiboroll under a patch of oak forest (*Quercus macrocarpa* with *Fraxinus pennsylvanica*) lay only 50 m (!) from a Pachic Udic Haploboroll under tall-grass prairie (cf. Heidt, 1977; Fröhling, 1983).

The main focus lay on a large-scale investigation of *climatic sequences* of soils, primarily in regions with an ustic soil moisture regime with a simultaneous maximum of clay content in the subsoil. Beginning from the abovementioned biosequence in the vicinity of Detroit Lakes, Minnesota, the selected soils included Udic Borolls in the eastern part of North Dakota and Typic Borolls in the vicinity of Bismarck (cf. McClelland et al., 1959; Mogen et al., 1959); further Aridic Borolls were seen in the northern short-grass prairie near Harlem, on the one hand, and Ustollic Haplargids in Montana's southern short-grass prairie in the vicinity of Billings, on the other (cf. Figure 2). From southern Minnesota and southern South Dakota to south of Sheridan, Wyoming, climatic sequences ranging from Udolls via Udic Ustolls to Ustollic Haplargids or Ustollic Camborthids were investigated. In the central part, beginning from the abovementioned biosequences in Iowa, Udolls were studied in the vicinity of Lincoln, Nebraska, and then Udic Ustolls, Typic Ustolls, and Aridic Ustolls in southern Nebraska and as far as northeastern Colorado.

The soil series selected for this purpose were ones with as widespread a distribution as possible; sampling was done, where possible, in the immediate proximity of the typifying pedon. Soil series descriptions with the analytical data of the U.S. Soil Survey are available for (almost) all of the selected soils; the majority of them are shown in color photographs in the book by Aandahl (1982). The present-day environmental factors were summarized by Weaver (1968).

In contrast to the northern and central Great Plains, the soils of the southern Great Plains, particularly the widespread Paleustolls and Paleustalfs (cf. Figure 2), often have their origin as far back as the mid-Pleistocene, some possibly in the late Tertiary. There is no causal relationship linking them to present-day soil-forming factors, in particular to the present climate and present potential natural vegetation (cf., however, Sections III.B.2 and IV). Therefore, no pedons were sampled in the southern Great Plains. Some climatic data, and thus an approximate soil-water balance, are summarized in Figure 6 in temperature–evapotranspiration–rainfall graphs, which are used in a simplified form by the Soil Survey Staff (1975). The northern climatic sequence of soils mentioned above is characterized by the upper row of graphs: Detroit Lakes, located in the deciduous forest at the boundary of the prairie, represents the abovementioned biosequence; Jamestown, in the ustic soil moisture regime, represents the Udic Borolls; Bismarck and Glendive, the latter already in the short-grass prairie, represent the Typic Borolls; and Harlem, the Aridic Borolls. The abovementioned southern climatic sequence of soils is characterized by the middle row of graphs: Washington, on the boundary between deciduous forest and forest–prairie, is still typical of the abovementioned climatic sequences in Iowa; Lincoln lies near the boundary between the udic and the ustic soil moisture regimes, i.e., between Udolls and Udic Ustolls; Red Cloud at the boundary between Udic and Typic Ustolls; McCook at the boundary between Typic and Aridic Ustolls; and Fort Morgan near the boundary between the ustic and aridic soil moisture regimes (cf. also Figure 2).

In accordance with the approach outlined above, one main emphasis was on *micromorphological* investigations of the soils mentioned, because they allow us to demonstrate clay illuviation, particularly in soils developed in loess and loess-derived sediments (cf. Section II.A). Approximately 60 thin sections of the soil sequences mentioned for the European portions of the USSR were investigated. In the southeastern part of Central Europe, especially in Hungary and the autonomous region of Vojvodina (Yugoslavia) and, for comparison, Dobrogea (Romania), around 100 thin sections of soils from near the boundary between the udic and ustic soil moisture regimes and particularly in the ustic soil moisture regime were evaluated (Bronger, 1976, pp. 36–61, 80–107). A good 300 thin sections were studied of the soils of the biosequences and climatic sequences in the central and northern Great Plains of the United States and the forest–prairie regions to the east of the Great Plains in Iowa and Minnesota.

Preliminary findings on some of these soils were reported previously (Bronger, 1978). In summary, there was a very conspicuous clay maximum in the subsoil in quite a number of the soils in the Great Plains, especially in the short-grass prairie, though by no means in all of them. It turned out that this clay maximum could not be explained by clay illuviation, or only to a very slight degree. On the other hand, pedogenetic clay formation from primary silicates ($>2\ \mu m$) is supposed to be minimal in Mollisols because of the low intensity of weathering, especially in Ustolls (summarized most recently by Allen and Fanning, 1983, pp. 168, 175; Anderson, 1987, p. 66). For this reason, in the abovementioned paper

(Bronger, 1978), two other possible reasons were considered. One was a *lithologic discontinuity* in the parent material, i.e., primarily in loess and loess-derived sediments. On the other hand, some of the soils that were investigated, particularly in the short-grass prairie, must be considered *relict soils,* i.e., of *polygenetic* origin; the subsoils were probably at least partially the result of much older soil-forming processes.

To test these two causal complexes the investigations had to include not only physical and chemical soil properties but also the content of *primary and secondary minerals,* especially *silicate clays.* For this purpose 16 soil profiles were selected from the soils that were investigated micromorphologically, and the pH and the $CaCO_3$ content of these were determined; additionally, the particle size distribution was determined by means of sieve and pipette analysis. Of these 16 profiles, nine, all in the mixed and short-grass prairie, were chosen for a more detailed investigation (cf. Figures 7–15). For each of these soil profiles the sand, silt, and clay fractions were extracted quantitatively, following CBD treatment (Mehra and Jackson, 1960), to determine the content of primary minerals in each fraction and to be able to estimate the content of clay minerals in the fractions 2 to 0.2 μm and <0.2 μm semiquantitatively. The quantitative separation and determination of the relative proportions of each of the individual fractions are much more exact than the pipette method (for details see Bronger et al., 1966; 1976, especially Tables I–VIII).

One way to determine *original petrographic* homogeneity is to use resistant *index minerals* (e.g., Barshad, 1967). These should be present in equal amounts in all horizons and neither decomposed nor displaced by the pedogenetic weathering. Preferably such index minerals should occur in quantities large enough for precise determinations of amounts in counts of a few hundred grains. For this reason the stable heavy minerals, such as zircon, rutile, and anatase, are unsuitable, because in these soils, as in the loess soils of Central Europe as well, they occur only in very small quantities. *Quartz,* however, is a very suitable index mineral. It is resistant to pedochemical weathering, at least in temperate latitudes, and always occurs in large quantities. If the percentage of quartz by weight in the different soil horizons is approximately constant, it is very likely that the parent material was originally vertically homogeneous (for details see Bronger, 1976, pp. 25–30; Bronger et al., (1976; Section III.A.2 in contrast to III.B.2). The proportions of the coarse sand, fine sand, and silt fractions, i.e., of quartz (Kundler, 1959; Schlichting and Blume, 1961; Meyer et al., 1962), are less suitable as an index of homogeneity because slight physical breakdown in the sand and/or silt fractions can cause considerable shifts in the relative proportions. This was pointed out especially by Arnaud and Whiteside (1963).

To determine the mineral composition for the fractions >200 μm, 200 to 63 μm, and 63 to 20 μm, several slides were examined for each fraction in polarized light. At least 300 particles were examined per slide, and the percentage of each mineral was calculated from the total counts per sample. The proportions of quartz, feldspasr, phyllosilicates, and other heavy minerals in the fractions 20 to 6 μm and 6 to 2 μm were determined by phase-contrast microscopy: 700 to 1600 particles in the fraction 20 to 6 μm and 1000 to 1800 in the fraction 6 to 2 μm were examined per slide. The percentages of the minerals in each fraction were then multiplied by the weight percentages of each fraction to give the percentage by *weight* shown in Figures 7 to 15 (lower part, left side). It can be demonstrated that the percentages by weight differ little from the percentages by volume (cf. Bronger, 1976, pp. 24–25) even though the minerals have different densities. In the upper parts of Figures 7–15 the percentages by weight of the different minerals, added up for all fractions, are represented as cumulative curves.

The composition of the clay subfractions was determined by *semiquantitative estimation* on the basis of the areas under selected peaks. Other weighting factors were those recommended by Laves and Jáhn (1972).* Illites and vermiculites were given a weighting of 1, kaolinites and smectites 0.25. The results were compared with the cation-exchange capacity of each fraction to correct gross deviations. Each mineral name stands for a group of closely related clay minerals with slightly varying compositions. The percentages by *weight* shown in Figures. 7 to 15 (lower part, right side) are consequently *estimates* only, in contrast to the weight percentages of the fractions >2 μm. The cumulative curves for the clay minerals (Figures 7–15, upper part) are therefore represented by dotted lines. Nevertheless, the results allow us to draw conclusions regarding the two causal complexes responsible for the conspicuous clay maximum in the subsoil of several of the selected soils—lithologic discontinuity and/or extremely polygenetic development.

III. Results

The results of the investigations can only be summarized along the lines of the general approach (Sections I and II.A).

A. USSR and Southeastern Central Europe

1. Micromorphological Studies

The Light Gray Forest soil in the deciduous forest zone—originally predominantly *Quercus robur*—near Pushino/Oka between Moscow and Tula (cf. Figure 2) has an approximately 25-cm-thick Ah horizon under-

*For methodological details cf. Bronger et al. (1966, 1976) and Bronger (1969/1970, 1976). In these studies, however, peak heights were used rather than peak areas, which resulted in other weighting factors, especially for smectites.

lain by a 20-cm-thick transitional BA horizon.* The approximately 70-cm-thick Bt horizon under it still contains some krotovinas, such as are typical of forest–steppe and steppe soils (see below). Micromorphologically the A and BA horizons show a fine spongy fabric with many pores but nowhere an eluvial fabric, as would be characteristic of an E horizon. In contrast, the upper Bt horizon reveals many aggregates and, additionally, so-called "white powder," which consists of uncoated grains of quartz, feldspar, and (secondarily) heavy mineral grains of the silt fraction (cf. Kundler, 1959). Despite this segregation, the proportion of the illuviation argillans that are partially mixed into the fabric by bioturbation is slight, though sufficient for an argillic horizon. The closest equivalent to this soil would be a Boralfic Udic Argiboroll (cf. Figure 1). With approximately 550 mm of precipitation and a mean annual temperature of 4.5°C, this is still a udic soil moisture regime (cf. Figure 4: soil moisture regime between the stations of Moscow and Kursk).

The three *Typical Thick* Chernozems in the forest–steppe near Kursk are very similar in their soil horizon sequence except for the Ap horizon in the profile under cultivated land. An approximately 80 to 90-cm-thick Ah horizon is free of carbonates down to a depth of 70 to 80 cm, with a pH (0.1 N KCl) between 6.1 and 6.6, even in the profile under forest. In all three profiles this is followed by an A(k) horizon with bieloglaska and pseudomycelia as signs of intensive carbonate metabolism (cf. Figure 5). Material from the underlying loess has been mixed into this horizon by bioturbation (cf. Bronger, 1978, photo 1). This is evident macroscopically from the large number of krotovinas, which are also found in the loess. The Ah and the A(k) horizons of all three soils are characterized micromorphologically by a fine spongy fabric rich in aggregates and pores *without* any traces of illuviation argillans as a sign of *clay illuviation*. Interestingly, this is also true of the Chernozem under oak forest *(Quercus robur)*, which is, however, only about 200 to 300 years old; prior to that there was steppe vegetation here also (cf. Afanasyeva et al., 1974; Ivanova et al., 1971; Hubrich et al., 1976). The soil moisture regime (cf. Figure 4, station Kursk) is close to the boundary between udic and ustic. The nearest equivalent to these three soils would be a Typic Vermiboroll (close to the Pachic Udic *Haplo*boroll).

Under the very few patches of original old forest, however, (with *Quercus robur, Tilia cordata, Fraxinus exelsior, Acer platanoides,* and *A. tataricum),* even in the southernmost forest–steppe near Kharkov (Figure 1), we find instead of Chernozem a Dark Gray Forest soil (Figure 5). It has an at least 50-cm-thick Ah horizon on top of a Bt horizon that is as

*The genetic horizon designations follow the international convention (FAO–Unesco, 1974; Wilding et al., 1983; Soil Survey Staff, 1987). A BA horizon, for example, has characteristics of both an overlying A horizon and an underlying B horizon but is more similar to the B than to the A.

much as 70 cm thick and contains some krotovinas. Micromorphologically this soil has a fine spongy fabric rich in pores, even in the lower portion of its Bt horizon.* Because of its high humus content, in the Ah horizon the fabric cannot be distinguished from that of a Chernozem. In the Bt horizon thin seams of illuviation argillans surround only some aggregates (cf. Bronger, 1978, photos 3,4), signifying only *slight clay illuviation*. The soil temperature regime is mesic (bordering on frigid, cf. Figure 1). The figures for the soil moisture regime lie between those for Kursk and those for Dnepropetrowsk (cf. Figure 4); thus, this is just barely an ustic soil moisture regime. The closest equivalent to this soil would then be an Udic Argiustoll or Udic Haplustoll bordering on an Argiudoll or Hapludoll.

The *Ordinary Chernozem* north of Zaporozhye with an Ah horizon as much as 60 to 65 cm thick (up to 55 cm free of carbonates) and an A(k) horizon 20 cm thick also has a fine spongy fabric rich in aggregates and pores. No traces of the illuviation argillans that are characteristic of clay illuviation are to be seen. Thus, this soil most nearly corresponds to a Typic *Hapl*ustoll (Figures 1 and 5). The Southern Chernozem directly north of Melitopol (Figure 1) with an Ah horizon about 50 cm thick and an A(k) horizon about 10 to 15 cm thick and the Dark *Chestnut soil* near Askania Nova, with an Ah horizon as much as 40 cm thick and an A(k) horizon 15 cm thick (Figure 5) show features similar to those of the Ordinary Chernozem. Bioturbation decreases slowly toward the south (cf. Bronger, 1978, photo 2) with increasingly drier climate (cf. Figure 4). No signs of clay illuviation are to be seen. Thus, both soils correlate most nearly with Aridic *Hapl*ustolls. Farther to the east in Kasakhstan, the Dark Chestnut soils can already correspond to Ustollic Orthids (e.g., stations Uralsk and Pavlodar, cf. Figures 1, 3, and 4).

In summary, our micromorphological investigations confirm the Russian results (cf. Section II.B), according to which in the sequence of climaphytomorphic soils under steppes extending from the Typical Thick Chernozem with a soil moisture regime at the boundary between udic and ustic via the Ordinary and the Southern Chernozem to the Chestnut soils, *no clay illuviation* took place. The process of clay illuviation occurs *only in soils under forest*. In soils under original old forest patches in the southernmost forest–steppe, just barely in the ustic soil moisture regime, clay illuviation can also be observed micromorphologically, but only to a very slight degree.

In the *steppes* of Dobrogea Romania, where it is warmer, the Chernozems and Dark Chestnut soils in the driest parts also show no signs of clay illuviation. In the few edaphically caused *forest* patches, however, we do not find Chernozem but a Dark Gray Forest soil (an Udic Haplustoll or an Udic Argiustoll) in an ustic soil moisture regime. This soil has a thick Ah horizon (47 cm) with an AB horizon to 63 cm above a 45-cm-thick Bt horizon. The micromorphology is also similar to the corresponding soil near Kharkov: in the Ah horizon, which is highly characterized by bioturbation, an eluvial fabric is lacking; in the Bt horizon, which is free of carbonates, clay illuviation can be seen only in the upper part, in the form of seams of illuviation argillans in only a few of the conductive channels (for further details see Bronger, 1976, p. 82 ff). In the Danubian plain in Romania the so-called "Danube Chernozem" (Haplic Chernozem, probably comparable to a Typic *Hapl*ustoll) does not show any signs of clay illuviation (Schönhals et al., 1982). According to these authors, only in the *Degraded Chernozem* of the *forest–steppe* southeast of Bucuresti—still with an ustic soil moisture regime–are signs of very slight clay illuviation demonstrable. It is thus more nearly a Haplic than a Luvic Chernozem, i.e., most likely an Udic Haplustoll.

In the central dry portion of the Carpathian Basin ["Nagy (Great) Alföld," Hungary and the Voivodina region Yugoslavia], some of the Chernozems developed in loess far from the water table (Typic Haplustolls) are partly rich in primary carbonates, particularly in the innermost, driest part (cf. Bronger, 1976, pp. 48–61; Neigebauer et al., 1983). But also in the somewhat moister peripheral regions, still with an ustic soil moisture regime, of the potential forest–steppe (an *Aceri tatarico–Quercetum pubescenti–roboris* association), the many investigated Chernozems (Udic Haplustolls) did not show any signs of clay illuviation. Even the Degraded Chernozems at the boundary between the ustic and the udic soil moisture regimes (cf. *Klima Atlas von Ungarn,* 1967) of this region are free of illuviation argillans (cf. Bronger, 1976, pp. 45–61, 82–103). An important reason for this is probably the high degree of bioturbation, so that even the *Degraded Chernozems* of this region must be considered Udic Haplustolls. Even in the moister region here, already in the udic soil moisture regime but with a water deficit from June to November, two *Brown Forest soils* studied in more detail (Bronger et al., 1976) show only traces of illuviation argillans, mostly reworked into the fabric. Therefore, despite considerable pedochemical clay formation (cf. section III.A.2), their Bt horizons can*not* be designated *genetically* as *argillic* horizons (for further information see Bronger, 1976, pp. 36–43, 103–107).

In the western part of Central Europe, in the so-called "Central German dry region" ("Mitteldeutsches Trockengebiet"), i.e., in the Thuringian Basin, the *Chernozems* are *relict soils* formed in a warmer and drier period of the early Holocene, mainly in the Boreal and early Atlanticum. During the cooler and moister second half of the Holocene, the Chernozems mainly became degraded to Brown Forest Soil-Lessivés (Mollic Hapludalfs). The latter already have a pronounced argillic horizon (Rau, 1965; Catt, 1989, Figure 1A). Where early neolithic agriculture prevented forest from spreading, however, the Chernozems are hardly degraded, and above all, they show *no* signs of clay illuviation (Rau, 1965).

2. Mineralogical and Clay Mineralogical Investigations

Recent (clay) mineralogical investigations on Gray Forest soils and a "Podzolized Chernozem" on loess in the Ukrainian forest–steppe (Breus et al., 1984) showed that the parent material had a pronounced lithologic discontinuity. For this reason it is hard to follow the logic of their statement that "maximum weathering and alteration of primary minerals are observed in the upper humic–eluvial horizons," especially since the clay maximum in these soils is always in the central portion of the profile (40 to 80 cm). The dominant clay mineral in the total clay fraction is illite ("hydromicas"). In contrast, in an Ordinary Chernozem in the central Ukraine, smectites are the dominant clay minerals, particularly in the fine clay fractions <0.2 μm (Tributh, 1971). The C horizon was not, however, included in the investigation in this case. For the abovementioned Chernozems (Typic and Udic Haplustolls) of the Danubian plain in Romania (Schönhals et al., 1982) there are very detailed clay mineralogical investigations, but the content of primary minerals (> 2 μm) was not part of the study. Therefore, the question of sedimentary homogeneity could not be answered (cf. section II.D).

The analysis of the content of primary minerals (>2 μm) and of clay minerals in the two abovementioned Brown Forest soils from the western edge of the central Carpathian Basin, with an udic soil moisture regime close to the boundary between udic and ustic, show that the percentage by weight of quartz in the different soil horizons is approximately constant. From this we can conclude that the loess as parent material was originally for the most part petrographically homogeneous. The pedochemical formation of illites and particularly of smectites in the fine-clay fraction <0.2 μm is not greatest in the A horizon but in the upper Bt horizon (cf. Bronger et al., 1976, Tables 3 and 4). Since, on the other hand, only traces of illuviation argillans can be found in these Bt horizons (cf. section III.A.1), the *maximum* amount of *weathering* and *clay mineral formation* occurs here and not in the topsoil as in more humid regions. The clay content is about 10% higher in the Bt horizon than in the A horizon in one of these Brown Forest soils and not quite 20% in the other. Therefore, on the basis of this criterion, they do not have an argillic horizon. On the other hand, the pedochemical clay formation is so high (15%) in both of these Brown Forest soils that they should not be classified as Inceptisols. There would be good reasons for classifying them as Mollic Hapludalfs, a point that is discussed elsewhere (Bronger and Bruhn, 1990; cf. also section IV). The large proportion of smectites ($\geqslant$50%) is not responsible for the almost total lack of illuviation argillans in this case (cf. sections II.A and IV). A Brown Forest soil-Lessivé (Typic Udalf) barely 40 km away from one of the two Brown Forest soils, but with 60 mm more annual precipitation, already shows well developed illuviation argillians (cf. Bronger, 1976, photos 6 and 7) in an argillic horizon.

B. Central and Northern Great Plains and Adjacent Forest–Prairie Regions

1. Micromorphological Studies

In biosequences of soils lying in close proximity to each other in the forest–prairie of southeastern Minnesota and central Iowa, the author observed the influence of the soil-forming factor vegetation on the process of clay illuviation. In soils under original prairie, i.e., in Udolls in central Iowa, clay illuviation is minimal. The widespread "Tama" soil, e.g., has a fine spongy fabric rich in aggregates (cf. Bronger, 1978; photos 6 and 7, in contrast to photo 5) not only in the Ah horizon (0 to 40 cm) and the AB horizon (40 to 60 cm) but also in the Bt horizon([4]) (60 to 140 cm, sometimes even deeper). Only the lower part is somewhat less rich in pores. Illuviation argillans are found only in the central part of the Bt horizon and only in very small amounts. In its lower part only traces are visible in the fabric, but these are old and show decreased birefringence. Because of the very slight clay illuviation (much less than 1% illuviation argillans, cf. Section II.A) and the lack of an eluvial fabric, this soil is more similar to a *Hapl*udoll than to an *Argi*doll. The associated Mollic Hapludalf ("Downs" soil) already shows more illuviation argillans, especially in the Bt horizon, but the proportion in the entire Bt horizon is still less than 1%. Only the associated Typic Hapludalf ("Fayette" soil), which was in all likelihood formed under a former patch of forest, has well developed illuviation argillans, at least in the lower Bt horizon. In most parts they are birefringent, and they are found to a certain extent in conductive channels (cf. Bronger, 1978, photo 5). The large increase in clay content from 19 to 21% in the Ah horizon to 32% in the Bt horizon (Roggenbau, 1983) cannot by any means be attributed to the process of clay illuviation alone. It is more likely that here, as in the abovementioned Udoll, the *maximum* amount of *weathering,* and thus *clay formation,* occurs in the *Bt horizon,* because it is moister, and not in the A horizon, which often dries out highly in summer (cf. Sections III.A.2, III.B.2, and IV).

In an equivalent biosequence in the extreme southeastern part of Minnesota, the "Prairie soil" ("Port Byron soil") shows only traces of clay illuviation; it is also classified as a Typic Hapludoll. In contrast to the above mentioned sequence in central Iowa, the "transitional soil", or Mollic Hapludalf ("Mt. Caroll" soil), already has noticeably more illuviation argillans in its middle and lower Bt horizon, similar to the Typic Hapludalf ("Seaton" soil). But here too the increase in clay content from 17 to 18% in the Ah horizon to 25% in the Bt horizon, or, respectively, from 16% to 27% (assuming there is sedimentary homogeneity), must be attributed primarily to the greater moisture content of the subsoil. In this area forest predominates because the climate is moister, and presumably also because of the more prominent relief. Küchler's general map (1964) shows this forest as maple–basswood forest *(Acer–Tilia).* In Marschner's

(1930; cf. also Wright et al., 1963) much more detailed map of Minnesota this was, however, a macromosaic of forest, "brush prairie," and prairie; i.e., according to Knapp's (1965) definition, following Walter (1968), it was a forest–prairie. The dominant soil is therefore a Typic Hapludalf (cf. Figure 2). In central Iowa the dominant vegetation according to Küchler (1964) is bluestem prairie *(Andropogon–Panicum–Sorghastrum),* which according to Oschwald et al. (1965) even covered most of Iowa with the exception of the gallery forests. Part of the prairie in eastern Iowa may have been covered with deciduous forest during the early and mid-Holocene (approximately 6200–2800 years B.P.), according to Hallberg (cited by Miller, 1974, p. 255ff). Farther northwest in Iowa, the prairie had begun to displace the oak–elm forest around 9000 years B.P. The period from 7700 to 3200 B.P. was the warmest and driest part of the Holocene, according to van Zant's (1979) pollen analyses. Thereafter, oaks returned to small areas. The dominant soils are Udolls (cf. Figure 2).

In the biosequence at the relatively sharp boundary between forest and prairie near Detroit Lakes, Minnesota (cf. Figure 2), the widespread "Barnes" soil of the prairie—still in an udic soil moisture regime (Figure 6)—is classified as an Udic *Haplo*boroll. Even the "transitional soil," or Mollic Eutroboralf ("Waukon" soil), shows only traces of illuviation argillans in the Bt horizon. In the biosequence just south of Devils Lake, North Dakota, in an ustic soil moisture regime (cf. Section II.D; Figures 2 and 6: Jamestown), the Boralfic Udic Argiboroll, or Albic Argiboroll ("Kelvin" or "Tonka" soil) under forest, shows a typical eluvial structure in the E horizon (formerly A2 horizon). In the Bt horizon the proportion of illuviation argillans is very small, on the whole, and for the most part they have been worked into the fabric by the soil fauna. The upper Bt horizon has a pronounced lithologic discontinuity, according to the particle size distribution, as does the B horizon of the Pachic Udic *Haplo*boroll ("Svea" soil) under original tall-grass prairie (Fröhling, 1983) only 50 m away from it. But this does not allow us to identify an argillic horizon on the basis of differing clay content.

The micromorphological investigations included climatic sequences of soils from the tall-grass prairie in the eastern part of North Dakota to the short-grass prairie of Montana and correspondingly from the eastern part of South Dakota to Wyoming and from eastern Nebraska to the northeastern part of Colorado (cf. Section II.D), always in the ustic soil moisture regime. For lack of space, the results are only briefly summarized here. In addition to Udic, Typic, and Aridic *Argi*borolls, or Udic, Typic, and Aridic *Argi*ustolls ranging to Ustollic Hapl*argids,* the study included *Haplo*borolls, *Hapl*ustoll, and Ustollic Camb*orthids* for comparison.

Neither in the Udic nor in the Typic Argiborolls do we find signs of clay illuviation in the form of illuviation argillans. In the Bt horizons of the two Typic Argiborolls ("Morton" and "Williams" soil around Bismarck), small amounts have a higher birefringence. This is certainly a result of the high clay content, approximately 50%, which consists predominantly of smectites (cf. Section III.B.2). The clay content of the A horizons is, however, just 20% lower. Micromorphologically, they, like the Haploborolls, show only in parts a fine spongy fabric rich in aggregates and pores, but the fabric is less aggregated than in the loess soils of the abovementioned more southerly climatic sequences. In the Aridic Argiustolls of the short-grass prairie in the north—in the "Scobey" soil near Harlem and particularly in the "Shaak" soil south of Billings, Montana (cf. Figure 2)—the differences in the clay content in the A versus the Bt horizon are much, or even very much, greater (cf. Section III.B.2; Figures 10 and 11). The "Shaak" soil is therefore classified as an Abruptic (Aridic) Argiboroll.

Two pedons of the "Hesper" soil not far from the "Shaak" soil lie at a lower elevation, i.e., with a higher mean annual temperature, and are consequently classified as Ustollic Haplargids. They also show strikingly great differences in clay content in the A versus the B horizon (cf. Section III.B.2; Figure 9). The differences are expressed very strongly in the micromorphology, especially in the four last-mentioned soils. The (dark) gray material of the A horizon, which is rich in skeletal matter in the "Hesper" soil, is, on the whole, well aggregated and contains sharply delineated lumps of yellowish brown material from the Bt horizon. In other places a well-defined boundary has developed. In the upper and lower Bt horizon the yellowish to reddish brown material with a much higher clay content is partially—probably secondarily—aggregated. The existing conductive channels in other parts are completely free of illuviation argillans. They are found in only a few very small places, aged and secondarily mixed into the fabric. Oriented birefringence, however, is found in many parts of the fabric. The BC horizon again is richer in pores and aggregates. According to the micromorphological findings, the very large differences in clay content between the A and the Bt horizons cannot be explained by the process of clay illuviation, or only to a slight degree.

In the southern climatic sequence, in contrast to the northern one, the differences in clay content in the A versus the Bt horizon are already great in the Udic Argiustolls. In two pedons of the "Hastings" soil (Udic Argiustoll) and two of the "Crete" soil [Pachic (Udic) Argiustoll] and in a "Holder" soil (Udic Argiustoll) in eastern Nebraska south of the Platte River, all developed in Peoria loess (cf. Figure 2), the clay content rises from 32 to 33% to 46% and 48%, respectively. Jordan (1967, p. 46) arrived at similar results, likewise in a "Hastings" soil, but the upper A horizon contained only about 23% clay. Micromorphologically not only many, or even most, parts of the A horizons show a well-developed fine spongy fabric rich in aggregates and pores. In parts of the Bt horizon as well there is still a spongy fabric (cf. Bronger, 1978, photo 8). In the Bt horizons illuviation argillans are found as a sign of clay illuviation only in traces on the surfaces of the less-rounded aggregates. In most parts of the Bt horizons they are completely lacking. At any rate, here too the much higher clay content in the Bt horizon compared with the A horizons, cannot be explained by clay illuviation, or only to a very slight degree. In comparison, in some C horizons small amounts of illuviation argillans are found, some in seams in voids, though with little birefringence. Other parts are already mixed into the fabric. Jordan (1967) already made this observation in the "Hastings" soil. The C horizons are, on the other hand, weakly developed Ck horizons because of the slight $CaCO_3$ content of the Peoria loess in Nebraska (under 10%, often under 5%: Maher, 1937), which is the parent material of the abovementioned soils (e.g., Hallberg, 1986).

Two examples each of the widespread Typic Argiustolls ("Holdrege" soils) in central Nebraska and of the Aridic Argiustolls ("Keith" soils) in western Nebraska and the extreme south(west)ern part of South Dakota have a well-developed fine spongy fabric rich in aggregates and pores not only in the A horizon. Some parts of the fabric of the Bt horizon in the "Holdrege" soils are slightly more dense, but the spongy fabric dominates. This is particularly prominent in the "Bt" horizons of the "Keith" soils, which are particularly rich in aggregates because of high biological activity (cf. Bronger, 1978, photo 9). Illuviation argillans as a sign of clay *illuviation* are completely lacking in these soils.

In contrast to the last-mentioned soils, two selected soils of the short-grass prairie in northeastern Colorado have great differences in clay content from the A to the Bt horizon, similar to those from Montana. Large parts of the Ah horizon of the selected Aridic Argiustoll ("Rago" soil, at the border with Nebraska) and the Aridic Paleustoll ("Weld" soil near Akron, Colorado) have a fine spongy fabric rich in aggregates and pores. The color of the matrix, which contains much more skeletal material than the subsoil, is (dark) brown. In striking contrast to this, the Btb horizon of the "Rago" soil generally has a very compact fabric with few pores. In some parts the material shows a more less strongly oriented birefringence, sometimes at the edge of a narrow conductive channel or crack. Illuviation argillans are lacking even in an aged form. In the "Weld" soil, many parts of the egg-yolk yellow to slightly reddish brown material of the Bt horizon have already obviously been transformed into a fabric rich in aggregates; parts are still birefringent (cf. Bronger, 1978, photo 10). Here too the Bt horizon is not an argillic horizon formed by clay illuviation but in all probability is an older, *relict* soil, i.e., a *polygenetic* soil. This is indicated in the "Rago" soil by the suffix "b", Btb referring to a buried soil.

2. Mineralogical and Clay Mineralogical Investigations

To test the two other possible reasons for the (much) higher clay content in the Bt horizons, compared with the A horizons—a *lithologic discontinuity* in the parent material or *polygenetic origin* of some soils, particularly in the short-grass prairie—not only soil chemical analyses but especially (clay) mineralogical investigations were necessary. Of 16 selected soil profiles whose particle size distribution was determined by sieve and pipette analysis, nine were chosen for a more detailed investigation. For this it was necessary to extract the sand, silt, and clay fractions quantitatively (cf. Section II.D). The results are shown in Figures 7 to 15. In some cases they deviated obviously from the results of the pipette analysis or, more specifically, corrected them. For instance, in the Typic Argiboroll ("Williams" soil, cf. Figure 15), the clay content does not increase from the lower A horizon to the Bt horizon. The clay content is highest in the

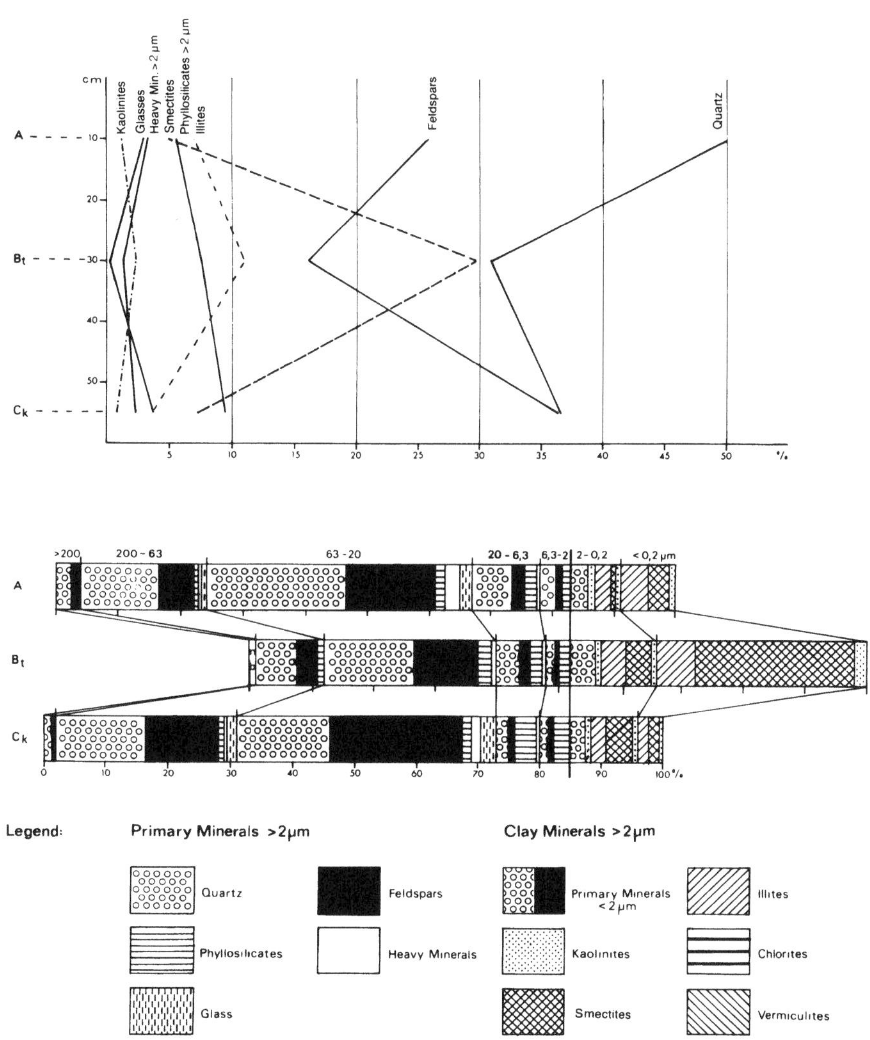

Figure 7. Mineralogical and clay mineralogical composition of an Aridic Paleustoll (Weld series, Colorado).

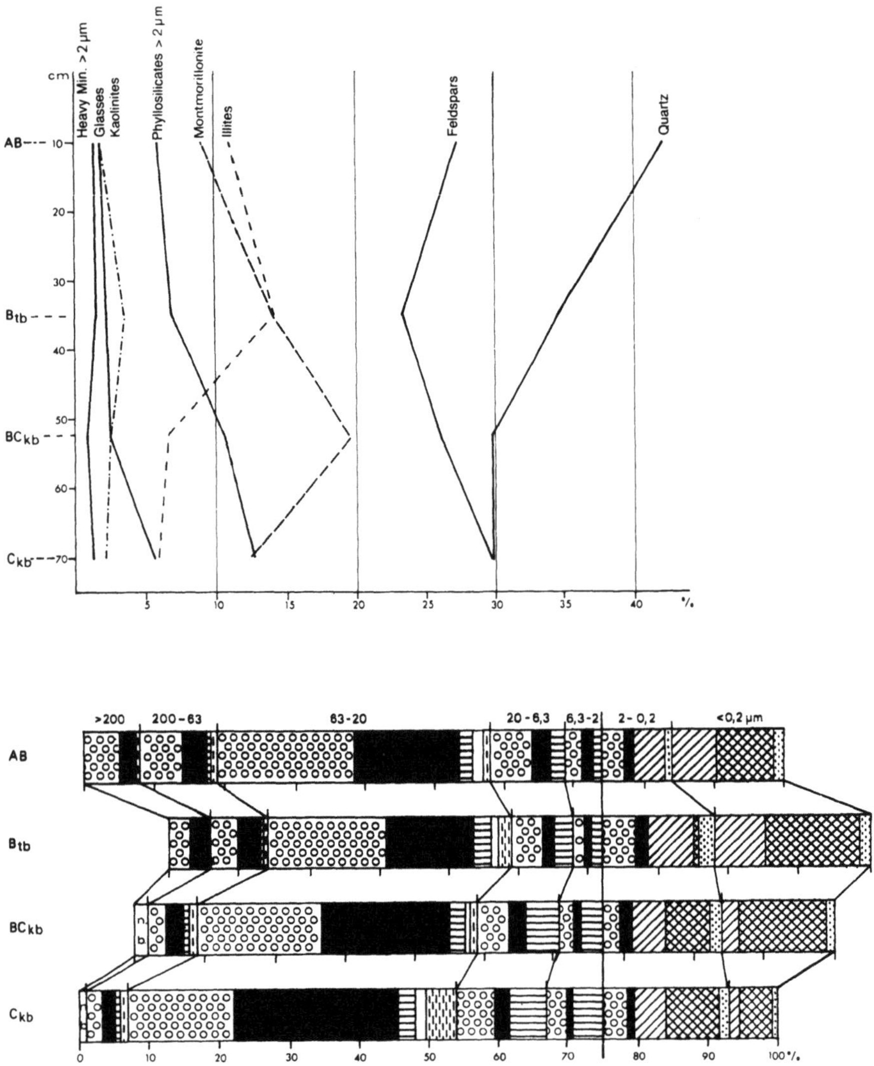

Figure 8. Mineralogical and clay mineralogical composition of an Aridic Argiustoll (Rago series, Colorado).

BCk and CBk horizon, also in contrast to the pipette analysis. In the Typic Argiustoll ("Reliance" soil, cf. Figure 14) as well, there is no recognizable increase in clay from the Ah to the Bt horizon, also in contrast to the pipette analysis, but there is an increase particularly from the lower Bt to the CBk horizon, especially in coarse and medium clay (2 to 0.2 µm). From this fact alone we can deduce that there was a *lithologic discontinuity* (cf. below). In the Aridic Argiustoll ("Keith" soil, cf. Figure

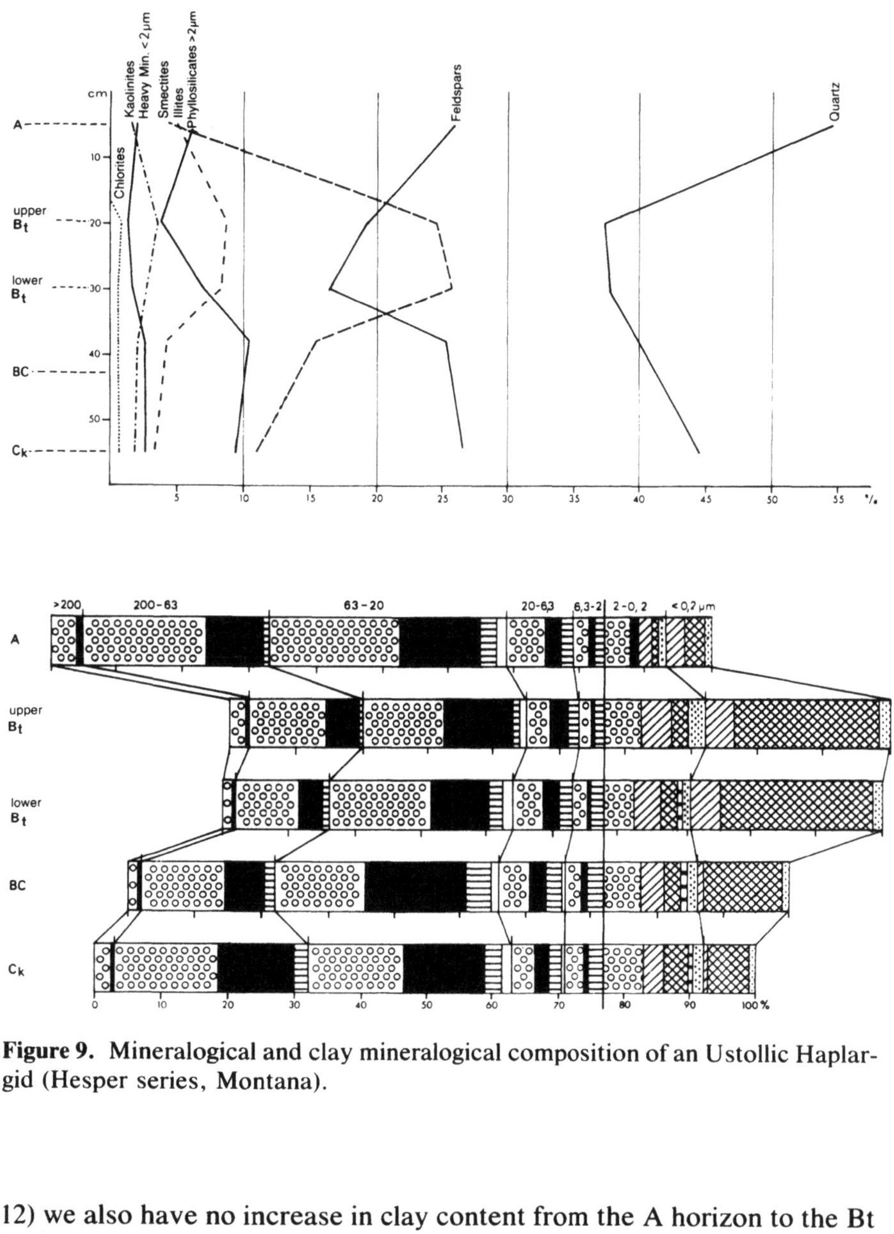

Figure 9. Mineralogical and clay mineralogical composition of an Ustollic Haplargid (Hesper series, Montana).

12) we also have no increase in clay content from the A horizon to the Bt horizon. Thus, all three soils must be designated as *Hapl*ustolls and not as *Argi*ustolls.

The cation-exchange capacity (CEC) of the nine soils lies between 47 and 65 me/100 g in coarse and medium clay (2 to 0.2 μm). In the A horizons of the "Weld," "Rago," and "Hesper" soils, in comparison, it is only around 34 to 38 me/100 g, and in the "Shaak" soil even as low as 27 me/100 g. In the fine clay fraction (<0.2 μm), the CEC is between 58 and

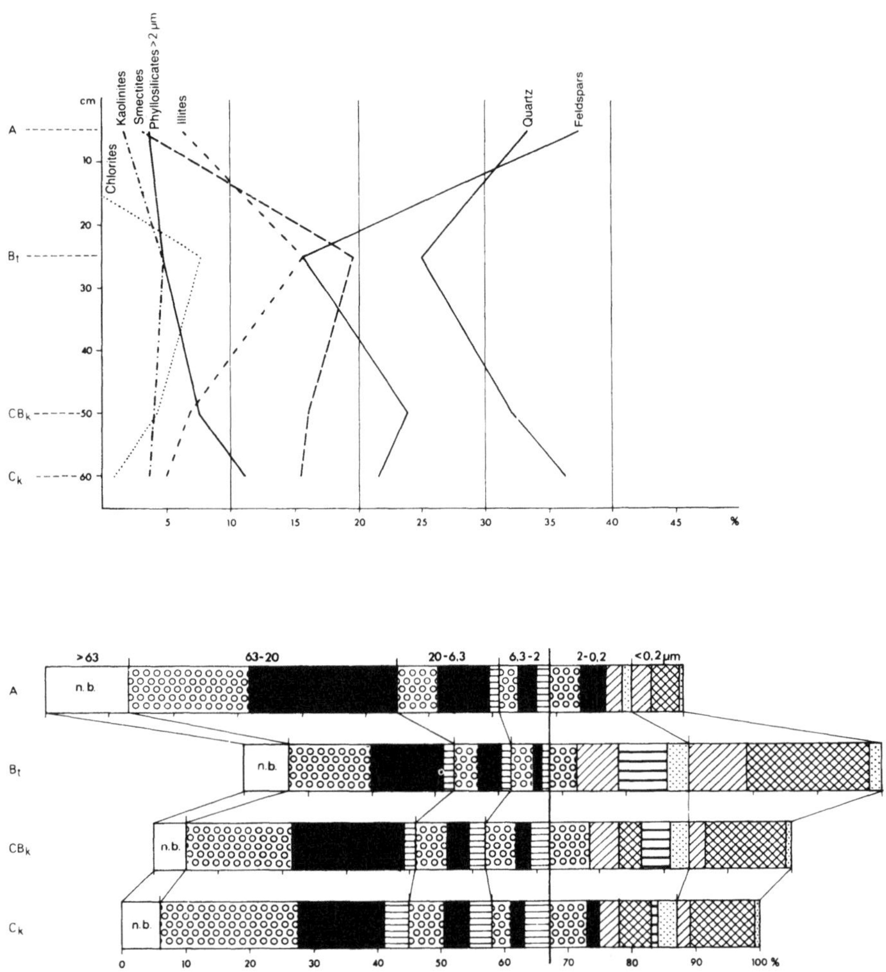

Figure 10. Mineralogical and clay mineralogical composition of an Abruptic (Aridic) Argiboroll (Shaak series, Montana.

81 me/100 g; only in the A horizons of the "Weld," "Rago," and "Hesper" soils is it between 45 and 51 me/100 g. On the whole, these figures correlate well with the smectite proportions of the clay subfractions (cf. Figures 7–15).

Among the exchangeable cations, *sodium* is of no significance in any of the soils investigated, even in the BC horizons that already have a higher pH, rising to 8.9 (H_2O) or 8.0 (0.1 N KCl).

The analyses of the Fe_o and Fe_d content of the "Weld, "Rago," and "Hesper" soils with particularly high increases in clay content in the subsoil (cf. Figures 7–9) do show maxima in the Bt horizon in each case. This

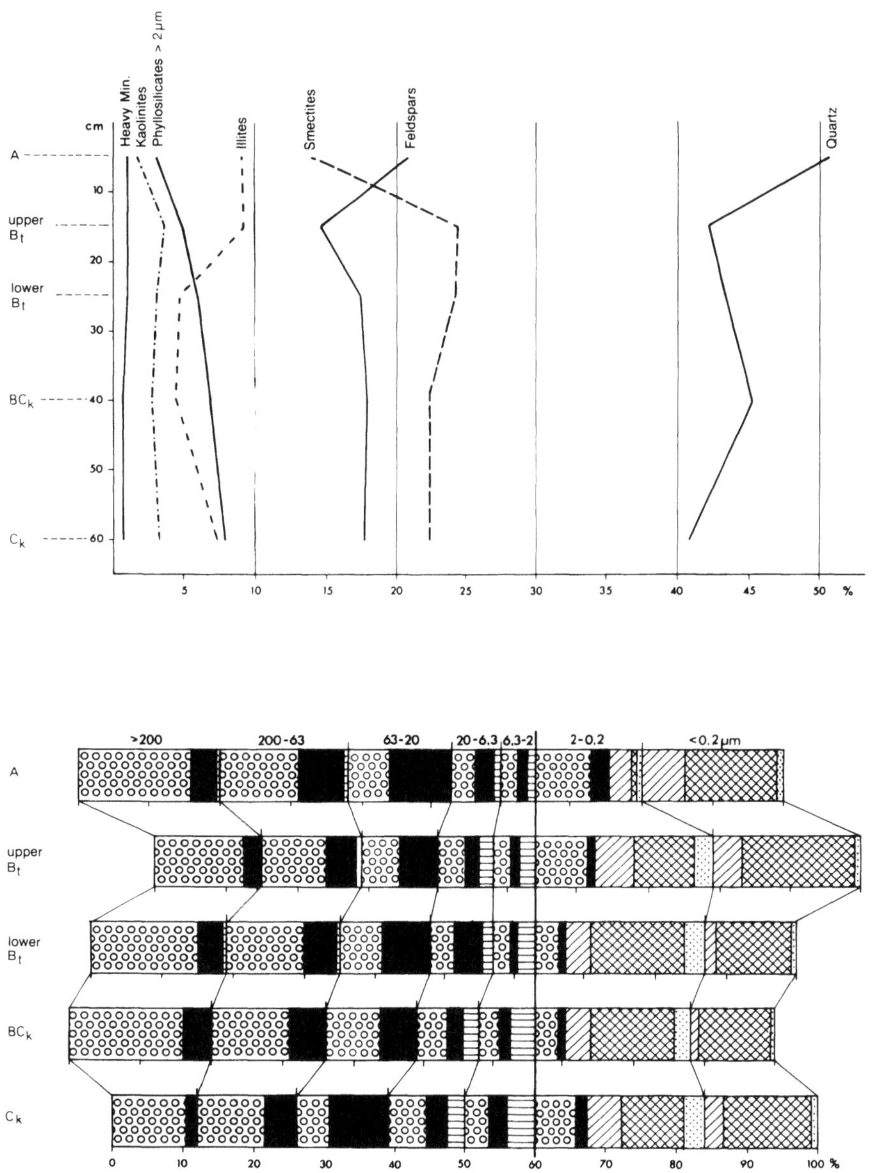

Figure 11. Mineralogical and clay mineralogical composition of an Aridic Argiboroll (Scobey series, Montana).

is not true, however, when they are converted to $Fe_{o,d}$ content per gram of clay. Thus, they are also no indication of clay illuviation.

Investigations of the content of primary minerals in the three silt fractions, and in three soils also the fractions 63 to 200 µm and >200 µm, and the clay mineralogical investigations yielded the following results.

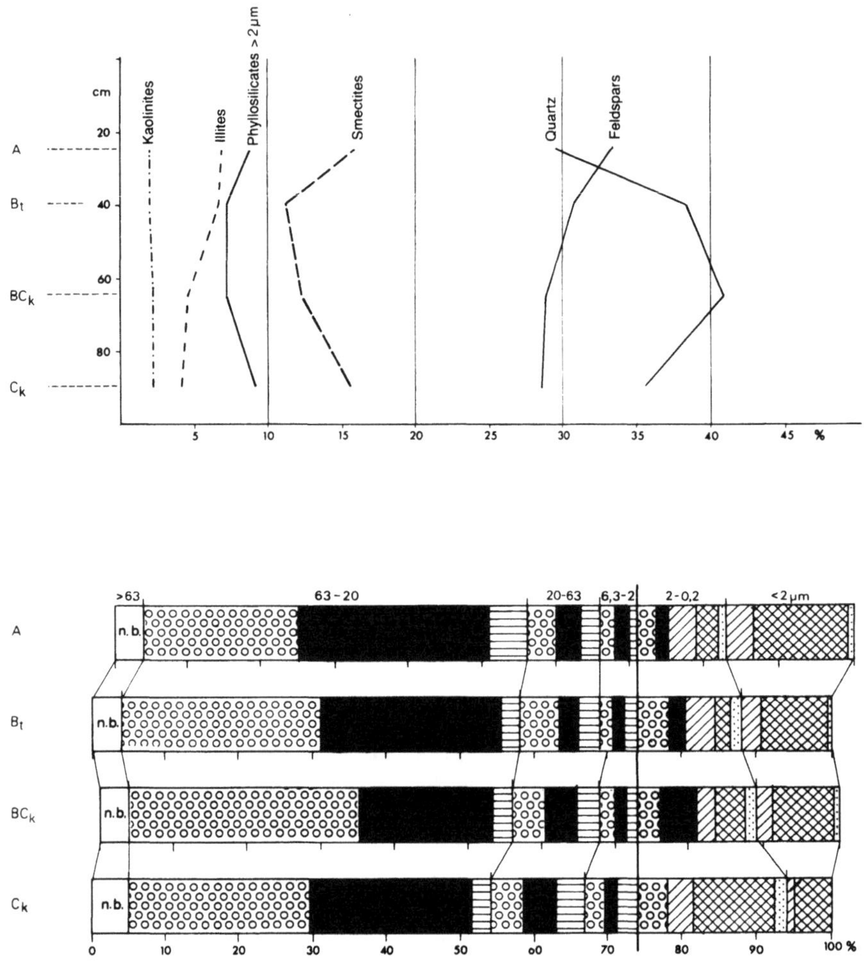

Figure 12. Mineralogical and clay mineralogical composition of an Aridic Argiustoll (Keith series, Nebraska).

The only soil we can assume was probably originally approximately homogeneous petrographically is the Typic Argiustoll ("Holdrege" soil, Figure 13), recognizable from the slight variation in the percentage by weight of quartz in the different soil horizons. We should stress that we can only determine that it is likely that a given soil was originally petrographically homogeneous. We cannot prove that this is the case, as we often read in the literature. Only an original petrographic discontinuity can be proven, often on the basis of the particle size distribution (cf. above). Assuming original petrographic homogeneity, the higher clay content in the Bt horizon, compared with the A horizon, described in Section III.B.1, cannot be attributed to clay illuviation. Here in the ustic

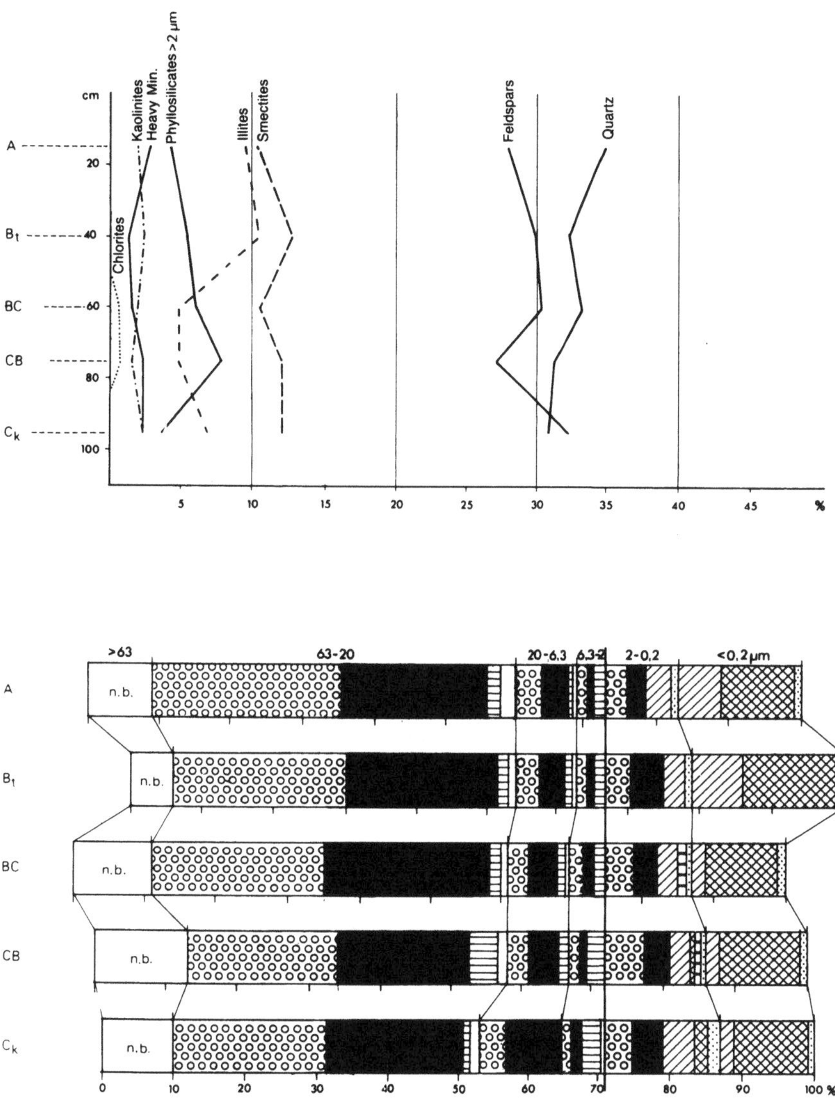

Figure 13. Mineralogical and clay mineralogical composition of a Typic Argiustoll (Holdrege series, Nebraska).

soil moisture regime—even more than in the udic soil moisture regime close to the boundary between udic and ustic (cf. Sections III.A.2 and III.B.1)—the maximum amount of weathering, and with it clay formation, takes place in the Bt horizon, because it is moister, and not in the A horizon.

In the "Keith" soil (Aridic Argiustoll, cf. Figure 12), in which no lithologic discontinuity was recognizable on the basis of the particle size distri-

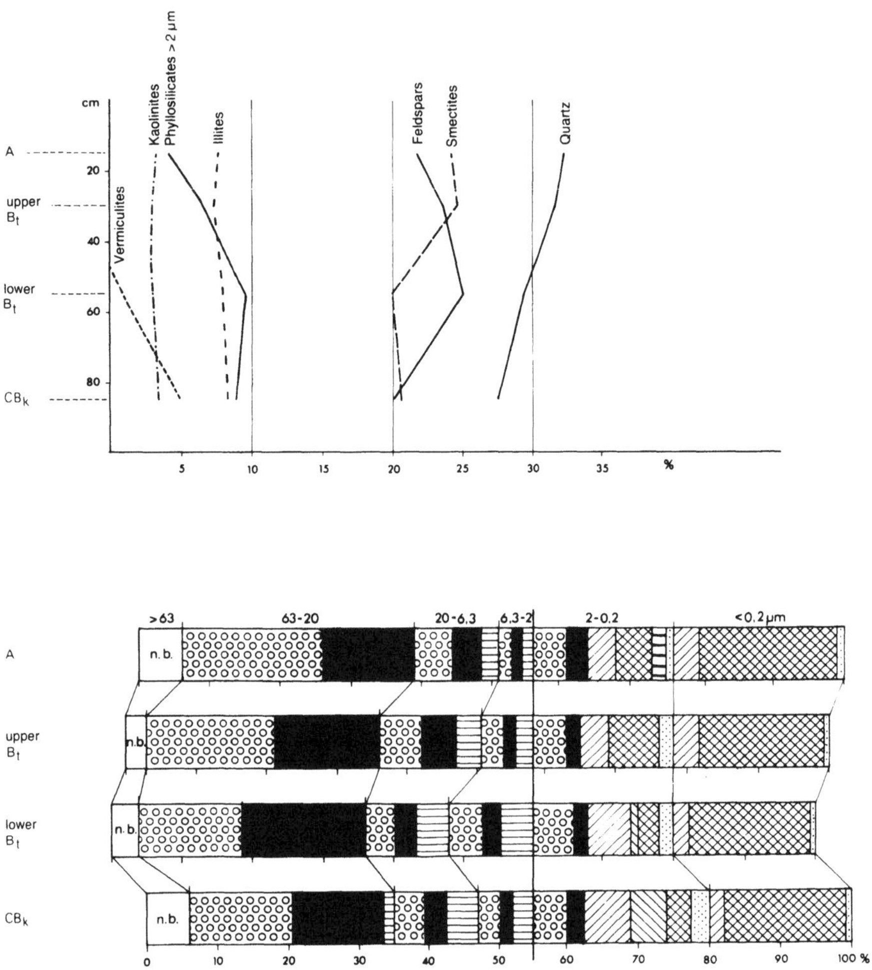

Figure 14. Mineralogical and clay mineralogical composition of a Typic Argiustoll (Reliance series, South Dakota).

bution, the great decrease in the quartz content from the BCk to the A horizon is evidence of a vertical petrographic discontinuity. In the "Williams" soil (Typic Argiboroll, cf. Figure 15), not only the particle size distribution (see above) but also the great variations in the quartz content, particularly in the lower Bt versus the A horizon, and the abrupt increase in the proportion of feldspars, particularly in the upper part of the soil profile, show that there was a petrographic discontinuity in the present material.

According to the (clay) mineralogical investigations, the large increase in clay content from the A to the Bt horizon in the selected soils of the dry western short-grass prairie, namely, the Aridic Paleustoll ("Weld"

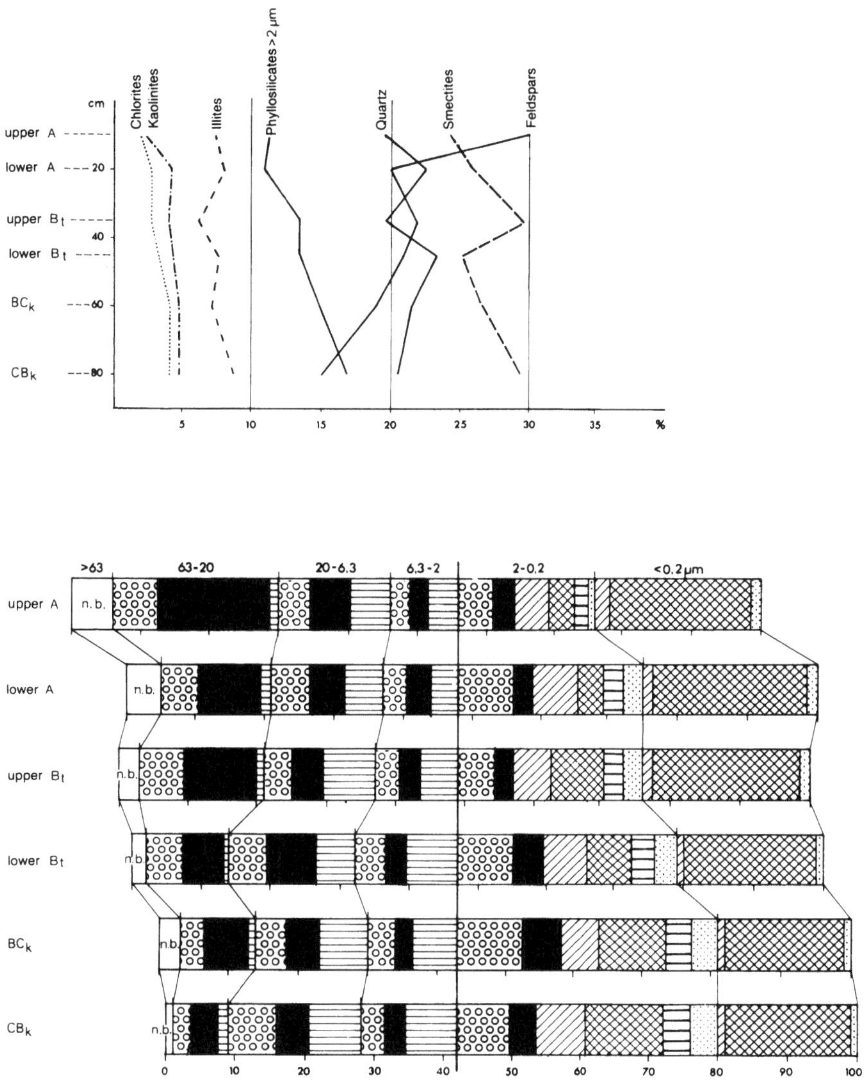

Figure 15. Mineralogical and clay mineralogical composition of a Typic Argiboroll (Williams series, North Dakota).

soil, Figure 7), the Aridic Argiustoll ("Rago" soil, Figure 8), the Ustollic Haplargid ("Hesper" soil, Figure 9), the Abruptic (Aridic) Argiboroll ("Shaak" soil, Figure 10), and the Aridic Argiboroll ("Scobey" soil, Figure 11), must also be attributed to an *original petrographic discontinuity* and not to clay illuviation. This is demonstrated by the great or even very great increase in quartz, which is resistant to weathering, from the Bt to the A horizon. In addition, there is a very great increase in feldspars, which weather poorly in this dry climate, in the Aridic Paleustoll (Figure

7), in the Ustollic Haplargid (Figure 9), and the Abruptic (Aridic) Argiboroll (Figure 10). On the whole, we can conclude that in all likelihood the *topsoil* of each of these soils developed on *more recent eolian* deposits of more or less sandy *loess* or loess-derived material. We do not consider it conceivable that the proportion of coarse silt and sand increases to this extent from the Bt to the A horizon as a result of argillipedoturbation (Hole, 1961), as in Vertisols, although micromorphological stress cutans, probably resulting from pressure caused by swelling, are frequently observed (see above). They are, however, restricted entirely to the subsoils.

Between the subsoil and Ck horizon a petrographic discontinuity can be identified, albeit on the whole relatively slight, in the Aridic Paleustoll (Figure 7), the Ustollic Haplargid (Figure 9), and in the Abruptic Aridic Argiboroll (Figure 10) on the basis of a decrease in the quartz content and in the Aridic Argiustoll (Figure 8) on the basis of an increase in the proportion of quartz. The micromorphological investigations (see above) have shown that the subsoils were more intensively weathered, in strong contrast to the A horizons. Despite a certain degree of petrographic discontinuity, we can conclude that the subsoil shows a much higher degree of weathering intensity, expressed primarily in a high degree of fine clay formation (<0.2 μm). This is associated with a considerable decrease in the feldspar content; the decrease in phyllosilicates plays a subordinate role quantitatively (cf. Figures 7–10). The most plausible interpretation is thus that these soils are relict, *polygenetic formations*. Their subsoils were at least partially the result of much older soil-forming processes, which possibly date back to the last interglacial period, the Sangamon interglaciation, or even further (cf. Section IV). This is also conceivable, considering that the loess cover tapers off toward the west (summarized by Thorp and Smith, 1952). Even though a greater age is suggested by the typologic designation Paleustoll ("Weld" soil), in any case the terms *Argi*ustoll, *Argi*boroll, and Hapl*argid* as soil types are genetically not tenable.

IV. Discussion

The results of the micromorphological and (clay) mineralogical investigations show that it is not sufficient merely to identify an increase in clay *content* of $\geq 20\%$ to designate a horizon *genetically* as an argillic horizon resulting from the process of clay illuviation. For instance, the soils of the European part of the USSR investigated here (cf. Section III.A.1; Figure 5) have a higher clay content not only in the lower part of the soil. In some of these soils the clay content of the underlying loess is even higher (cf. Afanasyeva et al., 1974), so that for this reason alone a lithologic discontinuity must be inferred. We found an extreme example that demonstrates that the clay content alone can actually be misleading in a loess soil at 2100-m altitude in the Kashmir Valley, India. The soil moisture regime is probably udic, whereas Srinagar, 20 km farther north but only 1600 m in altitude, already has a xeric soil moisture regime (Bronger et al., 1987; Figure 9). The soil lies under a *Pinus wallichiana* forest with a dense herb cover consisting of grasses and low herbs. The clay content of the mollic epipedon (0 to 30 cm), at 26.5%, is only about 20% lower than that of the Bt horizon (30% to 32%, cf. Bronger and Heinkele, 1985). The A horizon has in most parts a well-developed fine spongy fabric rich in aggregates and pores without any signs of clay illuviation. Contrasting with this, in the Bt horizon the proportion of illuviation argillans, mostly well laminated, is extremely high, higher than we have found in any other loess soil in any of the regions described in this chapter. Particularly in the upper and middle part of the Bt horizon, illuviation argillans are found not only in thick seams lining conductive channels (cf. Bronger et al., 1987, Figure 6), but large amounts are already mixed into the fabric, though still highly birefringent. In the lower, dense Bt horizon they are found almost exclusively in conductive channels. In places they are microlaminated, and occasionally they completely fill these channels. This extremely high degree of clay illuviation in this Typic Argiudoll close to an Ultic Argixeroll hardly shows up in the curve representing clay content (cf. Bronger and Heinkele, 1985, Figure 4). A possible reason for this is an original lithologic discontinuity.

In loess regions with an ustic soil moisture regime, and perhaps in the Udolls, the maximum of clay formation generally is found in the (upper) Bt horizon and not in the A horizon. This was already shown by Bourne and Whiteside (1962) in a "Medical Chernozem" from the tall-grass prairie. We were able to verify this on the basis of weathering balances for two Brown Forest soils on the border from the udic to the ustic soil moisture regime in southeast Central Europe, where only traces of illuviation argillans were observable (cf. Section III.A.2), and probably in a Typic "Argi"ustoll ("Holdrege" soil) in central Nebraska (cf. Sections III.B.1–2; Figure 13). This would only explain a ratio of clay in the Bt to that of the Ah horizon of 1.2 or less but not of 1.4 to 1.5 and more, as in the Udic Argiustolls in eastern Nebraska. These differences can only be explained to a slight degree by clay illuviation (cf. Section III.B.1). Therefore, we must assume that there is a lithologic discontinuity here, especially because these soils developed in Peoria loess and are thus probably Holocene formations. This too needs to be demonstrated by (clay) mineralogical studies.

As the clay mineraological investigations showed, in the soils of the mixed and short-grass prairie, smectites generally dominate over illites, particularly in the fine-clay fraction (<0.2 μm, cf. Figures 7–15). Ruhe (1984a,b) came to basically very similar conclusions with regard to Peoria loess and soils on a traverse extending from the upper Mississippi River through Iowa, southern Nebraska, and northwestern Kansas to the Colorado border. Here montmorillonites dominate the clay minerals (<2 μm). They account for 58% to 79%, compared with 15% to 24% for illites. Vermiculites (0 to 10%) and kaolinites (3% to 11%) play a subordinate role. This study included in particular several profiles of the "Fayette" soil (Typical Hapludalf, see above). Fenton (1983, p. 158) reported a dominance of montmorillonites in the clay fraction throughout a "Tama" soil [Typic Argi(Hapl?)udoll] in central Iowa. Both soils were described above micromorphologically (Section III.B.1) as the end members of a small-scale biosequence. Thus, a dominance of smectites or montmorillonites has been well demonstrated for the entire region from the forested areas along the upper Mississippi River to the short-grass prairie of the central and northern Great Plains. Therefore, the conclusion of Nettleton et al. (1969, cf. Section II.A) that the missing illuviation argillans ("clay skins") were never formed because of significant shrink–swell potentials cannot be applied to the area we are investigating at any rate. Illuviation argillans as an unequivocal sign of clay illuviation are observed as a general rule only in loess soils under forest and still in the forest–prairie as well, that is, in the udic soil moisture regime up to close to the udic–ustic boundary. But they are observed only in exceptional cases in the ustic soil moisture regime, and where they are observed, only to a very slight extent.

In the end, the above assumption (Section II.C) that clay illuviation also takes place in the ustic soil moisture regime leaves questions open. Why do we find, particularly in the tall-grass prairie of the central (and northern) Great Plains, not only the above-described Udic *Argi*ustills but also the Udic *Hapl*ustolls that we also investigated micromorphologically, such as the widespread "Moody" soil? Or why are there in addition to the *Argi*borolls also *Haplo*borolls, such as the "Kranzburg" soil or the "Barnes" soil? For this reason too, the most likely explanation for the much higher clay content in the Bt horizon (cf. Section III.B.1) of, e.g., the abovementioned Udic Argiustolls in eastern Nebraska—all developed on Peoria loess—is an original lithologic discontinuity.

The inference that the soils investigated in the short-grass prairie are additionally in all likelihood *relict, polygenetic soils* is to be tested in further investigations. Thermoluminescence dating to provide information on the approximate age of these soils is being undertaken.

V. Summary and Conclusions

The argillic horizon as a main diagnostic horizon in the U.S. *Soil Taxonomy* is considered genetically to be an illuvial horizon in which layer-lattice silicate clays have accumulated by illuviation to a significant extent.

In the loess soils of the climatic sequence from forest–steppe to dry steppe in Eastern Europe, which were formed during the Holocene period, clay illuviation can only be observed micromorphologically in soils under old forest patches in the forest steppe (at the border between the udic and the ustic soil moisture regimes) but not under steppe vegetation. The same holds in southeastern Central Europe. Even two Brown Forest soils formed in homogeneous loess, still in an udic soil moisture regime, show only traces of illuviation argillans in the Bt horizon. On the other hand, the maximum amount of weathering and clay formation occurs here and not in the A horizon.

In the central and northern parts of the transition from forest to prairie and in the prairies especially in the central and northern part of the Great Plains of the United States, not only Udolls, but also Borolls, Ustolls, and even Aridisols with a $\geq$20% higher clay content in subsoil are widespread. This Bt horizon is referred to genetically as a horizon of clay illuviation, or an argillic horizon. But from the micromorphology of about 300 thin sections studied, we must conclude that the clay maximum in more than 20 main and widespread soil series on uneroded stable surfaces of Typic and Aridic Argiborolls, Udic, Typic, and Aridic Argiustolls, and Ustollic Haplargids (all free of Na^+) cannot be explained by clay illuviation, or only to a very slight extent. Argiudolls in the former tall-grass prairie show very small amounts of illuviation argillans. Only the Typic Hapludalf associated with the Udolls, e.g., in the forest–prairie in Iowa, shows the micromorphological features of clay illuviation.

Additional investigations of the quantitative composition of the primary and clay minerals, including the fine-clay fraction ($<0.2\ \mu m$), of nine selected soils show that the cause of the clay maximum in the subsoil must most likely be sought in a lithologic discontinuity in the parent material (mostly loess or loess-derived sediments). This is demonstrated especially by the extremely variable content of quartz, which is a stable mineral in the temperate climatic zone (examples in Figures 7 to 15). At least four soils of the short-grass prairie with a much higher clay content in the B horizons are *polygenetic* soils or *paleosols*, in the sense that the subsoils are the result of much older soil-forming processes probably occurring during a period with a moister climate. The genetic classification of the soils investigated here as *Argi*borolls, *Argi*ustolls, and Hapl*argids* must therefore be reconsidered. The micromorphological and (clay) mineralogical investigations show that it is not sufficient merely to identify an increase in clay *content* of $\geq$20% to designate a horizon *genetically* as an argillic horizon resulting from the process of clay illuviation.

Acknowledgments

I am very grateful to A.R. Aandahl, C.R. Berdanier, C.S. Holzhey, and W.D. Nettleton, USDA, SCS National Soil Survey Laboratory, Lincoln, Nebraska; D. Lewis, State University of Nebraska, Lincoln, Nebraska; K.W. Flach, USDA, SCS, Washington, D.C.; T.E. Fenton, G.A. Miller,

F.F. Riecken, and N. Wollenhaupt, Iowa State University, Ames, Iowa; R.H. Rust, University of Minnesota, St. Paul, Minnesota; B. Worcester and C. Heidt, North Dakota State University, Fargo, and W. Freymiller and L. Joos, SCS, Bismarck, North Dakota; C.A. Mogen, SCS, Livingston, Montana; J. Stephens, SCS, Buffalo, and W. Elwonger, SCS, Sundance, Wyoming; D. Malo, South Dakota State University, Brookings, R. Radeke, SCS, Rapid City, and R. Schlepp, SCS, Kadoka, South Dakota; and R. Pollock, SCS, Aurora, Nebraska for guidance in the field and stimulating discussions in the summers of 1976, 1978, and 1980. I would especially like to thank Mrs. R. Giese, Kiel, who did most of the soil physical, chemical, and clay mineralogical work, as well as E. Kalk, Kiel, who determined the mineral composition of the sand and silt fractions of the selected soils, and the Deutsche Forschungsgemeinschaft for supporting field and laboratory work (Grants Br 303/12–14, 16, 17). My sincere thanks also go to Mrs. K.D. Dege, Kiel, for translating this chapter.

References

Aandahl, A.R. 1972. *Soils of the Great Plains.* Map 1:2.5 Mio., Lincoln, Nebr.

Aandahl, A.R. 1982. *Soils of the Great Plains. Land use, crops and grasses.* University of Nebraska Press, Lincoln.

Afanasyeva, E.A., N.I. Gorbunov, I.I. Karmanov, I.P. Kokovina, E.T. Muzychkin, G.S. Grin, V.D. Kissel, N.I. Polupan, V.F. Ivanov, R.T. Molchanov, and A.G. Nazarov. 1974. *Guide to soil excursion: The East-European plain. Forest–steppe and steppe zones. Tour 1.* Nauka, Moscow.

Akhtyrtsev, B.P. 1979. Genesis of Gray Forest soils. *Sov. Soil Sci.* 11:521–531.

Allan, R.J., and F.D. Hole. 1968. Clay accumulation in some hapludalfs as related to calcerous till and incorporated loess on drumlins in Wisconsin. *Soil Sci. Soc. Am. Proc.* 32:403–408.

Allen, B.L., and D.S. Fanning. 1983. Composition and soil genesis. pp. 141–192. In: L.P. Wilding, N.E. Smeck, and G.F. Hall (eds.), *Pedogenesis and soil taxonomy (I. Concepts and interactions).* Elsevier, New York.

Anderson, D.W. 1987. Pedogenesis in the grassland and adjacent forests of the Great Plains. *Adv. Soil Sci.* 7:53–93.

Andreyeva, N.P., and V.V. Redkov. 1983. Chernozems and some features of their humus, northern Kazakhstan. *Sov. Soil Sci.* 2:30–37.

Arnaud, R.J.S., and E.P. Whiteside. 1963. Physical breakdown in relation to soil development. *J. Soil Sci.* 14:267–281.

Arneman, H.F. 1963. Soils of Minnesota. Extension Bulletin-278:1–8. Agricultural Ext. Serv., University of Minnesota and USDA.

Arnold, R.W. 1965. Multiple working hypothesis in soil genesis. *Soil Sci. Soc. Am. Proc.* 29:717–724.

Arnold, R.W., and F.F. Riecken. 1964. Grainy gray ped coatings in brunizem soils. *Proc. Iowa Acad. Sci.* 71:350–360.

Barshad, I. 1967. Chemistry of soil development. pp. 1–70. F.E. Bear (ed.), *Chemistry of the soil.* Reinhold, New York.

Bartelli, L.J., and R.T. Odell. 1960. Laboratory studies and genesis of a clay enriched horizon in the lowest part of the solum of some brunizem and Gray–Brown–podzolic soils in Illinois. *Soil Sci. Soc. Am. Proc.* 24:391–395.

Borovskiy, V.M., and U.U. Uspanov. 1972. Soils of Kazakhstan and ways of using them in the economy. *Sov. Soil Sci.* 4:19–33.

Bourne, W.C., and E.P. Whiteside. 1962. A study of the morphology and pedogenesis of a medial chernozem developed in loess. *Soil Sci. Soc. Am. Proc.* 26:484–490.

Breburda, J. 1966. Entstehung von Böden mit Tondurchschlämmung in der Sowjetunion. *Z. Pflanzenernahr. Düng. Bodenkd.* 111:65–72.

Breburda, J., G. Heide, and H. Zakosek. 1972. Die wichtigsten Böden der europäischen Sowjetunion. *Fortschr. Geol. Rheinl. Westfal.* 21:109–132.

Breus, N.M., E.P. Shevyakov, D.I. Kovalishin, and M.B. Kutsykovich. 1984. Mineralogical composition of the Gray Forest soils of the Ukrainian forest–steppe. *Sov. Soil. Sci.* 3:81–93.

Brewer, R. 1964. *Fabric and mineral analysis of soils.* John Wiley & Sons, New York.

Bronger, A. 1966. Lösse, ihre Verbraunungszonen und fossilen Böden—ein Beitrag zur Gliederung des oberen Pleistozäns von Südbaden. *Schr. Geograph. Inst. Univ. Kiel* 25/2:1–113.

Bronger, A. 1969/1970. Zur Mikromorphologie und zum Tonmineralbestand quartärer Lößböden in Südbaden. *Geoderma* 3:281–320.

Bronger, A. 1976. Zur quartären Klima- und Landschaftsgeschichte des Karpatenbeckens auf paläopedologischer und bodengeographischer Grundlage. *Kieler. Geogr. Schr.* 45:1–268.

Bronger, A. 1977. Die Sequenzen Braunerde (-Lessivé)—erdiger Braunlehm (-Leissivé). Tonverlagerung als Klimaindikator? *Soil Micromorphology, Proc. 5th Int. Working Meeting on Soil Micromorphology, Granada, Spain* 1:303–323.

Bronger, A. 1978. Climatic sequences of steppe soils from Eastern Europe and the USA with emphasis on the genesis of the "argillic horizon." *Catena* 5:33–51.

Bronger, A. 1979. Zur Struktur und Genese der Bodendecke der Great Plains der USA—Ein Überblick. *Kieler. Geogr. Schr.* 50:471–490.

Bronger, A. 1980. Zur neuen "Soil Taxonomy" der USA aus bodengeographischer Sicht. *Peterm. Geogr. Mitt.* 124:253–262.

Bronger, A. 1986. Argillic horizons in modern loess soils in an ustic soil moisture regime in the temperate climatic zone? *Trans. XIII. Congr Int. Soil Sci. Soc. Hamburg* III:1068–1069.

Bronger, A., and N. Bruhn. 1990. Clay illuvation in semiarid–tropical (SAT) Alfisols; a first approach to a new concept. *Developments in Soil Science* 19 (Soil Micromorphology: a basic and applied science): 175–181 Elsever, Amsterdam.

Bronger, A., and T. Heinkele. 1985. Tonverlagerung in Böden mit einem "xeric" Bodenwasserhaushalt? *Mitt. Dtsch. Bodenkundl. Ges.* 43:719–724.

Bronger, A., E. Kalk, and D. Schröder 1976. Über Glimmer- und Feldspatverwitterung sowie Entstehung und Umwandlung von Tonmineralen in rezenten und fossilen Lößböden. *Geoderma* 16:21–54.

Bronger, A., R.K. Pant, and A.K. Singhvi. 1987. Pleistocene climatic changes and landscape evolution in the Kashmir Basin, India: Paleopedologic and chronostratigraphic studies. *Quatern. Res.* 27:167–181.

Bullock, P., N. Federoff, A. Jongerius, G. Stoops, T. Tursina, and U. Babel. 1985. *Handbook for soil thin section description*. Waine Research Publications, Wolverhampton.

Bullock, P., and M.L. Thompson. 1985. Micromorphology of Alfisols. pp. 17–47. In: L.A. Douglas and M.L. Thompson (eds.), *Soil Micromorphology and Soil Classification* (SSSA Special Publication No. 15), Madison/Wisc.

Catt, J.A. 1989. Relict Properties in Soils of the Central and North-West European Temperate Region. In: A. Bronger and J.A. Catt (eds.), *Paleopedology, Nature and Application of Paleosols, Catena* Suppl. 16:41–58.

Cline, A.J., and D.D. Johnson. 1963. Threads of genesis in the seventh approximation. *Soil Sci. Soc. Am. Proc.* 27:220–223.

Culver, J.R., and F. Gray. 1968. Morphology and genesis of some grayish clay pan soils of Oklahoma. *Soil Sci. Soc. Am. Proc.* 32:851–857.

Dankert, W.N., and J.V. Drew. 1970. Pedogenetic distribution of zinc in Mollisols and associated entisols in Nebraska. *Soil Sci. Soc. Am. Proc.* 34:916–919.

Docucaëv, V.V. 1883. *Russian chernozem*. Israel Program for Scientific Translation, Jerusalem, 1967.

Durasov, A.M., and Y.N. Almayev. 1970. Genetic peculiarities of typical and leached chernozems in the high trans-Volga region (based on the group composition of humus). *Sov. Soil Sci.* 2:641–647.

Ehwald, E. 1984. V.V. Dokucaevs "Russkij Cernozëm" und seine Bedeutung für die Entwicklung der Bodenkunde und Geoökologie. *Peterm. Geogr. Mitt.* 128:1–11.

FAO–Unesco. 1974. *Soil Map of the World*, 1:5 Mio. Vol. I: Legend. Paris.

FAO–Unesco. 1977. *Soil Map of the World*, 1:5 Mio, Vols. V-2 and VIII-1. Paris.

Fenneman, N.M. 1931. *Physiography of the western United States*. McGraw-Hill, New York.

Fenneman, N.M. 1938. *Physiography of the eastern United States*. McGraw-Hill, New York.

Fenton, T.E. 1983. Mollisols. In: Wilding, L.P., N.E. Smeck, and G.F. Hall (Eds.): *Pedogenesis and Soil Taxonomy. II. The Soil Orders*. Amsterdam: Elsevier, 125–163.

Fiziko-Geograficheskiy Atlas Mira [*Physicogeographical World Atlas*]. 1964. Moscow. Translation of the legend matter and explanatory text in *Soviet Geography: Review and Translation* Vol. IV, Nos 5–6:1–403, 1965, American Geographical Society, New York.

Florov, N. 1929. Zur Frage der Degradierung der dunkelfarbigen Böden von Nordamerika. *Bodenkundl. Forsch. Beih. Mitt. Int. Bodenkundl. Ges.* 1:200–224.

Fridland, V.M. 1967. Structure of the soil mantle in the principal soil zones and subzones of western USSR. *Soviet Soil Sci.* 5:589–597.

Fridland, V.M. 1974. Structure of the soil mantle. *Geoderma* 12:35–41.

Fridland, V.M. 1981. Principles of construction of the profile-genetic component of a basic soil classification. *Sov. Soil Sci.* 3:73–88.

Fridland, V.M., and A.A. Erokhina. 1963. Comparative genetic characteristics of soils of North America, USSR and western Europe. *Vopr. Genet. Pochv.* 216–266. Published in English by Amerind. Publ. Co, New Delhi, 1976.

Fröhling, J. 1983. *Untersuchungen von Bodensequenzen der nördlichen und mittleren Langgras-Prärie unter besonderer Berücksichtigung der Genese des "albic horizon."* Diplomarbeit, Kiel.

Fullerton, D.S., and R.B. Colton. 1986. Stratigraphy and correlation on the Montana plains. pp. 69–82. In: V. Sibrava, D.Q. Bowen, and G.M. Richmond (eds.), *Quaternary Glaciations in the Northern Hemisphere*. Pergamon Press, Oxford.

Gavrilyuk, F.Y. 1979. Concerning the soil classification problem. *Sov. Soil Sci.* 11:446–448.

Gerasimov. I.P. 1969. Revision of the genetic principles of Dokucaev's soil science in the new American classification of soils and in works on compilation of the New World soil map. *Sov. Soil Sci.* 6:511–524.

Gerasimov, I.P. 1975. Experiment in genetic diagnosis of the soils of the USSR on the basis of elementary soil processes. *Sov. Soil Sci.* 7:257–263.

Gerasimov, I.P. 1980. FAO–Unesco world soils units in the light of the concept of elementary soil processes (ESP). *Sov. Soil Sci.* 12:148–151.

Gile, L.H. 1970. Soils of the Rio Grande Valley border in southern New Mexico. *Soil Sci. Soc. Am. Proc.* 34:465–472.

Glazovskaya, M.A. 1983. *Soils of the World. Vol. I: Soil Families and Soil Types*. Balkema, Rotterdam.

Glazovskaya, M.A. 1984. *Soils of the World. Vol. II: Soil Geography*. Balkema, Rotterdam.

Goss, D.W., S.J. Smith, and B.A. Stewart. 1973a. Movement of added clay through calcareous material. *Geoderma* 9:97–103.

Goss, D.W., S.J. Smith, B.A. Stewart, and O.R. Jones. 1973b. Fate of suspended sediment during basin recharge. *Water Resources Res.* 9(3):668–675.

Haase, G., and R. Schmidt. 1970. Die Struktur der Bodendecke und ihre Kennzeichnung. *Albrecht Thaer. Arch.* 14:399–412.

Hallberg, G.R. 1986. Pre-Wisconsin glacial stratigraphy of the central plains region in Iowa, Nebraska, Kansas and Missouri. pp. 11–15. In: V. Sibrava, D.Q. Bowen, and G.M. Richmond (eds.), *Quaternary glaciations in the Northern Hemisphere (Report of the International Geological Correlation Programme Project 24*. Pergamon Press, Oxford.

Hallberg, G.R., and T.J. Kemmis. 1986. Stratigraphy and correlation of the glacial deposits of the Des Moines and James lobes and adjacent areas in North Dakota, South Dakota, Minnesota and Iowa. pp. 65–68. V. Sibrava, D.Q. Bowen, and G.M. Richmond (eds.), *Quaternary glaciations in the Northern Hemisphere (Report of the International Geological Correlation Programme Project 24)*. Pergamon Press, Oxford.

Heidt, C.J. 1977. *Interactions between soils and natural vegetation*. M.Sc. thesis, North Dakota State University, Fargo.

Hole, F.D. 1961. A classification of pedoturbation and some other processes and factors of soil formation in relation to isotropism and anisotropism. *Soil Sci.* 91:375–377.

Holliday, V.T. 1985. Morphology of late Holocene soils at the Lubbock Lake archeological site, Texas. *Soil Sci. Soc. Am. J.* 49:938–946.

Holliday, V.T. 1985b. Early Holocene soils at the Lubbock Lake archeological site, Texas. *Catena* 12:61–78.

Holzhey, C.S., W.D. Nettleton, and R.D. Yeck. 1973. Microfabric of some argillic horizons in udic, xeric and torric soil environments of the United States. pp. 747–760. *Soil Microscopy: Proc. Fourth Int. Working Meeting on Soil Microscopy, Ontario.*

Hubrich, H., G. Markuse, P. Rytz, and H. Schlüter. 1976. Stationäre Geographische Forschungen im Zentralen Schwarzerdegebiet der UdSSR. *Petermanns Geogr. Mitt.* 120:37–42.

Hunt, C.B. 1974. *Natural regions of the United States and Canada*. W. H. Freeman & Co., San Francisco.
Ivanova, E.N., and N.N. Rozov (eds.). 1967. *Classification and determination of soil types, Nos. 1–5*. Israel Programme for Scientific Translation, Jerusalem.
Ivanova, E.N., P.A. Letunov, N.N. Rozov, V.M. Fridland, J. Shashko, and S.A. Shuvalov. 1962. *Soil-geographical zoning of the USSR in relation to the agricultural usage of lands*. Israel Programme for Scientific Translation, Jerusalem.
Ivanova, Y.N., Y.V. Lobova, N.A. Nogina, N.N. Rozov, V.M. Fridland, and S.A. Shuvalov. 1969. Present status of the doctrine of soil genesis in the USSR I. *Sov. Soil Sci.* 1:265–277.
Ivanova, Y.N., Y.V. Lobova, N.A. Nogina, N.N. Rozov, V.M. Fridland, and S.A. Shuvalov. 1970. Development and current status of research on the genesis of soils in the USSR II. *Sov. Soil Sci.* 2:129–144.
Jenny, H. 1941. *Factors of soil formation*. McGraw-Hill, New York.
Jenny, H. 1980. The soil resource. Origin and behavior. *Ecol. Stud.* 37:1–377.
Jordan, R.H. 1867. Formation and transfer of clay in Hastings silt loam, a Udic Arginstoll. M.Sc. thesis, University of Nebraska, Lincoln.
Kauritschewa, I.S., and I.D. Gromyko. 1974. *Atlas Poch CCCP [Soil Atlas of the USSR]*. Kolos, Moscow.
Klima Atlas von Ungarn. 1967. Volume II. Akadémiai Kiadó, Budapest.
Knapp, R. 1965. *Die Vegetation von Nord- und Mittelamerika und der Hawaii-Inseln*. G. Fischer, Stuttgart.
Köppen, W. 1939. Klimakunde von Rußland in Europa und Asien. pp. 1–96. In: W. Köppen and R. Geiger (eds.), *Handbuch der Klimatologie*. (Bd. III, Teil N 2: Tabellen,) Borntraeger, Berlin.
Kovda, V.A. 1975. *Soil map of the world*. Compiled by G.V. Dobrovsky, L.A. Dvokhov, V.A. Kovda, E.V. Lobova, B.G. Rozanov, N.N. Rozov, E.M. Samoilova, M.N. Stroquanova, and V.D. Vasilievskaya, 1:10 Mio. Moscow.
Kovda, V.A., B.G. Rozanonv, and Y.M. Samoylova. 1969. Soil map of the world. *Sov. Soil Sci.* 6:1–9.
Kubiena, W.L. 1956. Mikromorphologie, Systematik und Entwicklung der rezenten und fossilen Lößböden. *Eiszeitalt. Gegenw.* 7:102–112.
Küchler, A.W. 1964. *Manual to accompany the map potential natural vegetation of the conterminous United States*. Am. Geogr. Soc., Spec. Publ. No. 36.
Küchler, A.W. 1972. The oscillations of the mixed prairie in Kansas. *Erdkunde* 26:120–129.
Kundler, P. 1959. Zur Kenntnis der Rasenpodsole und Grauen Waldböden Mittelrußlands im Vergleich mit den Sols lessivés des westlichen Europas. *Z. Pflanzenernähr. Düng. Bodenkd.* 86:16–36.
Laves, D., and G. Jähn. 1972. Zur quantitativen röntgenographischen Bodenton-Mineralanalyse. *Arch. Acker. Pflanzenbau Bodenkd.* 16:735–739.
Lebedeva, I.I. 1969. Nature of the illuvial horizon of light-gray forest soils and moraine loams. *Sov. Soil Sci.* 6:10–19.
Liverovskiy, Y.A. 1969. Some unresolved problems in classification and systematization of USSR soils. *Sov. Soil Sci.* 1:106–116.
Maher, J.C. 1937. Calcium carbonate content of the Peorian loess of Nebraska. Masters Thesis, University of Nebraska, Lincoln.
Marschner, F.J. 1930. *The original vegetation of Minnesota* (Map 1:500,000). Compiled from the U.S. General Land Office notes, U.S. Dept. Agriculture.

Published by North Central Forest Exp. Station, Forest Service, USDA, St. Paul, 1974.

Mather, J.R. (ed.). 1963. *Average climatic water balance data of the continents. Part III. USSR.* C.W. Thornthwaite Assoc. Lab. Climatol. Publ. Vol. XVI, No. 2:276–378, Centerton NJ.

Mather, J.R. (ed.). 1964. *Average climatic water balance data of the continents. Part VII. United States.* C.W. Thornthwaite Assoc. Lab. Climatol. Publ. Vol. XVII, No. 3:417–615, Centerton, NJ.

McClelland, J.E., C.A. Mogen, C.A. Johnson, F.W. Schroer, and J.S. Allen. 1959. Chernozems and associated soils of eastern North Dakota: Some properties and topographic relationships. *Soil Sci. Soc. Am. Proc.* 23:51–56.

McKeague, J.A. 1983. Clay skins and argillic horizons. pp. 367–387. In: P. Bullock and C.P. Murphy (eds.), *Soil micromorphology. Vol. 2, Soil Genesis.* AB Academic Publishers, Berkhamsted.

McKeague, J.A., R.K. Guertin, F. Pagé, and K.W.G. Valentine. 1978. Micromorphological evidence of illuvial clay in horizons designated Bt in the field. *Can. J. Soil Sci.* 58:179–186.

McKeague, J.A., R.K. Guertin, K.W.G. Valentine, J. Belisle, G.A. Bourbeau, A. Howell, W. Michalyna, L. Hopkins, F. Page, and L.M. Bresson. 1980. Estimating illuvial clay in soils by micromorphology. *Soil Sci.* 129:386–388.

McKeague, J.A., C. Wang, G.J. Ross, C.J. Acton, R.E. Smith, D.W. Anderson, W.W. Pettapiece, and T.M. Lord. 1981. Evaluation of criteria for argillic horizons (Bt) of soils in Canada. *Geoderma* 25:63–74.

Mehra, D.P., and M.L. Jackson. 1960. Iron oxide removal from soils and clays by a dithionite-citrate-bicarbonate system buffered with sodium bicarbonate. *Clays Clay Miner.* 7, 317–327

Meyer, B., E. Kalk, and H. Fölster. 1962. Parabraunerden aus primär carbonathaltigem Würmöß in Niedersachsen. I. Profilbilanz der ersten Folge bodengenetischer Teilprozesse: Entkalkung, Verbraunung, Mineralverwitterung. *Z. Pflanzenernähr. Düng. Bodenkd.* 99:37–54.

Miller, G.A. 1974. *Soil parent material, stratigraphy and soil development, Cedar County, Iowa.* Ph.D. Thesis, University of Iowa, Ames.

Mogen, C.A., J.E. McClelland, J.S. Allen, and F.W. Schroer. 1959. Chestnut, chernozem and associated soils of western North Dakota. *Soil Sci. Soc. Am. Proc.* 23:56–60.

Murphy, C.P., and R.A. Kemp. 1987. Micromorphology and the argillic horizon—a reappraisal. pp. 257–261. In: *Soil Micromorphology: Proc. VIIth Int. Working Meeting on Soil Micromorphology, Paris, 1985.*

Neigebauer, V., N. Vuchich, and N. Milkovich. 1983. Yugoslavian chernozem. *Sov. Soil Sci.* 15.5:10–20.

Nettleton, W.D., K.W. Flach, and B.R. Brasher. 1969. Argillic horizons without clay skins. *Soil Sci. Soc. Am. Proc.* 33:121–125.

Oschwald, W.R., F.F. Riecken, W.H. Dideriksen, F. Scholtes, and W. Schaller. 1965. *Principal soils of Iowa.* Dept. of Agronomy Spec. Rep. 42:1–77. Iowa State University, Ames.

Pawluk, S., and L. Bal. 1985. Micromorphology of selected mollic epipedons. pp. 63–83. In: L.A. Douglas and M.L. Thompson (eds.), *Soil micromorphology and soil classification.* SSSA Special Publication No. 15, Madison.

Ponomareva, V.V. 1974. Genesis of the humus profile of chernozem. *Sov. Soil Sci.* 6:393–402.

Rau, D. 1965. Untersuchungen zur Morphologie und Genese der Lößböden des Thüringer Beckens. *Abh. Zentr. Geol. Inst.* 4:1–71.
Roggenbau, M. 1983. *Landschaftsentwicklung aufgrund von Biosequenzen von Böden in der zentralen Waldprärieregion der USA*. Thesis, University of Kiel.
Rozanov, B.G. 1982. General world classification. *Sov. Soil Sci.* 4:71–78.
Rozov, N.N., and Y.N. Ivanova. 1967a. Classification of the soils of the USSR. Principles and a systematic list of soil groups. *Sov. Soil Sci.* 147–156.
Rozov, N.N., and E.N. Ivanova. 1967b. Classification of the soils of the USSR. Genetic Description and Identification of the principal subdivisions. *Sov. Soil Sci.* 288–300.
Rubilin, E.V., M.G. Kozoreva, and A.I. Zubkov. 1974. On the age of the humus in the chernozems of the European USSR. *Trans. 10th Int. Cong. Soil Science, Moscow*. VI(II):347–353.
Rubiln, E.V., and M.G. Kozyreva. 1980. Age of some chestnut soils of the European USSR. *Sov. Soil Sci* 12:1–9.
Ruhe, V.V. 1984a. Clay-mineral regions in Peoria loess, Mississippi River basin. *J. Geol.* 92:339–343.
Ruhe, V.V. 1984b. Loess-derived soils, Mississippi Valley region: I. Soil-sedimentation system. II. Soil–climate system. *Soil Sci. Soc. Am. J.* 48:859–867.
Rust, R.H. 1983. Alfisols. pp. 253–281. In: L.P. Wilding, N.E. Smeck, and G.F. Hall (eds.), *Pedogenesis and soil taxonomy. II. The soil orders*. Elsevier, Amsterdam.
Samoylova, Y.M., and V.I. Makeyeva. 1979. Chernozem–meadow soils and their identification. *Sov. Soil Sci.* 11:635–640.
Schlichting, E., and H.P. Blume. 1961. Art und Ausmaß der Veränderungen des Tonmineralbestandes typischer Böden aus jungpleistozänem Geschiebemergel und ihrer Horizonte. *Z. Pflanzenernähr. Düng. Bodenk.* 95:227–239.
Schönhals, E., G. Mavrocordat, H. Tributh, C. Chirita, S. Zanelli, and V. Mehedinti. 1982. Pedogenetische Untersuchungen an Tschernosemen und Parabraunerden der Donau–Ebene Rumäniens und des nördlichen Oberrheintieflands (vorderpfälzer Tiefland). *Gieß. Abhandl. Agrar. Wirtschaftsforsch. Eur. Ostens.* 115:1–157.
Schroeder, D. 1984. *Soils—facts and concepts*. International Potash Institute, Bern.
Severson, R.C., and H.F. Arneman. 1973. Soil characteristics of the forest–prairie ecotone in northwestern Minnesota. *Soil Sci. Soc. Am. Proc.* 37:593–599.
Shantz, H.L., and R. Zon. 1924. Natural vegetation. In: *Atlas of American agriculture*. U.S.D.A., Washington DC.
Shubina, I.G., A.S. Konovalova, and V.V. Grigor'yev. 1979. Separation (identification) of virgin and cultivated Sod-Podzolic, light gray and gray soils. *Sov. Soil Sci.* 11:433–445.
Smith, G.D. 1983. Historical background of soil taxonomy. pp. 23–49. In: L.P. Wilding, N.E. Smeck, and G.F. Hall (eds.), *Pedogenesis and soil taxonomy, I. Concepts and interactions*. Elsevier, Amsterdam.
Smith, G.D. 1986. *The Guy Smith interviews: Rationale for concepts in Soil Taxonomy*. SMSS Techn. Monograph No. 11 Washington.
Smith, G.D., W.H. Allaway, and F.F. Riecken. 1950. Prairie soils of the upper Missippi Valley. *Adv. Agron.* 2:157–205.
Smolikova, L. 1967. Polygenese der fossilen Lößböden der Tschechoslowakei im Lichte mikromorphologischer Untersuchungen. *Geoderma* 1:315–324.

Soil Conservation Service. 1967. Distribution of principal kinds of soils: Orders, suborders and great groups. In: *National Atlas of the United States of America*. Geological Survey, U.S. Dept. Interior, Washington, DC.

Soil Survey Staff. 1975. *Soil taxonomy. A basic system of soil classification for making and interpreting soil surveys. Agr. handbook No. 436*. U.S. Govt. Printing Office, Washington, DC.

Soil Survey Staff. 1987. *Keys to Soil Taxonomy*, (third printing) *SMSS Monograph No. 6*. Cornell University Ithaca, NY.

Stremme, H. 1930. Böden der feuchttrockenen gemäßigten Regionen: 1. Die Steppenschwarzerden. 2. Die Prairieböden. pp. 257–294. In: E. Blanck (ed.), *Handbuch der Bodenlehre*, Bd. III, Springer Berlin.

Thorp, J., and K.T.U. Smith. 1952. *Pleistocene eolian deposits of the United States, Alaska and parts of Canada*. Map 1:2.5 Mio. Geological Society of America, New York.

Tonkonogov, V.D., and N.Y Rubilina. 1980. Nature of the pale yellow and whitish eluvial horizons of Sod-Podzolic soils. *Sov. Soil Sci.* 12:379–389.

Tributh, H. 1971. Die tonmineralogische Zusammensetzung eines "Gewöhnlichen Tschernosems" bei Novo-Moskovsk/Ukraine. *Giess. Abh. Z. Agrar. Wirtschaftsord. Eur. Ostens* 56:27–45.

USDA (ed.) 1960. *Soils of the north central region of the United States*. USDA, Madison.

Valkov, V.F., and V.S. Kryshchenko. 1973. Argillization in the chernozems and chestnut soils of the northern Caucasus. *Sov. Soil Sci.* 5:385–391.

van Zant, K. 1979. Late glacial and postglacial pollen and plant microfossils from Lake Okoboji, northwestern Iowa. *Quatern. Res.* 12:358–380.

Volkovinster, V.I. 1969. Soil formation in the steppe basins of southern Siberia. *Sov. Soil Sci.* 1:383–391.

Volobuyev, V.R. 1976. Comparative investigations of the soil units identified in some publications on the soils of the world. *Sov. Soil Sci.* 8:48–61.

Volobuyev, V.R. 1980. Tentative comparative examination of soil classifications. *Sov. Soil Sci.* 12:158–166.

Walter, H. 1968. *Die Vegetation der Erde in Öko-physiologischer Betrachtung. Bd. II: Die gemäßigten und arktischen Zonen*, 2nd ed. G. Fischer, Jena.

Walter, H. 1974. *Die Vegetation Osteuropas, Nord- und Zentralasiens*. Fischer, Stuttgart.

Weaver, J.E. 1968. *Prairie plants and their environment. A fifty-year study in the midwest*. University of Nebraska Press, Lincoln.

Wernstedt, F.L. 1972. *World climatic data*. Climatic Data Press, Lemont, PA.

Westin, F.C., and D.D. Malo. 1978. *Soils of South Dakota (with a soil map 1:500,000)*. Plant Science Dept., South Dakota University, Brookings.

White, E.M. 1953. *The genesis and morphology of some transitional brunizem–gray brown podzolic soils*. Ph.D. dissertation, Iowa State University, Ames.

White, E.M., and F.F. Riecken. 1955. Brunizem–gray brown podzolic biosequences. *Soil Sci. Soc. Am. Proc.* 19:504–509.

Wilding, L.P., N.E. Smeck, and G.F. Hall. 1983. Glossary of Horizon designations. pp. 383–387. In: L.P. Wilding, N.E. Smeck, and G.F. Hall (eds.), *Pedogenesis and soil taxonomy, Vol. II*. Elsevier, Amsterdam.

Wright, H.E., F.C. Winter, and H.L. Patton. 1963. Two pollen diagramms from southeastern Minnesota: Problems in regional late- and post-glacial vegetational history. *Geol. Soc. Am. Bull.* 74:1371–1396.

Yarilova, Y.A. 1972. Comparative characteristics of fabric components in some chernozems of the USSR. pp. 357–369. In: S. Kowalinski (ed.), *Proceedings of the Third International Working Meeting on Soil Micromorphology 1969*. Panstwowe Wydawnictwo Nankowe, Warszawa.

Yarilova, Y.A. 1974. Micromorphology of chernozems. In: *Chernozemy SSSR [Chernozems of the USSR]*, part 1. Kolos, Moscow.

Yarilova, Y.A., and V.V. Bystritskaya. 1976. Morphological and micromorphological structure of the soils of the Khomutovskaya steppe in the Sea of Azov region. *Sov. Soil Sci.* 8:268–278.

Yarilova, Y.A., and N.V. Rubilina. 1975. Micromorphology of Sod-Podzolic soil on a moraine and a mantle loam. *Sov. Soil Sci.* 7:281–290.

Zachariae, G. 1964. Welche Bedeutung haben Enchytraeen im Waldboden? pp.57–68. In: A. Jongerius, (ed.), *Soil micromorphology*. Elsevier, Amsterdam.

Zolotun, V.P. 1974. Origin of loess deposits in the southern part of the Ukraine. *Sov. Soil Sci.* 6:1–12.

Myths and Scientific Realities of Agroforestry as a Strategy for Sustainable Management for Soils in the Tropics

R. Lal

Contents

I. Introduction

Tropical rainforest (TRF), a very valuable resource of genetic diversity and a regulator of many processes that impinge on global climate, is rapidly being depleted. Estimates of TRF conversion rate vary widely and range from 6 to 30 million hectares annually depending on what is consid-

Advances in Soil Science, Volume 15

ered as TRF and what is meant by conversion. Regardless of the accuracy of data, the current rate of deforestation of TRF is higher than ever before. In fact, rapid deforestation of TRF is one of the major environmental issues of modern times. Important among principal causes of deforestation are small-scale farming and shifting cultivation, large-scale commercial agriculture and plantation establishment, pasture development, and population resettlement schemes. The continuous need for conversion of TRF is fueled by the lack of sustainable land use and farming systems. There are two strategies to reduce demand for conversion of TRF. One is to increase production substantially from existing lands by adapting technologies that sustain economic production without causing degradation of soil and water resources. The second strategy to reduce pressure on TRF would be to restore the productivity of degraded soils and bring those back into production. For either of these strategies, trees and tree-based farming systems may be the key to achieving the goal of sustainability.

II. Farming Systems Prevalent in the Tropical Rainforest Zone

A farming system is a resource management strategy. It involves management of trees, crops, animals, labor, capital, and land. An ideal system is the one that achieves economic and sustained production while preserving the resource base and maintaining high environmental quality.

A. Predominant Farming Systems in the Tropics

A farming system may be predominantly crop-based, tree-based, or animal-based (Figure 1). Cropping systems can be an important component of the farming system.

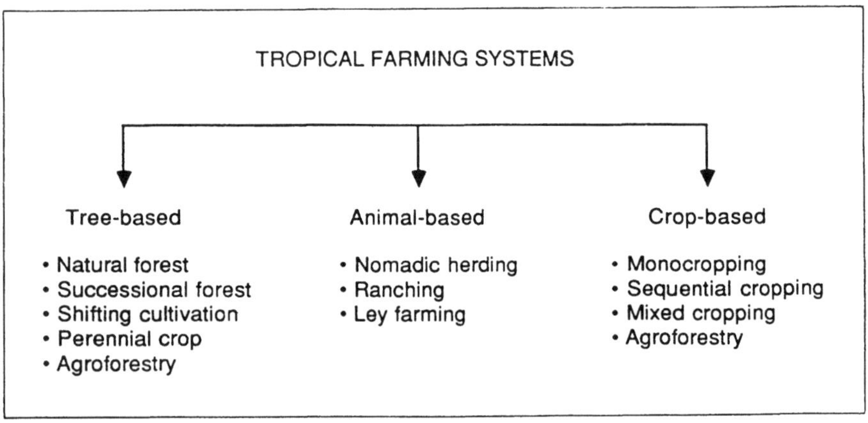

Figure 1. Major types of farming systems.

The traditional, extensive farming systems involve harvesting nonwood forest products (gathering and hunting) from primary or successional forest, shifting cultivation, and nomadic herding. Usually, traditional farming systems are subsistence, labor intensive, but based on low purchased input, complex, diverse, and, therefore, difficult to define and classify. These systems are subsistence and, with increasing population pressure, have proven unsustainable. Although a wide range of traditional farming systems are practiced throughout the humid tropical regions, there are distinct differences among the regions regarding the predominant systems of agricultural production.

In tropical Asia, systems based on intensive wetland rice cultivation have for centuries met the food needs of the world's most densely populated area. Although tree-crop plantations (rubber, oil palm, and tea) on weathered acid soils constitute a major portion of the export earnings of the agricultural sector, shifting cultivation is practiced on low-fertility and marginal soils of Thailand, Sumatra, the outer islands of Indonesia, and the highlands of northeastern India.

In the humid regions of tropical Africa, food crops are traditionally grown by shifting cultivation and related bush-fallow systems. Food needs are met primarily by the systems based on cultivation of root crops (*Manihoc* and *Dioscorea* sp.) and plantain (*Musa* sp.) grown in rotation with bush fallow. Such systems require little or no purchased inputs but have become increasingly inefficient because of shortening of the fallow period as a result of higher land pressure, i.e., increasing population and food needs.

In tropical America extensive grazing of pasture is an important form of current land use. In addition, food crops are also grown by shifting cultivation and related bush-fallow systems.

B. Characteristics of Traditional Farming Systems

Traditional upland farming systems in regions of shifting cultivation have the following general characteristics:

1. Traditional farming systems range in diversity from tree crops and nomadic herding to semipermanent and permanent agriculture. A range of crops are grown simultaneously on the same piece of land. Mixed cropping is common.
2. In shifting cultivation, forest follow of 7 to 15 years is widely used for soil fertility restoration and pest control. The duration of cropping and forest fallow varies among soils, vegetation types, population densities, and social structures.
3. Organic farming, involving mulch and household waste, is practiced around homesteads as kitchen gardens or compound farms. A large number of crop species including plantain and vegetables are grown in association, along with the raising of some animals.

4. Farm size is often very small, one that can be managed by family labor, and often ranges from less than 1 ha to 5 ha.
5. Traditional farming systems are mostly subsistence. Although some cash is generated by the sale of surplus produce, the economic status of the farm household is characterized by the lack of security to take risks.

C. Classification of Traditional Farming Systems

Traditional systems are classified on the basis of the following:

1. *Type of rotation*. The traditional rotation involves alteration between cropping for a few years and fallowing for a lengthy period to restore soil productivity (Plate 1). Cultivation is periodically shifted to areas covered by natural vegetation. Systems are classified according to the nature of fallow vegetation, e.g., forest fallow, bush fallow, savanna fallow, or grass fallow. In the humid tropics, the predominant rotations are forest and bush-fallow systems (FAO/SIDA, 1974).
2. *Cropping intensity*. Shifting cultivation systems are also classified on the basis of cropping intensity. Cropping intensity is related to the land use factor (L). The latter is defined as the ratio of cropping period plus fallow period to cropping period. Another variant of the land use factor is termed as the cultivation factor (R). The latter is defined as the ratio

Plate 1. Shifting cultivation and related bush-fallow systems have been extensively practiced in the tropics.

of cropping period to cropping period plus fallow period, expressed in percent.

3. *Types of crops*. Some important farming systems in the humid tropics are those based on tree crops. These systems involve growing food crop annuals in association with perennials. Commonly grown perennials include cocoa, oil palm, plantain, coffee, and rubber. Food crops such as upland rice (*Oryza sativa*), maize (*Zea mays*), beans (*Phaseolus* spp.), vegetables, etc. are grown in between the trees for the first 2 to 3 years.
4. *Association of crops and animals*. In some ecological regions, raising livestock is an important component of the traditional farming system. In humid tropics of Africa, small ruminants are commonly raised on farm households.

III. Need for Introducing Permanently Sustainable Farming Systems

Most traditional farming systems are subsistence; e.g., sale of crops or animals accounts for less than 25% of the farm produce (Ruthenberg, 1980). In comparison, systems based on trees or perennial crops and mixed farming, because they are larger and employ labor, are at least partly commercialized. Partly commercialized systems are those that trade between 25% and 50% of the farm produce. Because of market constraints, however, it is not always possible to introduce tree-crop-based systems.

Shifting cultivation and bush-fallow systems in the humid tropics are viable as long as there is enough land to enable 10 to 20 years fallow. If not, soil degradation and productivity decline set in rapidly, and stability of the system is lost. In fact, attempts at introducing some form of permanent cultivation in the humid tropics have persistently met some of the most troublesome ecological and social problems of acceptance. Introduction of permanent farming and continuous cultivation in regions of TRF has been successful only in exceptional situations.

A. Causes of Success of Traditional Farming Systems

Viable and intensive upland farming systems, such as those found in eastern Nigeria and Sri Lanka, owe their traditional success to the following:

1. Reduction of the fallow period to the minimum and the ability to cultivate on a permanent basis.
2. Minimum dependence on chemical fertilizers and other fossil-fuel-based inputs. Soil fertility is maintained through application of organic manure and household wastes.

3. Maintenance of a high level of soil organic matter content, which helps preserve soil structure and prevent erosion.
4. Increased farm production to raise the economic status from subsistence to a partly/fully commercialized farm and to minimize the need for clearing new land.
5. Integration of trees with grain crops.

The compound farms are usually multistorey crops involving coconuts (*Cocos nucifera*), cocoa (*Theobroma cacao*), banana (*Musa* spp.), coffee (*Coffea* spp.), cocoyam (*Xanthomonas* spp.), and cassava (*Manihot esculenta*). Cattle and small ruminants also graze in between these diverse crops.

B. Some Alternative Farming Systems

Some cropping/farming systems that have been tried as alternative to shifting cultivation and bush-fallow rotation are the following.

1. Large-Scale Mechanized Farming

Mechanized farming has been tried widely in Africa, but with limited success. Major physical obstacles to intensive cropping of maize, cowpea, groundnut, etc. by motorized farm operations in tropical Africa are severe soil compaction, accelerated erosion, and low soil fertility (Wood, 1950; de Wilde, 1967). Lal (1985) reported significant compaction of the watershed in general and headlands in particular. The crop stand, growth, and yield at the boundaries near the turning points of farm equipment were noticeably poor. Infiltration rates on compacted soils were 5 to 10 times less than the initial rate on an uncompacted soil. In the quest to shorten the turn-around time to grow a second crop, farmers/managers often try to speed up the harvesting procedure. Mechanized harvesting is particularly damaging if the soil is wet.

2. Ranching and Pasture Development

Conversion of TRF for pasture development can be successful only when the initial soil fertility is high and the stocking rate has been kept low. Severe economic and environmental problems have arisen wherever these conditions are not met. Excessive and uncontrolled grazing can cause at least as many problems as mechanized farming. Excessive grazing depletes the vegetation cover, changes the species composition, exposes soil to high-intensity rains, compacts the surface soil layer, and decreases the infiltration rate of the surface soil layer as a result of trampling (Pereira and Hosegood, 1961; Lundgren and Lundgren, 1972). Dunne (1979) observed that the sediment yield from some Kenyan catchments was related exponentially to the annual runoff. The exponent of the mean annual runoff was highest for the grazed land.

3. Forestry

It is generally believed that the establishment of forest plantations and tree crops produces an ecologically stable system (Plate 2). For example, Lugo and Liegel (1987) studied the impact of exotic plantation species on soil fertility and soil productivity in Puerto Rico. They observed that forest plantations had no harmful effects on the site's potential to support native species and did not deplete soil nutrients. However, the belief in restorative effects of forest plantations may be an overgeneralization because, with improper management, severe soil compaction and accelerated soil erosion and land degradation are also observed under some plantations. Plantations also place heavy demands for plant nutrients on the soil (Lundgren, 1978). Undoubtedly with good management, the chances of attaining ecologically compatible systems are better with establishment of tree crops. Tree crops protect the soil against raindrop impact and high insolation by their continuous vegetation cover.

In some agricultural production systems, tree crops are important components. These include the following:

1. Fruit and nut orchards and plantations yielding oil, rubber, gums, tannins, and drugs.
2. Forest plantations providing timber and related forest products.
3. Voluntary trees and shrubs maintained in natural or planted fallows to recycle nutrients for use by subsequent arable crops.
4. Trees grown in association with food crops in compound farms and as plantation crops associated with animals.
5. The Taungya system, or agrisilviculture, in which food crops are associated with forest trees for the first 2 to 3 years.
6. Forest trees as components of integrated land development plans.

Different types of forestry systems include those discussed in the following paragraphs.

a. Trees in Monocultures and Plantations

Many tree species are most profitably grown in pure stands, e.g., coconut (*Cocos nucifera*), oil palm (*Elaeis guineensis*), mango (*Mangifera indica*), papaya (*Carica papaya*), bananas and plantains (*Musa* spp.), citrus fruits (*Citrus* spp.), cocoa (*Theobroma cacao*), coffee (*Coffea* spp.), and Guaraner. Sometimes, however, it is more profitable to interplant arable crops, particularly prior to canopy development. For example, annual staples such as yams (*Dioscorea* spp.), maize (*Zea mays*), cocoyams (*Xanthomonas* spp.), and cassava (*Manihot esculenta*), intercropped with oil palms, are found to stimulate the latter's growth during the first few years but suppress it later on (Sparnaaij, 1957) (Plate 3).

Intercropping of annuals in large plantations is less practical unless carried out by smallholders under supervision (to prevent damage to trees).

a

b

c

Plate 2. Plantations of perennial crops (a) oil palm, (b) banana, (c) coffee, and others are supposedly ecologically stable systems.

Perennial crops that fruit annually, such as banana,are more frequently intercropped with annual crops or with cocoa at the early stages of plantation establishment than with tall woody tree crops. Smallholders commonly intercrop oil palms with rubber using wider spacings than for pure stands (Ochse, 1961; WAIFOR, 1960). In Nigeria and Ghana, shade-tolerant staples, such as cocoyam, are often grown along the outer margins of cocoa plantations. The land between coconuts is often planted with staples such as cassava or used to produce fodder for grazing animals. This is possible because the relatively open canopies of coconut plantations allow more light penetration than do those of oil palm or rubber plantations.

b. Agrisilviculture or the Taungya System

Agrisilviculture is the practice of growing forest trees in combination with agricultural crops. In the traditional Taungya system, parcels of forested land are allocated to farmers who clear and cultivate them for a short period during which new forest trees are planted and tended (King, 1968). This system, which in most respects is similar to a traditional long-term bush-fallow slash-and-burn approach, has the double advantage of decreasing costs of forest establishment while producing food. It is, however, wasteful of those trees that are either too young to be used or whose

Plate 3. Hedgerows of perennials are pruned just prior to sowing a food crop annual. Prunings are used as mulch.

value has not yet been recognized. This is an area where interdisciplinary research could help the smallholder considerably. *Gmelina arborea*, for example, has great potential for use in the Taungya system.

c. Trees and Woody Shrubs in Natural and Planted Fallows

The main trees used in traditional farming systems in Africa include those that become dominant during a bush-fallow period and those that are regularly planted. The former include the following (Okigbo and Lal, 1977):

1. Dominant species: *Alchornea cordifolia*, *Acioa baterii*, and *Anthonotha macrophylla*.
2. Abundant species: *Harungana madagascariensis*, *Dialium guineense*, and *Cnestis ferruginea*.
3. Common species: *Monodora tenuifolia*, *Uvaria chamae*, *Cola digitata*, *Berlinia grandifolia*, *Afzelia bella* var. *bella*, *Myrianthus arborea*, and *Rothmania* spp.

Some tree species recommended for planted fallowing include *Acioa baterii*, *Anthonatha macrophylla*, *Gliricidia sepium*, and sometimes *Dialium guineense* and *Nuclea latifolia* (Okigbo, 1976). The integrated management of such planted fallow species has hardly been researched. There has been an increasing interest in the use of *Leucena leucocephala* in the Philippines and Hawaii for nitrogen fixation, browse, and wood. However, leucaena has been seriously attacked by the insect psyllid.

d. Trees Grown in Association with Arable Crops on Compound Farms

On compound farms and home gardens, complex mixtures of tree crops, lianas, herbaceous climbers, and annuals are grown together, the diversity index approximating that of a natural tropical rain forest. Member species include the following (Okigbo, 1976):

1. Fruit trees: oil palm (*Elaeis guineensis*), mangoes (*Mangifera indica),* citrus (*Citrus* spp.), African oil bean (*Pentaclothra macrophylla*), and bitter cola (*Garcinia cola*).
2. Leaf vegetables: *Pterocarpus soyauxii* and *Afzelia bella* var. *bella.*
3. Boundary and staking plants: *Newbouldia laevis* and *Dracaona arborea.*
4. Browse plants: *Ficus* spp. and *Ricinodendron heudelotii.*
5. Lianas: *Tetracarpidum conophorum,* which produces an edible nut, and *Landolphia owariensis,* which produces an edible fruit.
6. Drug plants: *Neem* (*Azadirachta* spp.).

C. Agroforestry Systems

The concept of mixtures to create diversity is appropriately extended by growing woody and herbaceous perennials in association with seasonal annuals. The root system of perennials is believed to be deeper and the canopy higher than those of annuals, and the appropriate mixtures can be managed to optimize the utilization of both above- and below-ground resources in both space and time (Nwoboshi, 1974; King, 1979; Budowski, 1982; Kang et al., 1981; Wilson and Lal, 1986). This belief regarding the apparent compatability of mixtures of deep-rooted perennials with shallow-rooted annuals has led to many popular myths. Some of the most common are that nutrients recycled by deep-rooted woody shrubs and perennials can be used by annuals; there is little or no competition for water because of the differences in depth of active root zone; leguminous woody shrubs and trees can fix nitrogen that can be utilized by annuals through appropriate management techniques; some shrubs also produce allelopathic compounds that can supress weeds. In theory, these benefits can occur. Consequently, agroforestry systems have recently received considerable attention in the literature for economic, ecological, and environmental concerns (MacDonald, 1979; Steppler and Nair, 1987; Gholz, 1987; MacDicken and Vergara, 1990). In reality, however, there are major obstacles to harnessing the potential benefits of such a system. For acid soils, there may be no nutrients available to recycle. Woody perennials may cause excessive shading and compete for water and nutrients with shallow-rooted annuals. Woody perennials may also have adverse allelopathic effects on food crops and may even host pests that cause severe damage to the associate crop. Some of the adverse effects can no doubt be minimized by intensive management. However, agroforestry systems are being recommended without verification of the actual benefits and without fine-tuning the management packages.

Some examples of these mixtures include growing maize in association with *Leucaena leucocephala* or grazed pastures of *Panicum maximum* plus *Centrosema pubescens* with *Gliricidia sepium*. Suitable combinations of species of perennials and annuals and cultural practices of their management are different for different soils and environments. Although leguminous shrubs and trees fix some atmospheric nitrogen, their deep root system utilizes water and nutrient reserves in the subsoil. The taproot system creates macrochannels and voids to facilitate water transmission. The foliage prunings applied as mulch to the soil surface provide nutrients for the shallow-rooted annuals grown in association with the shrubs.

Agroforestry can be defined as "a system combining agricultural and tree crops of varying longevity (ranging from annual through biannual and perennial plants), arranged either temporally (crop rotation) or spatially (intercropping), to maximize and sustain agricultural yield" (Vergara, 1982). It is a landuse that involves deliberate retention, introduction, or mixture of trees or other woody perennials in crop/animal production fields to benefit from the resultant ecological and economic interactions (MacDicken and Vergare, 1990). Livestock may be an important component of agroforestry systems. There is a wide range of agroforestry systems (Nair, 1990). Different types of agroforestry systems with potential in the humid tropics are briefly outlined in Figure 2.

Rotational agroforestry is the traditional shifting cultivation in which trees and woody species of natural regeneration are rotated (5 to 40 years) with the cultivation of annuals (1 to 3 years). In spite of the extensive land use, the system has a long-term sustainability built into it (Nye and Greenland, 1960). Improvements in rotational agroforestry or shifting cultivation can be made by substituting improved over natural fallows such as multispecies tree gardens. These improved fallows may also serve as woodlots, home gardens, and compound farms. In comparison with traditional shifting cultivation based on natural fallows, tree species composition can be controlled in improved planted fallow.

Intercropping systems imply the continuous presence of both annual and perennial groups of plants on the same site at the same time. This spatially arranged system of growing food crop annuals and perennials enables continuous production of food and tree products with a minimum necessity of restorative idle fallows. Some of the examples of intercropping systems of agroforestry include alley cropping, boundary planting of trees and woody hedges, and a corridor system of land use developed by I.N.E.A.C. in the Belgian Congo (Jurion and Henry, 1969).

Given a compatible association of trees and annuals, agroforestry systems are likely to sustain economic productivity (rather than maximum productivity) without causing severe degradation of soils and environment. Because of the fragile nature of most upland soils of the tropics, some degradation is inevitable with any cultivation system. The rate of

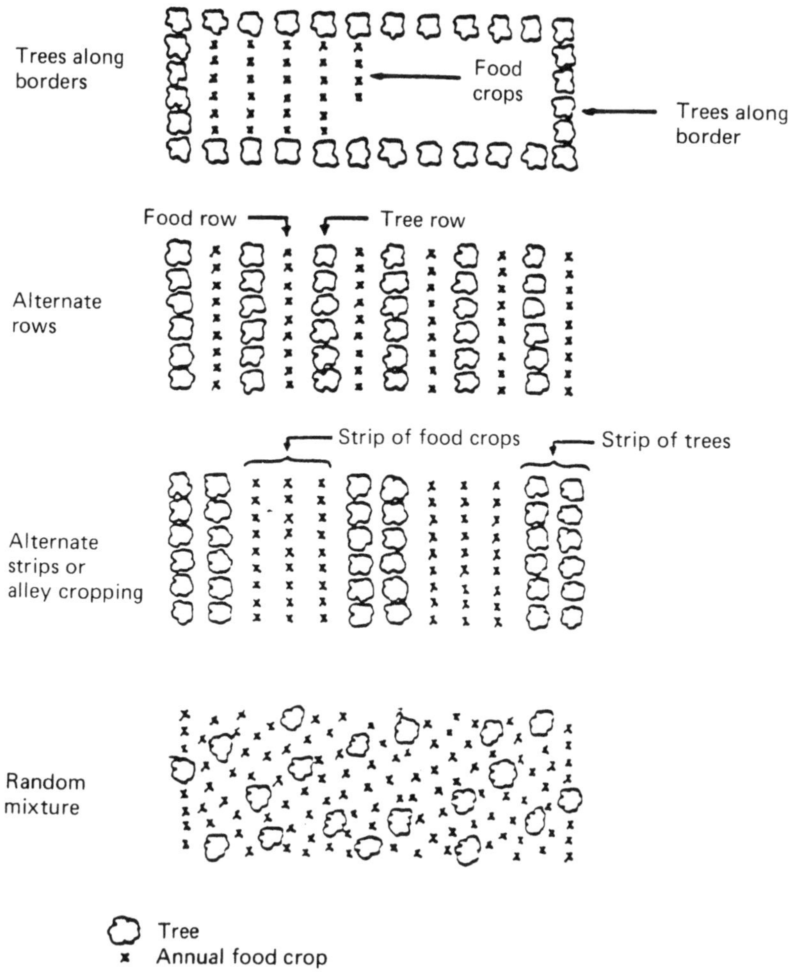

Figure 2. Integral agroforestry systems under various spatial arrangements (Vergara, 1981. Reprinted by permission from the East-West Center).

such degradation, however, is lower with agroforestry than with annual crop rotations. Soil organic matter, pH, soil structure, infiltration rate, cation-exchange capacity, and the base saturation with different cations are maintained at more favorable levels in agroforestry systems for the following reasons: reduced losses to runoff and soil erosion, efficient nutrient recycling, biological nitrogen fixation by leguminous trees, a favorable soil temperature regime, prevention of ultradesiccation, and improved macroporosity because of root and other biochannels (Lal, 1989b). It is important to note, however, that trees have both positive and negative effects on soils (Lundgren, 1978; Budowski, 1982; Nair, 1984,

1987; Young, 1986). Among negative effects are those caused by competition for limited resources (nutrients, water, and light) and alleleopathic effects. The latter are important in some tree species. Although growth characteristics of shrubs and their demand on soil moisture and nutrient reserves play an important role, management of the system is critical. Mismanaged trees can also cause severe erosion and land degradation.

There are successful examples of agroforestry systems in the humid tropics. Notable among these is the Pulpwood Project in the Philippines (Spears, 1987). The project is a traditional cash-crop-tree farming operation where trees are grown in discrete woodlots for production of pulpwood. The Paper Industries Corporation of Philippines (PICOP) has 3,800 participating farmers covering about 22,000 hectares. The key to its success was the high financial returns from tree growing. Adequate market incentives and security of land tenure were the basic factors responsible for an enthusiastic response by farmers.

Another example of the successful use of agroforestry in restoring eroded land is available from Rwanda. At Nyabisindu, a complex system combining trees, animals, and crops was developed building on community's existing knowledge. Trees and hedges were used to establish erosion-control strips yielding fruit, wood, and fodder while providing a protective ground cover against soil erosion. Extensive use was also made of perennial crops to further stabilize the soil (Dover and Talbot, 1987; Lazier et al., 1983). In Amazonian Ecuador, a sustainable system has been developed to raise tropical sheep in association with cassava and contour strips of *Inga edulis*. The latter is a deep-rooted leguminous fuelwood tree. After the cassava is harvested, a perennial leguminous ground cover, *Desmoduim,* is planted between the trees. Sheep graze the desmodium while enriching the soil (Bishop, 1983).

A key to success of these projects was building on traditional knowledge, farmers' involvement in choice of species, and economic incentives of more income than from the traditional systems. Resource conservation and land restorative effects were the additional benefits built into the systems.

IV. Alley Cropping

Alley cropping is an agroforestry system in which food crops are grown in alleys formed by contour hedgerows of trees or shrubs (Plates 4 and 5). The hedgerows are preferably established from native trees or shrubs. These trees are periodically pruned to prevent shading of the food crops. Shrubs and trees act as a windbreak, facilitate nutrient recycling, suppress weed growth, decrease runoff, and reduce soil erosion. The prunings can be used as fodder for livestock and as a source of nitrogen-rich mulch.

Plate 4. Alley-cropping maize with *Leucaena*. *Top:* Prunings of *Leucaena* at about 1.2 m above ground to facilitate manual pruning. *Bottom:* Pruning of *Leucaena* at about 10–15 cm above ground level using a tractor-driven mower.

Plate 5. Alley-cropping cowpea with *Leucaena* pruned with a mechanized tractor-driven mower just prior to sowing.

Some tree species with potential use for nonacid tropical soils are *Leucaena leucocephala, Gliricidia sepium, Acioa baterii,* and *Alchornia cordifolia*. Other species that have a potential for acid soils in such a system include *Alchornea cordifolia, Acioa baterii, Anthonatha macrophylla, Harungana madagascariensis, Dialium guineense, Cnestis ferruginea, Flamingia congesta, Nuclea latifolia, Calliandra callothysrus, Samanea saman, Inga edulis,* and *Erythrina* sp. Detailed lists of tree species for different environments are given by Fassbender (1987), NAS (1979), and Gholz (1987). The hedgerows of *Leucaena, Gliricidia,* and *Cassia* can be established from seed (following appropriate treatment for breaking seed dormancy). Other species are established from seedlings or stem cuttings. Establishment from direct stem cuttings often results in a patchy stand with a high rate of seedling mortality. Seedlings established from stem cuttings are also easily uprooted.

Such agroforestry systems, when successfully established, have a potential to produce two or more outputs, e.g., food grains, fodder, mulch, fuelwood, and staking material. However, the beneficial effects of these systems depend on many factors, e.g., tree species, land area allocated to trees, hedgerow management, crop management, soil type, and the prevalent climate. The "neighborly effects" of crop mixtures are usually of three types: (1) indifferent effects in which yield of each component is proportional to the areas and resources allocated, (2) symbiotic effects in which two species grown together benefit from one another, and (3) competitive effects in which two species compete with one another for the limited resources. The competitive behavior may be expressed in terms of aggressive growth and canopy to smother the associated crops or through allelopathy in which chemical suppressants are involved to retard the growth and physiological functions of an associated crop. From an agricultural point of view, successful systems are those that have indifferent or symbiotic effects. Symbiotic effects of crop mixtures and pastures are well established (Donald, 1963; Okigbo and Greenland, 1978; Willey, 1979; Willey and Reddy, 1981). An objective of agroforestry research, therefore, is to select species and mixtures with desirable characteristics.

A. Growth and Yield of Food Crops

If the species are chosen with indifferent effects, yields of food and tree crops will be proportional to the area allocated to each component. In general, tree crops have a much larger feeding zone or "root room" than food-crop annuals. This implies that the relative proportion of area allocated to trees and shrubs should be decreased with decreasing seasonal/annual rainfall amount. That is why widely spaced and randomly distributed *Acacia albida* is likely to be more successful in semiarid and arid regions than closely spaced hedgerows of *Leucaena* or *Gliricidia*.

A few experiments conducted in Nigeria have demonstrated that if soil moisture is not a limiting factor, some yield increase is possible in maize grown in association with *Leucaena* or *Gliricidia* (Table 1; Kang et al., 1981) and cassava grown in association with *Acioa* (Table 2; Kang and Duguma, 1985). The effects are, however, not well defined. For example, the data in Table 1 show that addition of 80 kg/ha of fertilizer resulted in more yield increase than addition of *Leucaena* prunings. Furthermore, prunings had no effect on maize yield in the fourth year. The data in Table 2 also show that the increase in cassava yield was far greater with fertil-

Table 1. Main season grain yield of maize alley cropped with *Leucaena leucocephala* as affected by application of *Leucaena* prunings and nitrogen*

N rate (kg/ha)	*Leucaena* prunings	Yield (ton/ha) 1979	1980	1981	1982	1983
0	Removed	—	1.0	0.5	0.6	0.3
0	Retained	2.1	1.9	1.2	2.1	0.9
80	Retained	3.5	3.3	1.9	2.9	3.2
	LSD (0.05)	0.4	0.3	0.3	0.4	0.8

*From Kang et al. (1984).

Table 2. Effect of alley cropping with various woody species on cassava yield on an ultisol*

Woody species/treatment	Fresh cassava yield (ton/ha)	
	Stem	Tuber
Control		
No fertilizer	7.3	14.2
With fertilizer	10.4	16.0
Acioa baterii		
Prunings only	11.4	15.0
Fertilizer only	15.3	26.1
Pruning + fertilizer	15.0	16.0
Gmelina arborea		
Prunings only	4.1	5.4
Fertilizer only	2.3	3.5
Prunings + fertilizer	5.2	6.9
Cassia siamea		
Prunings only	7.0	13.2
Fertilizer only	6.7	10.7
Prunings + fertilizer	9.2	17.2

*From Kang and Reynolds (1986).

izer than with prunings of *Acioa baterii*. Furthermore, prunings applied with fertilizer had no effect whatsoever. Effects of prunings on increase in yield in the absence of fertilizers are obtained through contributions by perennials of some essential nutrient through recycling and biological fixation. The actual amount of plant nutrients made available depends on soil properties, climate, range of nutrients made available by perennial species through prunings, and biomass production (Table 3). Also in Nigeria, the data of Kang and Mulongoy (1987) reported in Figure 3 show that maize yield from applications of 40 kg or 80 kg of N per ha as urea was more than that from an equivalent application of N from either fresh or dry loppings of *Gliricidia*. In addition to large quantities of loppings required, the efficiency and timing of application are important considerations.

The total nutrient yield from researcher-managed plots can be high. The total nitrogen and potassium contained in the biomass may range from 40 to 200 kg/ha (Fassbender, 1987; Kang and Mulongoy, 1987; Atta-Krah and Sumberg, 1987). However, the entire amount of nutrients may not be available for plant growth. The losses of nitrogen by leaching and erosion, escape in gaseous forms into the atmosphere, and immobilization by organisms are possible causes of low efficiency. Furthermore, other essential nutrients (e.g., P and Zn) may become growth-limiting for seasonal crops.

Table 3. Estimated nutrient yield from hedgerow (4-m interrow spacing) prunings of four fallow species grown in alley cropping on a degraded alfisol in southern Nigeria*

Species	Biomass yield (ton/ha per year)	Nutrient yield (kg/ha per year)				
		N	P	K	Ca	Mg
Acioa baterii	3.0	41	4	20	15	5
Alchornea cordifolia	4.0	85	6	48	42	8
Gliricidia sepium	5.5	169	11	149	104	18
Leucaena leucocephala	7.4	247	20	184	98	16

*Kang and Wilson (1987).

Another example of the effects of alley cropping with *Leucaena* on yield of maize is available from the Philippines. The data by O'Sullivan (1985) presented in Table 4 indicate a yield increase by *Leucaena* of 0.7 ton/ha without fertilizer and 0.3 ton/ha with fertilizer application. Data from another experiment from the Philippines by Watson and Laquihon (1985) (Table 5) show that *Leucaena* hedgerows increased maize yield by

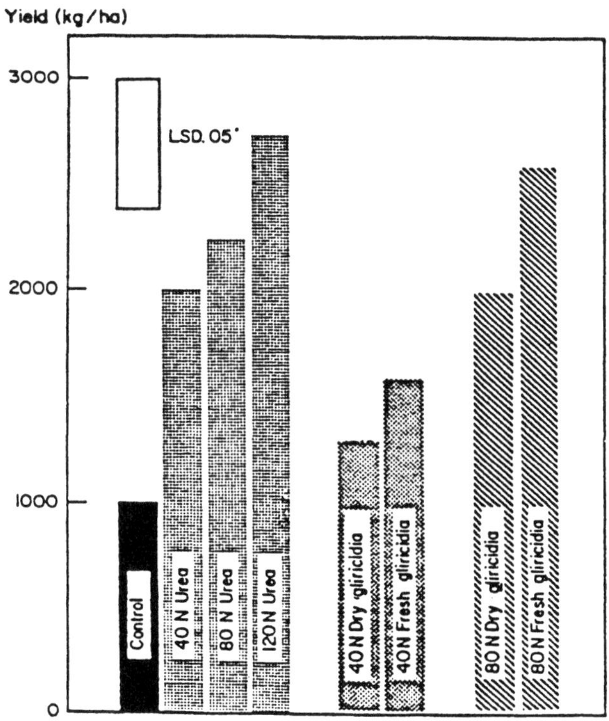

Figure 3. Effect of application of urea and *Gliricidia* loppings on grain yield of maize grown on psarmentic ustorthent (Kang and Mulongoy, 1987).

Table 4. Effect of alley cropping with *Leucaena leucocephala* on maize yield in Philippines*

Treatment	Maize yield (ton/ha)
Maize intercropped with *Leucaena*	
With fertilizer†	2.4
Without fertilizer	1.2
Maize alone	
With fertilizer†	2.1
Without fertilizer	0.5

*From O'Sullivan (1985).

†Fertilizer at 100 kg N and 22 kg P per hectare.

about 1.4 ton/ha with or without fertilizer. However, similar to the data shown in Figure 3, the nutrient use efficiency was much superior when it was applied as chemical fertilizers rather than as *Leucaena* fertilizers. In Hawaii, Guevara (1976) obtained 56% more yield of maize grown in association with *Leucaena* than without it.

Some interesting results of the effects of various tree species used as live stakes on fresh tuber yields of yam are reported from Ivory Coast by Budelman (1987) and Budelman and Pinners (1987). The data in Table 6 show that only the yam–*Gliricidia* association produced acceptable yields. Other associations suppressed growth and yield of yam, primarily because of a severe effect of shading. Shading effects of *Leucaena* can also be severe on maize, especially if pruning is not done frequently (Atta-Krah and Sumberg, 1987).

Experiments conducted on an acid tropical soil at Yurimaguas, Peru, have shown that crop yields tend to increase with wider spacing between rows of trees despite the fact that the areas with wider spacing have less mulch applied (Table 7; Tropsoils, 1987). The data in Figure 4 from Salazar and Palm (1987) show a drastic increase in maize grain yield with increase in distance from the hedgerow. Furthermore, except for cow-

Table 5. Effect of alley cropping and fertilizer applications on grain yield of maize in the Philippines*

Treatment	Grain yield (ton/ha)
No fertilizer	1.3
Leucaena hedgerows	2.7
Leucaena hedgerows + fertilizer†	3.7
Leucaena prunings equivalent to fertilizer rate	2.6
Fertilizers†	3.7

*From Watson and Laquihon (1985).

†Commercial fertilizer applied at a rate of 100 kg N and 50 kg P per hectare.

Table 6. Yam tuber yield grown in association with *Cassia siamea* and *Gliricidia sepium**

Treatment	Fresh tuber yield (ton/ha)			Average hulls per ha
	1985	1986	1987	
Monoculture	11.1	5.1	16.2	8,000
Grown with *Cassia*	6.0	4.1	10.1	6,000
Grown with *Gliricidia*	9.6	6.8	16.4	6,400

*Modified from Budelman and Pinners (1987) and Budelman (1987).

peas, grain yields of crops studied in this Yurimaquas experiment were generally low (Table 8). Furthermore, yields from alley-cropping plots were generally equal to or less than those from controls. For example, rich grain yields in rotations 4 and 6 were significantly greater in fertilized controls than in alley-cropping systems. Yields of cowpeas in cycles 2

Table 7. Grain yield of rice from an experiment conducted at Yurimaquas, Peru with three species of shrubs*

Species	Row from the shrub	Grain yield (kg/ha)
Inga	1	723
	2	1942
	3	2162
	4	1986
	5	1975
	6	2052
	8	1923
	10	1875
Erythrina	1	1582
	2	2014
	3	1888
	4	2059
	5	2056
	6	2044
	8	2107
	10	1997
Leucaena	1	1718
	2	2084
	3	1948
	4	2128
	5	2253
	6	2321
	8	2106
	10	2419

*From Tropsoils (1987).

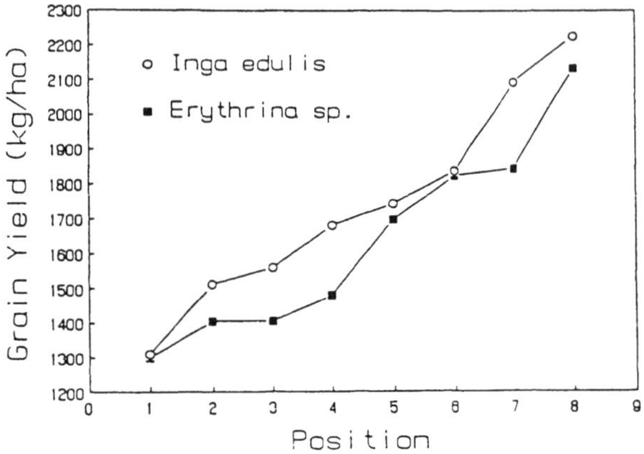

Figure 4. Corn yields at different distances from a hedge in an alley-cropping system (Salazar and Palm, 1987).

and 5 were highest in the nonfertilized control. In cycle 2, yields with *Inga* alley cropping were significantly lower than those from the nonfertilized controls. Corn yields were low regardless of the treatment. Szott (1987) concluded from these data that the inability to obtain higher yields in the alley-cropping treatments than in the nonfertilized control resulted from the shading and root competition for nutrients and water. Shade alone was not responsible for lower crop yields. Mulch from prunings,

Table 8. Mean grain and dry matter production in the cropping systems: Yurimaquas, 1984–1986*

	Cropping system (kg ha^{-1})†							
Cycle crop	Cc	Ie	Nc	Fc	Cc	Ie	Nc	Fc
	Grain‡				Dry matter			
1. Corn	634*a*		390*a*	369*a*	1762*b*		2268*b*	4339*a*
2. Cowpea	778*ab*	526*b*	1064*a*	972*ab*	1972*b*	1791*b*	2597*b*	4766*a*
3. Rice	231*a*	211*a*	488*a*	393*a*	1138*b*	1160*b*	1723*b*	3718*a*
4. Rice	156*c*	205*bc*	386*b*	905*a*	929*b*	1151*b*	2121*b*	5027*a*
5. Cowpea	415*a*	367*a*	527*a*	352*a*	1398*b*	1353*b*	1404*b*	3143*a*
6. Rice		386*b*	382*b*	1557*a*		1054*b*	1037*b*	4797*a*

*From Szott (1987).

†For grain or dry matter, means within a row that are followed by the same letter are not significantly different, based on results from the least-significant-difference test, $p = 0.05$. Cc, *Cajanus cajan* alley cropping; Ie, *Inga edulis* alley cropping; Nc, nonfertilized, nonmulched control; Fc, fertilized, nonmulched control.

‡Corn grain yield based on 15.5% moisture content; rice and cowpea grain yields based on 14% moisture content. *Inga* plots in cycle 1 and *Cajanus* plots in cycle 6 were not cropped.

despite relatively high N concentrations, immobilized N during the early stages of crop growth. The competition for N was especially severe in rice. In some cases, surface-applied prunings mulch physically impeded the emergence of crop seedlings. Severe reduction in crop yields with alley cropping and the fact that the overall yields were low on these acid soils raise the question of whether the "recycling" hypothesis is relevant to soils where subsoil is extremely low in K, Ca, Mg, and other cations. There must be something to recycle if recycling is going to be an advantage of this system.

Results of an experiment with alley cropping of maize and beans with *Gliricidia* on a relatively fertile oxisol in Costa Rica were similar to those obtained from an ultisol in Peru. Kass and Araya (1987) studied the effects of alley cropping with *Gliricidia* at three nitrogen levels of 0, 100, and 200 kg/ha per year. The data in Table 9 show that yields in alley-cropping systems were always lower than in the control plots. However, the application of mulch from *Gliricidia* prunings invariably increased the yields of maize and beans. Grain yield in mulch treatment was superior to that in alley-cropping treatments even at no input of chemical fertilizers. An important observation made by Kass (Kass and Araya, 1987) was

Table 9. Grain yields of maize and beans as affected by fertilizer, mulching, and alley cropping with *Gliricidia* on an oxic dystropept, Pital District, San Carlos, Costa Rica†

		Bean yield (kg/ha)		Maize yield (kg/ha)	
Treatment	N rate (kg/ha)	1985	1986	1985	1986
Control	0	740	1451	1799	635
	100	1213	1795	2355	310
	200	1214	1996	1717	390
	mean	1055	1957	1747	442
Mulch*	0	906	1689	2271	835
	100	1046	1874	2565	622
	200	1119	1965	2451	712
	mean	1023	1843	2429	723
Alley cropping, 6 m	0	758	1242	1543	468
	100	913	1410	1553	886
	200	916	1348	1109	524
	mean	862	1333	1402	627
Alley cropping, 9 m	0	853	1309	1610	654
	100	936	1607	1143	805
	200	1035	1443	1328	764
	mean	941	1453	1360	727

*Mulch applied at 20 ton/ha.

†Modified from Kass and Araya (1987).

that the positive response to alley cropping was associated with low inherent fertility.

Similar to acid soils, experiments conducted in India also showed a significant decrease in yield when grain crops were grown in association with tree crops (Singh, 1987). However, when *Leucaena* prunings were applied as mulch (fertilizer), there was an increase in grain yield (Table 10). For example, intercropping with *Leucaena* reduced sorghum grain yield by 24%, 11%, and 24% for nitrogen application rates of 0, 15, and 50 kg/ha, respectively. Addition of *Leucaena* prunings, on the other hand, increased sorghum grain yield by 72%, 31%, and 4% for nitrogen application rates of 0, 25, and 50 kg/ha. Addition of prunings was equivalent to a fertilizer rate of about 50 kg of N per hectare.

A very important question of sustainability can only be addressed by yield data from long-term experiments. There are only a few data from experiments that have been carried for long enough time to answer this question. Furthermore, agronomic experiments conducted on small plots do not provide adequate information on the economic feasibility of a highly labor-intensive system. Results of a field experiment conducted at the International Institute of Tropical Agriculture (IITA) for 6 consecutive years are shown in Table 11. Each crop of maize received chemical fertilizers at the rate of 100 kg N/ha, 26 kg P/ha, and 30 kg K/ha. Hedgerows were pruned three or four times during the growing season. Maize grain yields for the first 4 years from 1982 through 1985 were satisfactory and averaged about 4 ton/ha. For the last two years (1986 and 1987), maize yields were low and averaged slightly over 2 ton/ha. The low yields were partly caused by drought but also by depletion of soil fertility and degradation of soil properties caused by intensive cultivation. Furthermore, the ratio of average maize grain yield in agroforestry versus control was 0.98, 0.78, 0.95, 0.98, 1.0, and 0.97 for the years 1982, 1983, 1984, 1985, 1986, and 1987, respectively. Regardless of the management sys-

Table 10. Effects of *Leucaena* prunings and of intercropping sorghum with *Leucaena* on grain yield of sorghum*

	Grain yield (10^2 kg/ha) at different N rates (kg/ha)		
Treatment	0	25	50
1. Control	6.2	12.1	16.8
2. Addition of *Leucaena* prunings	10.7	15.8	17.4
3. Intercropping with *Leucaena*	4.7	10.8	12.7
4. Intercropping with *Leucaena* plus pruning	9.3	13.9	16.6
LSD (0.05)	1.8		

*From Singh (1987).

Table 11. Effects of agroforestry systems on relative grain yields of maize on a tropical alfisol*

Treatment	Relative grain yield for different years†					
	1982	1983	1984	1985	1986	1987
Plow-till	102	122	90	107	67	57
No-till	100	102	100	124	60	67
Leucaena, 4 m	92	82	92	119	52	50
Leucaena, 2 m	109	90	95	104	42	62
Gliricidia, 4 m	97	97	90	112	65	55
Gliricidia, 2 m	95	95	82	119	40	70

*Modified from Lal (1989b).

†Relative yield is calculated as a ratio of actual yield to average yield of all seasons and all treatments expresed in percent. Fertilizer applied at the rate of 100 kg N, 30 kg K, and 26 kg P per hectare.

tem, maize grain yield declined with cultivation duration at the rate of 340 kg/ha per year or 8.5% per year (Figure 5). The maximum yield reduction of 22% from alley cropping was observed in 1983. For the 6-year period as a whole, maize grain yield from alley-cropping systems was about 10% lower than that of the control. This reduction was almost proportional to the area planted to hedgerows. Maize grown in association with *Leucaena* or *Gliricidia* in this system is a typical example of the "indifferent effects."

In case of severe drought stress, however, *Leucaena* with its deep and extensive root system can agressively compete for soil moisture with maize. The data in Table 12 show the yield of maize that experienced severe drought within the first 6 weeks after sowing. In comparison with no-till, there was a reduction in grain yield by 17% by *Leucaena* and by 8% by *Gliricidia*. The harvest index was also reduced.

In addition to results from the small-scale plots, large-scale experi-

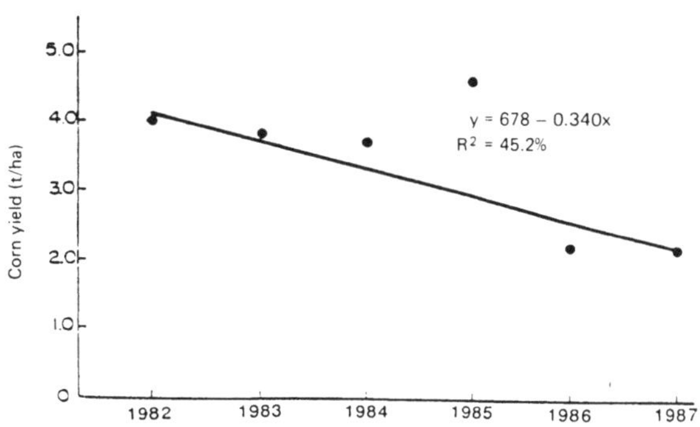

Figure 5. Decline in yield of corn with continuous cultivation (Lal, 1989b).

Table 12. Maize response to alley cropping with *Leucaena* and *Gliricidia* 6 years after establishing the hedgerows on an alfisol in Western Nigeria

Treatment	Grain yield (ton/ha)	Stover yield (ton/ha)	Harvest index (%)
Plowed control	2.3	2.6	46.9
No-till	2.7	2.9	48.2
Leucaena, 4 m	2.0	3.1	39.2
Leucaena, 2 m	2.5	3.0	45.4
Gliricidia, 4 m	2.2	3.1	41.5
Gliricidia, 2 m	2.8	3.1	47.5

ments involving plot size of 2 to 3 ha were also conducted at IITA. *Leucaena* hedgerows were established in 1984 on plots that were somewhat degraded from past management. Yield comparisons reported in Table 13 are among plots where maize was either alley cropped with *Leucaena* or grown in rotation with *Mucuna utilis*. Although the comparisons are difficult to make from a purely statistical point of view, the data presented indicate an interesting scenario. Even with 1 year of *Mucuna* fallowing, total grain yield of maize grown in rotation with *Mucuna* over the 5-year period was 33% more than that grown in association with *Leucaena*. A significant part of this difference was attributable to the loss of space to *Leucaena* hedgerows. Furthermore, mechanized prunings of *Leucaena* created soil compaction in a zone of about 50 cm on both sides of the row.

Cowpea was grown in the second season following maize. No fertilizers were applied to the cowpea. In contrast to maize, cowpea germination

Table 13. Maize grain yield in mechanized system with and without alley cropping with *Leucaena**

	Grain yield (ton/ha)	
Year	Maize alley cropped with *Leucaena* (plot 5)	Maize in rotation with *Mucuna* (plot 3)
1984	1.6	2.3
1985	1.3	1.6
1986	0.6	*Mucuna*
1987	1.2	2.4
1988	1.7	2.2
Total	6.4	8.5

*Unpublished data from IITA's West Bank experiment. *Leucaena* hedgerows were established in 1984.

and seedling establishment were significantly suppressed by *Leucaena*- and *Gliricidia*-based systems. Cowpea germination in agroforestry systems was often as low as 10% to 20% of the control. Patchy and low cowpea stands often necessitated reseeding for the second or even third crop. Crop stand generally improved, however, if cowpea was sown 2 to 3 weeks after the fresh prunings of *Leucaena* and *Gliricidia* were incorporated into the soil. Cowpea stand was always the lowest when sown 1 to 2 days after the *Leucaena* and *Gliricidia* prunings were turned under. Because the second season in subhumid regions of West Africa is short, and rains usually end abruptly toward the end of October, an early sowing is necessary to ensure some yields.

The low cowpea stand may result from allelopathic effects. Preliminary experiments have shown that leachate from freshly pruned leaves of *Leucaena* and *Gliricidia* inhibits cowpea germination. Atta-Krah and Sumberg (1987) also observed allelopathic effects of *Gliricidia* on some weeds. The most likely allelopathy to have occurred under these situations is the possibility of some biologically active substances being liberated from the freshly decomposed leaves and shoots. Such allelopathic mechanisms are common even among crop mixtures.

Grain yields of cowpea from 1982 through 1987 (Table 14) show that yields were greater in the control plots than in agroforestry systems. The ratio of average cowpea yield in control to alley-cropping systems was 1.33, 1.29, 1.35, 1.56, 3.56, and 1.31 for 1982, 1983, 1984, 1985, 1986, and 1987, respectively. Yield suppression was comparatively more by *Leucaena* than by *Gliricidia*. With some exceptions, yield suppression was more for 2-m than 4-m spacing. Over and above the reduction in crop stand, the possibility of competitive effects among legumes cannot be ruled out. Similar adverse effects are observed on ground nut when grown in association with *Acacia albida* in the Sahel. Cowpea may have been

Table 14. Effects of agroforestry systems on relative grain yields of cowpea on a tropical alfisol*

	Relative grain yield for different years†					
Treatment	1982	1983	1984	1985	1986	1987
Plow-till	128	78	79	77	176	65
No-till	270	147	212	139	177	38
Leucaena, 4 m	177	91	103	73	51	39
Leucaena, 2 m	129	57	89	28	26	42
Gliricidia, 4 m	168	106	119	105	80	37
Gliricidia, 2 m	124	95	120	72	41	40

*Modified from Lal (1989b).

†Relative yield is calculated as a ratio of actual yield to average yield of all seasons and all treatments, expressed in percent. No fertilizer was applied.

adversely affected by the competition between root systems for nutrients and between shoot systems for light. Specific mechanisms for these effects are not known. Competitive effects of perennial hedgerows are also reported for maize, sorghum, and castor bean in semiarid regions (Hoekstra, 1982; Singh et al., 1986; Randhawa and Venkateswarlu, 1986).

It is relevant to evaluate the graphic presentation of yield data over a period of 6 years to assess the long-term trends. Regardless of the systems, yield of maize and cowpea declined with the cultivation duration. The yield decline with time, however, was more consistent in cowpea than maize (Figure 6). The average grain yield declined at the rate of 96 kg/ha per year or 17% per year. Similar to the data on maize, the response of cowpea grain yield to alley cropping with *Leucaena* in a mechanized system involving plot size of 2 to 3 ha is shown in Table 15. With the exception of 1987, grain yield of cowpea grown in association with *Leucaena* was lower by about 40% than the yield of rotation-grown cowpea. The total yield over a period of 4 years was about 5% lower when cowpea was grown in association with *Leucaena* than without it (1664 kg versus 1742 kg).

Some attempts have been made to evaluate the economic feasibility of the alley-cropping system. Verinumba et al. (1984) and Ngambeki (1985) concluded that alley-cropping maize with *Leucaena* may be economical under severe cash constraints and where hired labor is available at relatively low cost. Kass and Araya (1987) concluded from experiments in Costa Rica that economic analysis of the data would not lead to a recommendation of alley cropping as a means of increasing food production in Costa Rica. It is apparent, therefore, that the available data on alley cropping do not yet support the idea that this system can sustain production without substantial input of chemical fertilizers. In fact, soil degradation and attendant yield reductions are observed even with application of fertilizers.

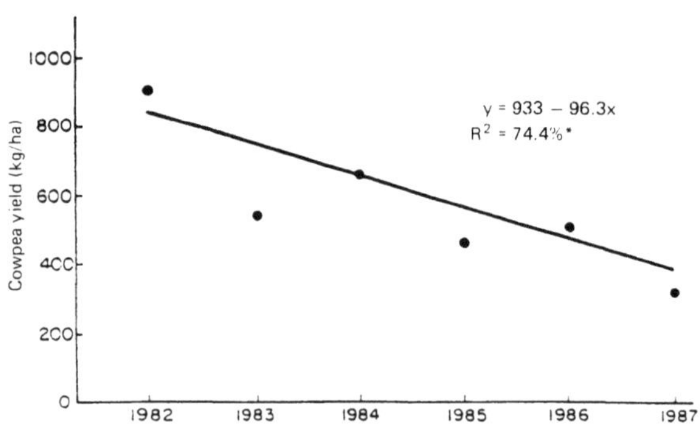

Figure 6. Decline in yield of cowpea with continuous cultivation (Lal, 1989b).

Table 15. Cowpea grain yield in mechanized system with and without alley cropping with *Leucaena**

	Cowpea grain yield (kg/ha)	
Year	Cowpea alley cropped with *Leucaena* (plot 5)	Cowpea following maize grown in rotation with *Mucuna* (plot 3)
1985	397	673
1986	310	*Mucuna*
1987	615	434
1988	342	635
Total	1664	1742

*Unpublished data from IITA's West Bank. *Leucaena* hedgerows were established in 1984.

B. Erosion Control

All other factors remaining the same, the soil erosion hazard is likely to be greater from crop-based than tree-based farming systems. Similarly, agroforestry-based systems are likely to cause more erosion than pure-stand forestry systems. The usefulness of an agroforestry system in erosion control and stabilizing uplands has been demonstrated for Java (Wiersum, 1981, 1984), the Philippines (Pacardo and Montecillo, 1983), Kenya (Raintree and Torres, 1986), Sumatra (Sukmana et al., 1985), and Nigeria (Lal, 1989b). The data in Table 16 from Nigeria show significant reductions in runoff and erosion by alley cropping, especially when *Leucaena* hedgerows are established 2 m apart. The data in Tables 17 through 20 indicate that, all other factors remaining the same, soil erosion from sloping land can be drastically reduced using agroforestry systems based

Table 16. Alley-cropping effects on runoff and soil erosion under maize–cowpea rotation measured in 1984*

	Runoff		
Treatment	Millimeters	Percentage of rainfall	Soil erosion (ton/ha per year)
Plow-till	232	17.1	14.9
No-till	6	0.4	0.03
Leucaena, 4 m	10	0.7	0.2
Leucaena, 2 m	13	1.0	0.1
Gliricidia, 4 m	20	1.5	1.7
Gliricidia, 2 m	38	2.8	3.3

*From Lal (1987a).

Table 17. Runoff control with alley cropping on an alfisol in Nigeria soon after pruning and sowing of maize (April–May, 1986; IITA, Ibadan, Nigeria)

Treatment	Runoff (mm)	Relative runoff
Plowed control	99.5	100
No-till	10.3	10
Leucaena, 4 m	26.6	27
Leucaena, 2 m	9.0	9
Gliricidia, 4 m	30.5	31
Gliricidia, 2 m	27.8	28

on successful establishment of perennial hedges on contour at close spacings. It is apparent that properly maintained hedgerows of *Leucaena* can effectively reduce runoff and control erosion in both manual and motorized systems of soil and crop management.

Integration of trees with annuals decreases runoff and soil erosion by one or all of the following mechanisms:

1. Reducing runoff velocity by creating a barrier formed by the closely spaced hedge of shrubs.
2. Decreasing runoff amount by allowing more time for water to infiltrate into the soil.
3. Minimizing raindrop impact and sheet erosion because of the protective ground cover provided by the shrub canopy and mulch available from leaf fall and prunings.
4. Curtailing sediment transport in overland flow, especially if the sediment-carrying capacity of water runoff is not filled.

Establishment of contour hedgerows or strips of *Leucaena* on steep lands in the Philippines and Indonesia have led to the formation of natural terraces. Terraces are formed by the entrapment of washed-off soil behind the hedges. Development of natural terraces is indicative of both strength and weakness of the system. The fact that sediments are trapped indicates the conservation effectiveness of contour hedgerows. On the other hand, the fact that sediments originated and moved (prior to being

Table 18. Runoff and soil erosion with mechanized alley cropping systems on an alfisol in western Nigeria*

Treatment	Runoff (mm)	Erosion (kg/ha)
Maize (with alleys)	1.4	2
Maize	8.4	4
Maize (after grazing)	10.2	100

*Rainfall of 5 June 1986 was 61 mm.

Table 19. Runoff loss from alley-cropping maize with *Leucaena* in a mechanized system for a 4-ha watershed versus manually cultivated 0.07-ha plot (first season, 1985, IITA, Ibadan)

Treatment	Total runoff (mm)	Relative runoff (mm)
Mechanized		
Alley cropping with *Leucaena*	38.4	662
No-till (with *Leucaena*)	109.8	1893
Manual		
Alley cropping with *Leucaena*	8.7	150
No-till (without *Leucaena*)	5.8	100

trapped) is indicative of the vulnerability of the ecosystem as a whole to the forces and agents of erosion. The ridges/terraces formed are also prone to break or leak by burrowing activities of rodents or by overflow. Erosion in a situation where the terraces break may be very serious indeed. Furthermore, *Leucaena* hedges can be maintained only if repeated cuttings are done at frequent intervals. This labor-intensive management, in addition to being a drudgery, can also be a serious drain on nutrients and water.

Erosion-preventive effects of *Leucaena* hedgerows must also be considered in evaluating the economic impact of an alley-cropping system. Failure to incorporate the effect of erosion's on-site costs on a farmer's land-use decision may result in misleading information. Ehui (1988) simulated soil erosion rates and maize grain yield over a 20-year period. He observed that maize grain yield declined very rapidly for the traditional cultivation practice. In contrast, a no-till system maintained high yields over the 20-year planning horizon. The yields in 4-m and 2-m spacing of *Leucaena* hedgerows were lower than that of no-till system because of the yield penalties related to land that is taken out of production. Using

Table 20. Soil erosion from alley-cropping maize with *Leucaena* in a mechanized system for a 4-ha watershed versus manually cultivated 0.07-ha plot (first season 1985; IITA, Ibadan).

Treatment	Soil erosion (kg/ha)	Relative erosion
Mechanized		
Alley cropping with *Leucaena*	76	33
No-till (without *Leucaena*)	10,736	4668
Manual		
Alley cropping with *Leucaena*	70	30
No-till (without *Leucaena*)	230	100

a discount rate of 10%, the present value of net return per hectare was in the following order: alley cropping 2 m with herbicides < alley cropping 4 m with herbicides < traditional system < alley cropping 2 m < no-till < alley cropping 4 m. The return per unit of labor, however, was in the order of alley cropping 2 m with herbicides < alley cropping 2 m < alley cropping 4 m with herbicides < alley cropping 4 m < traditional system < no-till system. It is interesting to note that the traditional system is quite economical in terms of returns per unit of labor.

C. Soil Properties

Soil properties, determinants of soil's productive potential and its ability to regulate the life-support processes, depend on soil management and the amount, quality/nature, and frequency of inputs. Presence of shrubs and deep-rooted perennials is likely to affect soil moisture and temperature regimes and nutrient uptake and depletion. Addition of mulch may alter soil organic matter content, structure, and activity of soil macro- and microfauna.

Experiments conducted at IITA showed that intensive cultivation for growing two sequential crops per year (corn and cowpea) for 6 years of 12 consecutive crops resulted in a decrease in soil organic matter content, total N, pH, and exchangeable bases in all systems including alley cropping and control. The relative rates of decline, however, were the most severe in control and less severe in alley-cropping systems. During the third and fourth years after establishing the *Leucaena* hedgerows, there was evidence of a slight increase (though not significant) in soil pH and exchangeable bases, indicating the possibilities of nutrient recycling by deep-rooted perennials. Five years of continuous cultivation resulted in severe reduction in soil content of organic carbon (Table 21) and total nitrogen (Table 22).

Table 21. Effects of continuous cultivation and of agroforestry and tillage systems on soil organic carbon contents (%) for two depths*

	1982		1986	
Treatment	0–5 cm	5–10 cm	0–5 cm	5–10 cm
Plow-till	1.70	1.12	0.42	0.28
No-till	2.50	1.41	1.08	0.52
Leucaena, 4 m	3.01	1.59	0.90	0.91
Leucaena, 2 m	2.35	1.10	0.71	0.65
Gliricidia, 4 m	2.26	1.53	0.63	0.60
Gliricidia, 2 m	2.38	1.47	0.62	0.61
LSD (0.05)	0.12		0.12	

*From Lal (1989b).

Table 22. Effects of continuous cultivation and of agroforestry and tillage systems on total nitrogen content (%) of soil for two soil depths*

	1982		1986	
Treatment	0–5 cm	5–10 cm	0–5 cm	5–10 cm
Plot-till	0.214	0.134	0.038	0.042
No-till	0.270	0.174	0.105	0.063
Leucaena, 4 m	0.397	0.188	0.103	0.090
Leucaena, 2 m	0.305	0.160	0.070	0.059
Gliricidia, 4 m	0.242	0.191	0.066	0.067
Gliricidia, 2 m	0.256	0.182	0.056	0.038
LSD (0.05)	0.01		0.01	

*From Lal (1989b).

The data in Table 21 show that soil organic carbon contents declined in all treatments with duration of cultivation. Mean organic carbon (average of all treatments) of 2.37% and 1.41%, respectively, for 0 to 5 and 5 to 10 cm depths in 1982 declined to 1.45% and 1.37% in 1983, 1.00% and 0.90% in 1984, 1.20% and 1.03% in 1985, and 0.73% and 0.60% in 1986. By the end of the fifth year, however, soil organic carbon contents of 0 to 5 and 5 to 10 cm depths for *Leucaena*- and *Gliricidia*-based systems were significantly more than those of the plow-till system. For 0 to 5 cm depth, the highest level of soil organic carbon content was observed for the no-till treatment. Soil nitrogen contents followed a trend similar to that of organic carbon (Table 22). Total nitrogen contents of 0 to 5 and 5 to 10 cm layers, respectively, declined from 0.281% and 0.172% in 1982 to 0.135% and 0.129% in 1983, 0.101% and 0.085% in 1984, 0.107% and 0.098% in 1985, and 0.073% and 0.059% in 1986. By the end of 1986, surface 0 to 5 cm layer of no-till and *Leucaena* 4-m treatments contained significantly more nitrogen than other treatments. The highest contents of organic matter and total nitrogen were observed for 5 to 10 cm layer of the *Leucaena* 4-m treatment. Soil pH declined with cultivation from an initial value of about 6.6 to 6.7 in 1982 to a final value of about 5.5 in 1986. By 1986, pH of 0 to 5 cm depth was 5.2 for plow-till, 5.7 for *Leucaena* 4-m, and *Leucaena* 2-m, 6.0 for *Gliricidia* 4-m and Gliricidia 2-m, and 6.1 for the no-till treatment.

D. Soil Productivity

Woody perennials and tree crops characteristically produce relatively large amounts of above- and below-ground biomass. Because of their perennial nature, there is a continuous addition of organic matter and biomass to the soil. Furthermore, tree crops influence the microclimatic factors such as soil and air temperatures, net radiation reaching the ground surface, evaporative demand, etc. (Lal, 1989a,b). Expectedly, soil and air temperatures are lower during the day in the vicinity of perennial hedges than farther away from them. Under these conditions, soil organic matter content is being continuously increased, activity of soil fauna enhanced, and soil structure improved. The relative magnitude of improvements in these factors depends on tree species, whether deciduous or not, duration trees have been grown, antecedent soil properties, and the prevalent climate.

In contrast, soil disturbance by plowing for cultivation of short-duration seasonals or annuals results in increase in soil and air temperatures, decrease in soil organic matter content, reduction in pH, and deterioration in soil physical and biological properties (Lal, 1983). Unless promptly and adequately replenished, nutrient depletion by crop removal, leaching, runoff, and erosion results in a rapid loss of soil fertility. The beneficial effects of growing tree crops and woody perennials in association with seasonals and annuals are notable mainly in terms of soil and water conservation. Deep-rooted perennials, shrubs and trees, decrease (but do not eliminate) the soil degradation risks of an intensive arable land use. Their effectiveness in eliminating erosion-caused degradation can be increased by proper management, including planting shrubs on the contour, contour plowing, and mulch farming. The rates of decline of soil organic matter content, CEC, total nitrogen, and soil structure are lower in tree-based than in plow-based arable land use systems. Consequently, soil productivity can be maintained for a relatively longer duration.

V. Transfer of Alley-Cropping Technologies

The implementable technology available to date from IITA, International Crops Research Institute for the Semiarid Tropics, International Council for Research on Agroforestry, International Livestock Center for Africa, and other national and international institutions indicates that a substantial increase in food production in tropical Africa can be achieved (Lal, 1987a). New resource management technologies include improved methods of land clearing, mulch farming, and conservation tillage, cover crops and planted fallows, and agroforestry or alley cropping. Planted fallows (such as growing *Mucuna utilis, Pueraria phaseoloides,* etc. for 1 year or more) and shrubs and trees are also important components of ecologically compatible and economically sustainable cropping systems (Lal et al., 1978, 1979; Juo and Lal, 1977; Okigbo and Lal, 1977; Sanchez and Salinas, 1981; Kang and Wilson, 1987; Wilson and Lal, 1986).

In addition to food production, successful transfer of improved technologies also has important environmental implications. First, increasing production from existing land could reduce the pressure on marginal lands. As a consequence, accelerated erosion and pollution of water by sediments and agricultural chemicals will be reduced. Above all, reduction in the need to bring new land under production would theoretically decrease the rate of deforestation. Conversion of tropical rainforest has been linked to an increase in atmospheric concentration of CO_2 (Bolin, 1977; Woodwell et al., 1978; Lovelock, 1983), other greenhouse gases (Bach, 1986), and the global warning trend (Revelle, 1982). Agroforestry and improved resource management technologies should be developed to reverse these trends. An important component of a potentially improved package of technologies is the agroforestry and alley-cropping systems.

Some on-station research has been conducted over the last decade regarding the applications of agroforestry systems in general and that of alley cropping in particular. We are getting data on potentials and restraints of alley-cropping systems. The ecological limits, soil types, and rainfall regimes of alley cropping with *Leucaena* and *Gliricidia* are being tested. The major thrust to date, however, has been toward conducting basic research and developing new component/subsystem technologies addressed to alleviating the resource-based physical constraints, e.g., low soil fertility, accelerated soil erosion, lack of fodder and fuelwood, shortage of staking material, nonavailability of fence posts, etc. In the past, there has been less emphasis on transferring the agronomic information available and on evaluation of new technologies under on-farm conditions. Transformation of low-output subsistence agriculture of the tropics into an economically viable enterprise can only be achieved if the proven technologies are transferred on a wide scale. What is, then, the way ahead?

The first step would be to conduct on-farm research in alley cropping and other agroforestry systems with full participation of the farmers concerned. It is apparent that the alley-cropping system is not economical with maize–cowpea rotation involving herbicides and other purchased inputs. The new research should be planned with native shrubs and traditional cropping systems.

Second, efforts should be made to transfer the best technological package known in a limited ecological region where it has been found successful—4-m alley cropping with *Leucaena* on alfisols with no herbicides and reduced input of fertilizers.

An important issue is the most cost-effective and efficient way to achieve this important task. An option to transfer technology is through developing cooperative demonstration networks. Regional networks can be organized on the basis of ecological regions. An ecological region would be delineated on the basis of similarities in soil, rainfall regime, and socioeconomic conditions.

Site selection for establishing a regional network must be based on a diagnostic survey of the farm community. These surveys are designed to (1) identify critical physical, institutional, social, and economic factors, (2) evaluate soil types and climax vegetation of the region, (3) assess representative farm size and existing farming systems, (4) develop linkages with the community leaders, and (5) organize a feedback mechanism from farmers. The diagnostic surveys must learn farmers' biases and opinions regarding what does and what does not work.

Diagnostic surveys should form the basis of establishing agronomic demonstrations. Field demonstration plots must be organized through the local extension service. To begin with, these demonstration plots should be conducted on a few benchmark locations for which agronomic details of alley cropping and other agroforestry techniques are known. Once the plots are fully operative, field days should be organized and informal and formal discussions be held with farmers and extension agents.

The majority of farmers in the humid tropics of Africa, Asia, and Latin America are smallholders who have neither the education nor available information services. They also do not possess the resources required to avail themselves the benefits of modern technologies, nor are they in a position to take risks. Furthermore, the majority of farmers in tropical Africa are women.

A major outcome of these activities is to provide feedback to researchers regarding the performance of technologies regarding practical problems that require further fine-tuning on research plots. Extension, research, and training must be closely linked.

VI. Research Priorities

The available data from research station experimentation indicate that our knowledge of biological processes in agroforestry systems is much less than that in agricultural and forestry-based systems. There is also a need to evaluate economic feasibility and social acceptability of the basic concepts, subsystems, and components developed. It is relevant to improve conceptual understanding of local agricultural problems and develop socially acceptable solutions. Furthermore, farming systems are site-specific, and the appropriate packages of technological components vary among soils and ecological regions. The final and workable package must be fine-tuned by local adaptive research. The following paragraphs address some important researchable topics.

Choice of compatible tree species. Appropriate tree species for agroforestry systems differ among soils and environments. For example *Leucaena leucocephala* cannot be grown on acid soils, and *Gliricidia sepium* has a high rate of seedling mortality. The most suitable tree or woody shrub should be multipurpose and meet diverse needs of the farm household. Field experiments should be conducted on growth patterns of suitable tree species, their canopy characteristics and rooting patterns, decomposition rates of their litter fall and prunings, water requirements, and possible adverse effects on crops grown in association with them.

Establishing a field nursery for evaluation of tree species is of a high priority.

Spatial and rotational sequence. An agroforestry system based on an intercropping system requires validation for proper spatial arrangement. The most compatible system is the one that optimizes the use of a limited resource without causing degradation of soil quality. Spatial arrangements in an intercropping system requiring validation include zonal systems such as alley cropping, contour strips, corridor system, etc. In a zonal system, spacing of perennial hedgerows depends on tree species, cropping system, slope angle, and slope length.

Shading is a major problem when growing food crops near trees. Research is needed to develop or identify the following: (1) systems for intercropping compatible species, e.g., *Cola nitida* and *Marantachloa* sp., (2) optimal designs for intercropping tree crops with annual staples, and (3) trees that perform best in pure stands; these could then be restricted to such locations as hedgerows and windbreaks.

Land development. Site preparation and method of land development are crucial to the success of an agroforestry system. Soil compaction caused by heavy machinery can cause severe problems of erosion and accelerated degradation (Lal, 1988).

Crop husbandry. Management systems of trees and food crops are different for agroforestry than for forestry or crop-based systems. Pruning of trees and hedgerows requires knowledge of the frequency and height of prunings so that trees do not compete with food crop annuals. Growing food crops in spatial arrangements with trees in a system that is regularly harvested for crops and tree products may put a heavy demand on nutrients. Suitable crop rotations must, therefore, be developed to exploit the desirable soil-enriching or restorative characteristics of the tree-crop component of an agroforestry system. Crop management practices with agroforestry systems are site specific and require local knowledge of fertilizer needs, weed control, interrow cultivation, cropping sequence and combinations, time of seeding, time of prunings, etc.

Animal husbandry. Agroforestry systems must, in some ecological regions, serve the need for mixed farming. Some trees are not good forage crops, and others may have toxic effects if grazed directly. The choice of appropriate species, therefore, is an important consideration. Living fences are widely used in tropical America (Budowski, 1982), but the relevant scientific knowledge regarding their management is scarce. Practically oriented research is needed toward choice of species, cost, method of establishment and planting density, when and how to place/install barbed wire, toxic effects, additional economic products, management, etc.

Economics. Unless a system is proven economical in the context of on-farm situations, it will not be adopted by the farming community. There is little economic validation of the agroforestry systems described herein.

Socioeconomic considerations should be of a high priority on any research agenda. Evaluation of social acceptability of these practices by farmers should be a major consideration in the process of technology generation.

Sociopolitical. Human ecology and sociology play an important role in acceptance and spread of technologies. We need to study the specific sociopolitical and institutional constraints and then design appropriate packages that alleviate those constraints. Agroforestry, similar to soil conservation measures, has some hidden benefits and involves investment for possible future payoffs. Public policies must, therefore, be developed to stimulate and sustain farmer's interest.

VII. Modus Operandi

In addition to the researchable topics outlined above, it is the approach to research that is extremely crucial.

The existing research networks (such as those organized by ICRAF and IITA) should address the issue of involving traditional cropping systems, native shrubs, and interests of the farming community. The research must be done under "on-farm" conditions with active participation by the farming community. The research plots should be managed by the farmer so that he/she understands the potential and restraints of these practices. A key consideration of the research program is that the farmers are partners at all levels from planning to implementation and execution. Researchers should, however, work very closely with the farming community and collect all data on biophysical and socioeconomic factors for interpretation of the results and drawing valid conclusions.

Farming systems research involves an interdisciplinary approach (1) to identify critical biophysical, institutional, and socioeconomic constraints, (2) to understand underlying processes and interactions among subsystems and environments, (3) to complement the commodity-based inputs with the resource-management components, (4) to validate prototype technology under on-farm conditions, (5) to appraise socioeconomic acceptability of the improved subsystems, and (6) to obtain researchable feedback from the farmers. These activities are carried out by a multidisciplinary team. The research team must comprise at least the following five disciplines: agronomist, forester, economist, sociologist/anthropologist, and extension specialist. The team should organize and conduct on-farm research considering the following steps.

Baseline surveys. Baseline diagnostic surveys must be conducted by the agronomist, economist, and forester to evaluate (1) existing farming systems in ecological regions of interest, (2) factors characterizing the biophysical and agroeconomic environment, (3) an availability of resources and their effects on productivity, and (4) feedback from farmers about what does and what does not work.

On-farm research. The on-farm research should be conducted in collaboration with a national research institution. Field experiments need to be established to evaluate different agroforestry systems and assess crop response to management. Field work should attempt to integrate components into viable packages and assess their applicability and social acceptability.

VIII. Linking Training with Research

In comparison with forestry and arable farming, agroforestry systems are complex and little understood. The lack of trained manpower is one of the reasons that the potentials of this system are not fully exploited. Few tropical foresters know about food crop annuals just as few agronomists know about tropical tree species. Consequently, there is a scarcity of trained personnel to improve and handle known agroforestry systems and develop new and innovative ones. There is also a conspicuous misunderstanding of the potentialities and constraints of agroforestry systems among policy makers. This ignorance has resulted in the scarcity or lack of funds allocated for research and development in agroforestry.

Needless to say, therefore, that a major constraint to the development of sustainable agroforestry systems is the lack of trained personnel to address two critically needed objectives: (1) conducting on-farm adaptive research and (2) transferring the knowledge that already exists. With these objectives in mind, it is apparent that research and training must be linked together and cannot be separated. Some of the ongoing research-cum-training network should be strengthened to involve researchers and extension specialists from different countries to address the critical issues objectively. The network, involving participants from Africa, Asia, and South America, should address the training needs at two levels: (1) advanced degree-related training and (2) on-the-job training to develop new skills.

The degree-related program could be linked with universities within the region, whereby students will complete their course requirement and return to their home country to conduct thesis research under on-farm conditions in the ecological region of their interest. It is also appropriate to exchange students among countries, e.g., students from Brazil conducting thesis research in Indonesia or Ivory Coast and vice versa. On completion of the field work students must return to their respective universities to analyze the data, write up the thesis, and complete the degree program.

On-the-job training for researchers and extension agents is equally important. Short refresher courses (2 to 4 weeks) should be offered within the tropics whereby a team of researchers would provide comprehensive training in crop management, forest management, economics, sociology, and extension techniques related to agroforestry. Lectures and workshops should be supplemented by field visits to where research is being conducted to obtain the first-hand information on topics of current interest. Each participant should receive a field manual specifically prepared on state-of-the-art of agroforestry. Resource persons to deliver these lectures should be drawn from different universities and research organizations.

IX. Potential of Agroforestry: Myth or Reality

A very important question of sustainability can only be addressed by evaluating agronomic yield data in relation to economic feasibility and soil properties. Furthermore, agronomic experiments conducted on small plots do not provide adequate information on economic feasibility of a highly labor-intensive system. Many agronomic experiments are conducted without the supporting data on soil analysis.

A. Agronomic Yield

Lal (1989b) reported that the grain yield declined with cultivation duration at the rate of 340 kg/ha per year for maize and 96 kg/ha per year for cowpea. The yield decline with time, however, was more consistent in cowpea than maize. For the 6-year period as a whole, maize grain yield from alley-cropping systems was about 10% lower than that of control. The yield decline in cowpea was at 17% per year.

Comparing the results of Lal's studies with those of others from elsewhere provides a useful evaluation of the alley-cropping system. The data by Kang et al. (1985) showed that applications of 80 kg/ha of N caused more drastic increase in maize yield than addition of *Leucaena* prunings. Furthermore, prunings had no effect on maize yield in the fourth consecutive year. The data on cassava yield from an experiment by Kang and Reynolds (1986) also showed that increase in cassava yield was more drastic with fertilizer than with prunings of *Acioa baterii*. Furthermore, prunings applied with fertilizer had no effect whatsoever.

Experiments conducted on an acid tropical soil at Yurimaguas, Peru, have shown that the inability to obtain higher yields in alley-cropping treatments than in nonfertilized control resulted from (1) shading, (2) root competition for nutrients and water, (3) immobilization by mulch, and (4) possible alleleopathic effects. These data cast considerable doubt on the relevance of the "recycling" hypothesis where subsoil is extremely low in nutrients and cations. There must be something to recycle if recycling is going to be an advantage of the alley-cropping system.

B. Economic Assessment

Economic evaluation is obviously an important tool to assess a technology. Although this labor-intensive technology may be economical under severe cash constraints and where hired labor is available at relatively low cost, the available data on alley cropping does not yet support the idea that the system can sustain production without substantial input of chemical fertilizers. In fact, soil degradation and attendant yield reductions are observed even with applications of fertilizers.

Erosion control is a definite advantage of closely spaced contour hedgerows of *Leucaena* or another shrub. However, erosion can also be controlled by no-till, cover crops, or grass strips (Lal, 1984). Nonetheless, erosion-preventive effects of *Leucaena* hedgerows must also be considered in evaluating the economic impact of an alley-cropping system. Ehui (1988) simulated soil erosion rates and maize grain yield over a 20-year period. Using a discount rate of 10%, the present value of net return per hectare was in the following order: alley cropping 2 m with herbicides < traditional system < alley cropping 2 m < no-till < alley cropping with 4 m. The return per unit of labor, however, was in the order of alley cropping 2 m with herbicides < alley cropping 2m < alley cropping 4 m with herbicides < alley cropping 4 m < traditional system < no-till system. It is interesting to note that according to Ehui's analysis, the traditional system is economical in terms of return per unit of labor.

C. Soil Properties

The data on soil properties presented by Lal (1989b) indicate that intensive cultivation for growing two sequential crops per year (corn and cowpea) for 6 years of 12 consecutive crops resulted in decrease in soil organic matter content, total N, pH, and exchangeable bases in all systems including alley cropping and control. This drastic decline in soil fertility was observed in relatively fertile alfisols. The relative rates of decline, however, were somewhat less in alley cropping than with plow-based control. These results are also supported by the data from acid tropical soils in Yurimaguas, Peru (Szott, 1987). Szott observed that there was significantly more P, K, and Ca + Mg in the upper 15 cm of soil with control treatments than with the alley-cropping treatments. Fertilized control significantly exceeded all other treatments in topsoil Ca + Mg. The pH values were also significantly greater in the fertilized control than in the other treatments.

D. Research Priorities

The available data from research station experimentation indicate that our knowledge of biophysical processes in alley cropping and other agroforestry systems is much less than that in agricultural and forestry-based systems. Considerable knowledge gaps still remain in the choice of appropriate species and their management. The latter is crucial. Most experiments have been conducted with intensive and monocropping of maize, cowpea, cassava, and rice. We must evaluate agronomic and economic aspects of agroforestry using traditional cropping systems, e.g., mixed cropping at lower plant population. We need more scientific data on interactions among species, specifically in relation to competition for water, nutrients, and light. There is a need to assess possible alleleopathic effects. We do not know the synchronization of nutrients released by the pruning and the period of maximum nutrient demand by the food crop.

In addition to understanding the biophysical processes, we must assess economic feasibility of the system in the context of the on-farm situations. Human ecology and sociology play an important role in acceptance and spread of technologies. We need to study the specific sociopolitical and institutional constraints.

Agroforestry has a potential to be a sustainable alternative to shifting cultivation. To date, however, we do not have a workable system for major soils and ecological regions of the tropics. Alley cropping has shown some advantages in relatively fertile alfisols but not in other soils and harsh environments. Although innovative agroforestry systems await development, scientific community must be objective in its approach. Researchers have yet to answer the question of sustainability on the basis of data that can withstand the scrutiny of systems performance indicators.

References

Atta-Krah, A.N., and J.E. Sumberg. 1987. Studies with *Gliricidia sepium* for crop/livestock production systems in West Africa. pp. 31–43. Special Publication 987-01, Nitrogen Fixing Tree Association/CATIE, Turrialba, Costa Rica.

Bach, W. 1986. Trace gases and their influence on climate. *Nat. Resources Dev.* 24:9–124.

Bishop, J.P. 1983. Tropical forest sheep on legume forage/fuel wood fallows. *Agroforestry Syst.* 1:79–84.

Bolin, B. 1977. Changes of land biota and their importance for the carbon cycle. *Science* 196:613–615.

Budelman, A. 1987. The above-ground structural compatability of *Glemingia macrophylla, Gliricidia sepium* and *Leucaena leucocephala* as live stakes for yams. pp. 82–89. Special Publication 87-01, Nitrogen Fixing Tree Association/ CATIE, Turrialba, Costa Rica.

Budelman, A., and Pinners, E.C.M. 1987. The value of *Cassia siamea* and *Gliricidia sepium* as *in situ* support systems in yam cultivation: Experiences from farm based trials. pp. 90–91. Special Publication 87-01, Nitrogen Fixing Tree Association/CATIE, Turrialba, Costa Rica.

Budowski, G. 1982. Applicability of agroforestry systems. pp. 6–12. In: L.H. MacDonald (ed.), *Agroforestry in the African humid tropics*. UNU, Tokyo.

Budowski, G. 1987a. Living fences: A widespread agroforestry practice in Central America. pp. 169–178. In: H.L. Holz (ed.), *Agroforestry: Realities, possibilities and potentials*. Martinus Nijhoff, Dordrecht.

Budowski, G. 1987b. The development of agroforestry in Central America. pp. 69–79. In: H.A. Stepler and P.K.R. Nair (eds.), *Agroforestry: A decade of development*. ICRAF, Nairobi, Kenya.

de Wilde, J.C. 1967. *Experiences with agricultural development in tropical Africa*, Vols. I and II. The John Hopkins Press, Baltimore.
Donald, C.M. 1963. Competition among crops and pastures. *Adv. Agron.* 15:1–118.
Dover, M., and L.M. Talbot. 1987. *To feed the earth: Agro-ecology for sustainable development.* World Resources Institute, Washington, DC.
Dunne, T. 1979. Sediment yield and landuse in tropical catchments. *J. Hydrol.* 42:281–300.
Ehui, S.K. 1988. Evaluating the impact of soil erosion and productivity loss on the relative profitability of alternative landuse systems in the tropics. Evidence from Western Nigeria. *Paper presented at the 2nd biennial meeting of the Rockefeller Foundation Social Science Research Program, CIP, Lima, Peru, 26–30 Sept. 1988.*
FAO/SIDA. 1974. Shifting cultivation and soil conservation in Africa. *FAO Soils Bull.* 24:1–248.
FAO/UNEP. 1983. *Guidelines for the control of soil degradation.* FAO, Rome.
Fassbender, H.W. 1987. *Modelos edafologicos de sistemas agroforestades.* GTZ/CATIE, Turrialba, Costa Rica.
Gholz, H.L. 1987. *Agroforestry: Realities, possibilities and potentials.* Martinus Nijhoff, The Hague.
Guevara, A.B. 1976. Management of *Leucaena leucocephala* for maximum yield and nitrogen combination to intercropped corn. Ph.D. Thesis, Univ. Hawaii, Honolulu.
Hoekstra, D.A. 1982. *Leucaena leucocephala* hedgerows intercropped with maize and beans. An ex-ante analysis of a candidate agroforestry landuse system for the semiarid areas in Machakes, Kenya *Agroforestry Syst.* 1:335–345.
Juo, A.S.R., and R. Lal. 1977. The effect of fallow and continuous cultivation on the chemical and physical properties of an alfisol in Western Nigeria. *Plant Soil* 47:567–584.
Jurion, F., and J. Henry. 1969. *Can primitive farming be modernized?* Hons Series, Brussels.
Kang, B.T., and B. Duguma. 1985. Nitrogen management in alley cropping systems. In: B.T. Kang and J. Van der Heide (eds.), *Nitrogen in farming systems in the humid and subhumid tropics.* Haren, Netherlands.
Kang, B.T., H. Grimme, and T.L. Lawson. 1985. Alley cropping sequentially cropped maize and cowpea with *Leucaena* on a sandy soil in southern Nigeria. *Plant Soil* 85:267–276.
Kang, B.T., and K. Mulongoy. 1987. *Gliricidia sepium* as a source of green manure in an alley cropping system. pp. 44–58. Special Publication 87-01, Nitrogen Fixing Tree Association/CATIE, Turrialba, Costa Rica.
Kang, B.T., and L. Reynolds. 1986. *Alley farming in the humid and subhumid tropics.* IITA, Ibadan, Nigeria.
Kang, B.T., and G.F. Wilson. 1987. The development of alley cropping as a promising agroforestry technology. pp. 227–243. In: H.A. Steppler and P.K.R. Nair (eds.), *Agroforestry: A decade of development.* ICRAF, Nairobi, Kenya.
Kang, B.T., G.F. Wilson, and T.L. Lawson. 1984. *Alley cropping: A stable alternative to shifting cultivation.* IITA, Ibadan, Nigeria.
Kang, B.T., G.F. Wilson, and L. Sipkens. 1981. Alley cropping maize and *Leucaena* in southern Nigeria. *Plant Soil* 63:165–179.
Kass, D.L., and J.F. Araya. 1987. Alley cropping with *Gliricidia sepium* on farm-

er's fields in Costa Rica. pp. 50–58. Special Publication 87-01, Nitrogen Fixing Tree Association/CATIE, Turrialba, Costa Rica.
King, K.F.S. 1968. Agrisilviculture, Bull. 1. University of Ibadan, Ibadan, Nigeria.
King, K.F.S. 1979. Agroforestry and the utilization of fragile ecosystems. *Forest Ecol. Manage.* 2:161–168.
Lal, R. 1983. No-till farming. IITA Monograph 2, Ibadan, Nigeria.
Lal, R. 1984. Soil erosion from tropical arable lands and its control. *Adv. Agron.* 37:183–248.
Lal, R. 1985. Mechanized tillage systems effects on physical properties of an alfisol in watersheds cropped to maize. *Soil Tillage Res.* 6:149–161.
Lal, R. 1987a. Managing soils of sub-Saharan Africa. *Science* 236:1069–1076.
Lal, R. 1987b. *Tropical ecology and physical edaphology.* J. Wiley & Sons, Chichester, UK.
Lal, R. 1987c. Effects of soil erosion on crop productivity. *CRC Crit. Rev. Plant Sci.* 5(4):303–368.
Lal, R. 1989c. Land degradation and impact on food and other resources. In: D. Pimentel and A. Hall (eds.), *Food and natural resources.* Academic Press, Orlando, FL: 85–140
Lal, R. 1989a. Conversion of tropical rainforest: Agronomic potential and ecological consequences. *Adv. Agron.* 39:173–264.
Lal, R. 1989b. Agroforestry systems and soil surface management of a tropical alfisol, I–V. *Agroforestry Syst.* 8:1–6, 7–29, 97–111, 113–132, 197–215, 217–238.
Lal, R., G.F. Wilson, and B.N. Okigbo. 1978. No-tillage farming after various grasses and leguminous cover crops in tropical alfisols. I. Crop performance. *Field Crops Res.* 1:71–84.
Lal, R., G.F. Wilson, and B.N. Okigbo. 1979. Changes in properties of an alfisol produced by various crop covers. *Soil Sci.* 127:377–382.
Lazier, J., A. Getahun, and M. Velez. 1983. The integration of livestock production in agroforestry. pp. 84–88. In L.H. MacDonald (ed.), *Agroforestry in the African humid tropics.* United Nations University, Tokyo.
Lieth, H. 1976. Biological productivity of tropical lands. *Unasylva* 28:24–31.
Lovelock, J.E. 1983. *Gaia, a new look at life on earth.* Oxford University Press, Oxford.
Lugo, A.E., and L.H. Liegel. 1987. Comparison of plantations and natural forests in Puerto Rico. pp. 41–44. *People and the tropical forest, MAB program.* 41–44.
Lundgren, B. 1978. *Soil conditions and nutrient cycling under natural and plantation forests in Tanzanian highlands.* Dept. of Soils, Swedish Univ. of Agric. Sciences, Uppsala, Sweden.
Lundgren, B., and L. Lundgren. 1972. Comparison of some soil properties in one forest and two grassland ecosystems on Mount Mer, Tanzania. *Geogr. Ann.* 54:227–240.
MacDicken, K.G., and N.T. Vergara. 1990. *Agroforestry: Classification and management.* John Wiley & Sons, New York.
MacDonald, M.H. 1979. *Agroforestry in the African humid tropics.* The United Nations University, Tokyo.
Nair, P.K.R. 1984. *Soil productivity aspects of agroforestry.* ICRAF, Nairobi, Kenya.

Nair, P.K.R. 1987. Soil productivity under agroforestry. pp. 21–30. In: H.L. Gholz (ed.), *Agroforestry: Realities, possibilities and potentials*. Marinus Nijhoff, Dordrecht.

Nair, P.K.R. 1990. Classification of agroforestry systems. In: K.G. MacDicken and N.T. Vergara (Eds) "Agroforestry: Classification and management", J. Wiley & Sons, New York: 31-57.

National Academy of Sciences. 1979. *Tropical legumes: Resources for the future*. NAS, Washington, DC.

Ngambeki, D.S. 1985. Economic evaluation of alley cropping *Leucaena* with maize–maize and maize–cowpea in southern Nigeria. *Agric. Syst.* 17:243–258.

Nwoboshi, L.C. 1974. Soil productivity aspects of agrisilviculture in the West African tropical moist forest zone. *Afr. Soils* 19:1–13.

Nye, P.H., and D.J. Greenland. 1960. *The soils under shifting cultivation*. Commonwealth Bureau of Soils. Tech. Commun. No. 51, Harpenden, England.

Ochse, J.J. 1961. *Tropical and subtropical agriculture*. Macmillan, New York.

Okigbo, B.N. 1976. Role of legumes in small holdings of the humid tropics. In: J.M. Vincent, (eds.), *Exploiting the legume–rhizobium symbiosis in tropical agriculture*. University of Hawaii, Honolulu.

Okigbo, B.N., and D.J. Greenland. 1978. Intercropping systems in tropical Africa. pp. 63–101. In: R.I. Papendick, P.A. Sanchez, and G.B. Triplett (eds.), *Multiple cropping*. ASA Special Publication 27, Madison, WI.

Okigbo, B.N., and R. Lal. 1977. Residue mulches, inter-cropping and agri-silviculture potential in tropical Africa. pp. 54–69. In S. Hill (ed.), IFOAM Conf., Montreal.

O'Sullivan, T.E. 1985. Farming systems and soil management: The Philippines/Australian Development Assistance Program experience. In: E.T. Craswell, J.V. Remenyi, and L.G. Nallana (eds.), *Soil Erosion Management. ACIAR Proc. Series* 6:77–81.

Pacardo, E.P., and L. Montecillo. 1983. *Effects of corn/ipil-ipil cropping systems on productivity and stability of upland agroecosystems*. Annual Report, UPLB-PCARRD Res. Project.

Pereira, H.C., and P.H. Hosegood. 1961. The productivity of semiarid thorn shrub country under intensive management. *Emp. J. Exp. Agric.* 3:89–98.

Raintree, J.B., and F. Torres. 1986. The agroforestry research in farming systems perspective: The ICRAF approach. In: *IARC's Workshop of FSR 17–21 Feb. 1986*. ICRISAT, Hyderabad, India.

Randhawa, N.S., and J. Venkateswarlu. 1986. Farming systems research: Indian experience. In: *IARC's Workshop on FSR, 17–21 Feb. 1986*. ICRISAT, Hyderabad, India.

Revelle, R. 1982. Carbon dioxide and world climate. *Sci. Am.* 247:33–41.

Ruthenberg, H. 1980. *Farming systems in the tropics*, 3rd ed. Oxford University Press, Oxford.

Salazar, A., and C.A. Palm. 1987. Screening of leguminous trees for alley cropping on acid soils of the humid tropics. In: Special Publication 87-01. Nitrogen Fixing Tree Association/CATIE, Turrialba, Costa Rica.

Sanchez, P.A., and J.G. Salinas. 1981. Low input technology for managing oxisols and ultisols in tropical America. *Adv. Agron.* 34:279–306.

Singh, G.B. 1987. Agroforestry in the Indian subcontinent: Past, present and future. pp. 117–140. In H.A. Steppler and P.K.R. Nair (eds.), *Agroforestry: A decade of development*. ICRAF, Nairobi, Kenya.

Singh, R.P., R.J. Van Den Beldt, D. Hocking, and G.R. Korwar. 1986. Alley-cropping in the semiarid regions of India. *Proc. Workshop on Alley Farming for Humid and Subhumid Regions of Tropical Africa,'' 10–14 March 1986*. IITA, Ibadan, Nigeria.
Sparnaaij, L.D. 1957. Mixed cropping in palm cultivation. *J. West Afr. Inst. Oil Palm Res.* 2:244–264.
Spears, J. 1987. A development-bank perspective. pp. 53–68. In: H.A. Steppler and P.K.R. Nair (eds.), *Agroforestry: A decade of development*. ICRAF, Nairobi, Kenya.
Steppler, H.A. and P.K.R. Nair. 1987. *Agroforestry: A decade of development*. ICRAF, Nairobi, Kenya.
Sukmana, S., H. Suwardjo, A. Abdurachman, and J. Dai. 1985. Prospects of *Flamingia congesta* for reclamation and conservation of volcanic skeletal soils. *Pembr. Pen. Tanah Dan Pipuk.* 4:50–54.
Szott, L.T. 1987. Improving the productivity of shifting cultivation in the Amazon Basin of Peru through the use of leguminous vegetation. Ph.D. Dissertation, North Carolina State Univ., Raleigh.
Tropsoils/NCSU. 1987. *Technical Report for 1986/87*. North Carolina State University, Raleigh.
Vergara, N.T. 1981. Integral agroforestry: a potential strategy for stabilizing shifting cultivation and sustaining productivity of natural environment. Working Paper, EPI, East-West Center, Honolulu, HI
Vergara, N.T. 1982. *New directions in agroforestry: The potential of tropical legume trees*. EPI, East–West Centre, Honolulu.
Vergara, N.T. 1987. Agroforestry: A sustainable landuse for fragile ecosystems in the humid tropics. pp. 7–20. In: H.L. Gholz (ed.), *Agroforestry: Realities, possibilities and potentials*. Martinus Nijhoff, Dordrecht.
Verinumbe, I., H.C. Knipscheer, and E.E. Enabor. 1984. The economic potential of leguminous tree crops in zero village cropping in Nigeria: a linear programming model. Agroforestry systems 2: 129-138.
WAIFOR. 1960. How to plant, care for and feed your oil palm. *Farmers Bull.* 2, Benin, Nigeria.
Watson, H.R., and A. Laquihon. 1985. Sloping agricultural land technology (SALT) as developed by the Mindanao Baptist Rural Life Center. In: *Workshop Proc. Protection and Amelioration Roles of Agroforestry. Institute of Forest Conservation, UPLB, Los Banos, Philippines: 4–11 Sept. 1985.*
Wiersum, K.F. 1981. *Observations on agroforestry in Java, Indonesia*. Agric. Univ. Wageningen.
Wiersum, K.F. 1984. Surface erosion under various tropical agroforestry systems. In: C.L. O'Loughlin and A.J. Pearce (eds.), *Symposium on effects of forest land use on erosion and slope stability*. University of Hawaii, Honolulu.
Willey, R.W. 1979. Intercropping: Its importance and research needs. I. Competition and yield advantages. *Field Crops Abstr.* 32:1–10.
Willey, R.W., and M.S. Reddy. 1981. A field technique for separating above- and below-ground interactions in intercroppings: An experiment with pearl millet groundnuts. *Exp. Agric.* 17:257–264.
Wilson, G.F., and R. Lal, 1986. New concepts of post-clearing land management in the tropics. pp. 371–382. In: R. Lal, P.A. Sanchez R. W. Cummings, Jr. (eds.), *Land clearing and development in the tropics*. A.A. Balkema, Rotterdam.

Wood, A. 1950. *The groundnut affair*. Bodley Head, London.

Woodwell, G.M., R.H. Whittaker, W.A. Reiners, E.G. Likens, C.C. Delwiche, and D.B. Botkins. 1978. The biota and the world carbon budget. *Science* 199:141–146.

Young, A. 1986. *Effects of trees on soils. Working paper*. ICRAF, Nairobi, Kenya.

Land Evaluation: From Intuition to Quantification

C.A. van Diepen, H. van Keulen, J. Wolf, and J.A.A. Berkhout

Contents

Advances in Soil Science, Volume 15

I. Introduction

According to Stewart (1968), land evaluation is "the assessment of the suitability of land for man's use in agriculture, forestry, engineering, hydrology, regional planning, recreation, etc." Many disciplines have contributions to make to land evaluation in its widest sense. The present review focuses on the role of soil science in land evaluation.

The task of soil science is to provide information on land use potential on the basis of soil factors, but in practice soil scientists often deal also with the integration of information from several other disciplines. In the course of time, "land evaluation" has taken many different forms indicated by various names such as land classification, land (use) capability classification, land, soil, or irrigation suitability classification, soil survey interpretation, soil survey application, soil evaluation, soil productivity rating, soil potential rating, productivity assessment, land assessment for. . ., ecological site classification, land judging, land appraisal, interpretative soil classification, forest site quality evaluation, land resource evaluation, and even land evaluation.

In this review land evaluation includes all methods to explain or predict the use potential of land. This definition embraces many different land evaluation methods from conventional common-sense approaches to computer simulations, includes comprehensive and partial, economic and ecological evaluations, and is independent of scale and choice of assessment criteria. Paraphrasing Aandahl (1958), we define the aim of land evaluation as providing information on the opportunities and constraints for the use of land as a basis for making decisions on its use and management.

This chapter examines the various approaches to land evaluation from a historical perspective. Emphasis is placed on evaluation for rural purposes and on the increasing tendency toward quantification since 1950.

II. Historical Perspective

Traditionally, soil science has developed along two main branches, soil fertility and soil geography, whose origins can be traced back to their founding fathers of the 19th century, Liebig and Dokuchaiev. Specialists of both branches have a role to play in answering the basic questions in land evaluation related to the production potential of soils. This calls for cooperation between soil fertility and soil survey specialists, which is seldom the case because of differences in approach to the study of soils and soil management problems (Buol et al., 1975). As a consequence, their evaluations are based on completely different soil factors. These differences have their roots in the past, and it is therefore of more than historic interest to examine the early approaches and philosophies in both branches, as recorded by Russell (1973), Soil Survey Staff (1951), and Simonson (1985).

A. Soil Fertility Research and Land Evaluation

Since the invention of artificial manure by Liebig in 1840, soil fertility research has been focusing on determining the influence of soil and other factors on plant growth and, in particular, the relationship between amount of an element applied and yield. One of the earliest leading concepts was Liebig's law of the minimum: plant growth is regulated by the factor present in minimum amount. Pot and field experiments were carried out to determine the effect of each single factor in isolation of all others. The approach was analytical in the sense that it tried to discover cause–effect relationships. Mitscherlich (1909), with his famous nonlinear equation

$$y = A \cdot (1 - e^{-cx})$$

where y is the yield obtained, A is the maximum yield, c is a proportionality factor, and x is the amount of the growth factor, introduced the use of mathematical equations to describe yield as a function of the supply of a plant nutrient. It formed the starting point for quantitative analyses of the relationships between growth and one or several interacting growth factors. This mathematical approach has been pursued until present, especially in Germany. Soil fertility research contributes to land evaluation through soil fertility evaluation, defined by Sanchez (1976) as "the process by which nutritional problems are diagnosed and fertilizer recommendations are made." Such recommendations pertain to a given crop and a given field and are made for one or at most of a few years. Soil fertility evaluations are based on transient properties of the topsoil mainly, and resampling is needed for each new evaluation. What holds for soil fertility research in relation to land evaluation also applies to research dealing with other soil management aspects such as tillage, compaction, irrigation, and drainage: it is oriented to experimental work and study of processes at a limited number of sites and deals with soil properties that can be changed by management.

B. Soil Geography and Land Evaluation

In soil geography the soil itself is the object of research. The focus is on the mapping of different kinds of soil and understanding of its properties. Making meaningful reliable soil maps became possible only after the Russian concept of soil was adopted. This concept was first formulated by Dokuchaiev in 1883 but spread through a book by Glinka in 1914, after which it was introduced in the U.S. Soil Survey by Marbut (1928).

In the United States, soil surveying and its application for practical purposes was promoted by the large soil conservation program during the 1930s. The experience gained from that work was recorded in the *Soil Survey Manual* (Soil Survey Staff, 1951). The basic concept is that soils are products of soil-forming processes over time under the influence of climate, plants, animals, parent rock material, and their place in the landscape. The genesis of a soil is reflected in its morphology in terms of observable soil characteristics. In soil surveys all soil characteristics are considered collectively, in terms of a complete integrated natural body, rather than individually. Soil map units are complex correlative entities that are distinguished on the basis of all permanent soil characteristics not affected by normal tillage or manuring. In a large part these diagnostic criteria are characteristics of the subsoil. This approach results in scientifically conceived soil maps that are intended to serve as a basis for multiple interpretations over several decades, "for purposes never dreamed of by the soil survey party doing the original fieldwork" (Soil Survey Staff, 1951). Interpretations include predictions on the adaptability of soils to various crops, on their behavior and productivity under different management systems, and on the yields of adapted crops under defined management practices.

The primary task of soil science in soil survey interpretation is one of synthesis rather than analysis. The approach is based on logical correlation, using empirical knowledge and sound judgment. The suitability of land for sustained use is judged from a broad view of all factors relating to the land and farming system simultaneously and from integration of all these factors.

In contrast to the soil map, interpretations are ephemeral, and reinterpretations are needed as changes occur in technology or in land use, but a resurvey of the soils is not needed. The American experience has greatly influenced later developments in soil survey and land evaluation in northwestern Europe, Canada, Australia, and in developing countries. Soil survey interpretation has been the forerunner of the land evaluation of the 1970s.

C. Development of Land Evaluation as a Discipline

The term "land evaluation" was used in 1950 at the Amsterdam Congress of the International Society of Soil Science, where Visser (1950a) presented a paper entitled "The trend of the development of land evaluation in the future." But until about 1970 "land classification" and "soil survey interpretation" were used instead of the term "land evaluation." Land classification was defined by Vink (1960) as "all those groupings of soils that are made from the point of view of the people that are using the soils in a practical sense." Land classification should not be confused with "soil classification," which refers to the scientific nomenclature of soils. The term "land evaluation" was revived in 1968 at the Symposium on

Land Evaluation in Canberra, organized by CSIRO. The broad definition by Stewart quoted in the introduction was given on that occasion, and in a similar connotation the term was propagated by FAO's *A framework for land evaluation* (FAO, 1976).

Visser (1950a) compared the American and the Western European approaches to land evaluation. In America, the soil map units representing natural genetic soil profile types served as the basis for evaluation. In Western Europe, land evaluation was approached by adopting schemes of appreciation by adjustment of figures. A final value was found by combining figures for each separate factor in accordance with its depressing effect on yield. Visser related these differences in approach to differences in aim of the land evaluation. The primary motive in a sparsely populated country like America was the search for fertile land. In Europe, however, where land is scarce, the aim was to identify deficient soil factors that could be remedied, in particular by manuring.

Visser proposed a working scheme in land evaluation research that embodied both the genetic profile approach and the factor approach, but he stressed that the nature of the problem to be solved should determine the choice of the method. The genetic soil profile types could be applied if only a few prominent factors were deficient but would not provide sufficient information for refined assessments in situations where combinations of deficient factors could not be shown on the soil map. However, such a combined approach was not elaborated, and the evolution of land evaluation since 1950 was one of separate developments in experimental soil science and in field-based soil science.

From 1950 to 1980, enormous efforts in soil mapping were undertaken in many countries, often in the form of systematic country-wide surveys. Such mapping activities were usually accompanied by some kind of land evaluation to show the usefulness of the soil map and to justify the continuation of the mapping program. As academic exercises such interpretations have contributed to methodological progress in land evaluation. In the same period, controlled experiments were executed on soil–crop response relations and on the behavior of soil water, but this was usually not considered land evaluation, and the results were often not considered applicable in survey-based land evaluation because the real world was of greater complexity than such research could describe.

For a long time soil surveyors have played the main role in land evaluation, although logically soil fertility specialists, ecologists, agronomists, economists, or systems analysts could have taken that position. Their dominant role is sometimes explained by the fact that they more than other people have participated in the field survey and, hence, have the most balanced insight in the area as a whole and thus have a direct knowledge of the land or "a feel of the land" (Brinkman and Smyth, 1973). However, that does not automatically apply to reevaluations on the basis of existing surveys. Another plausible reason is that soil maps provide adequate information for land use planning by their complete coverage of large areas with relevant factors not shown on other maps. Dent and Young (1981) state that a soil survey is usually at a scale that is suitable for land development projects for arable farming, because it shows the local pattern of variation in land of different qualities. Thus, it can distinguish between kinds of land, in contrast to a climatic map that is of a more global nature and does not differentiate within the zone of interest.

III. Major Systems and Trends since 1950

The major trends in land evaluation since 1950 have been a shift from broad to specific assessments, increasing use of nonsoil factors, and increasing quantification. This has led to a diversity in approaches, ranging from straightforward soil survey interpretations to more sophisticated multidisciplinary integrated regional studies and to application of simulation techniques. However, there has not been a regular succession of approaches. Conventional and modern procedures are employed side by side, depending on purpose, scale, and detail of study and background of the land evaluator. Many of the currently applied conventional procedures are based on methodologies of half a century ago. Usually, when a new, more quantitative land evaluation procedure was developed, the previous empirical approach was not abandoned, as it remained necessary as a first approximation (Vink, 1960). Such a mixed hierarchical application of qualitative and quantitative evaluation procedures has recently also been proposed by Bouma (1989b) for its cost effectiveness and working speed.

Modern land evaluation has gradually developed into an interdisciplinary field of study, aiming at the integration of knowledge of land resources and land use. However, these developments have not always been smooth, and confusion has plagued many land evaluation studies. According to Vink (1960), "land classification always has been, and often still is, a highly controversial subject." Kellogg (1951) attributed the confusion to the failure to distinguish between the influence of physical and economic factors on land use capability. This confusion was often compounded with those about definition and purpose. A common error was to make an interpretation for a narrow purpose, say the need for soil conservation, and then to use this for another purpose, such as prediction of soil productivity. Another cause of confusion was the failure to appreciate that interpretations are based on assumptions about the economic and technological conditions, the validity of which may be local and short-lived. And many assumptions were not stated explicitly.

In the following, some sources of confusion and attempts to clarify them are treated along with the major trends and systems since 1950.

A. Conventional Standard Evaluation Systems in the United States

The state of the art in the United States was presented at a meeting in 1956 in Cincinnati, with the papers published in the *Proceedings of the Soil Science Society of America* (Aandahl, 1958; Klingebiel, 1958; Krantz, 1958; Odell, 1958; Pierre, 1958; Stockstad, 1958).

1. Soil Survey Interpretation for Productivity Assessment

Aandahl (1958) started from the notion that soils are used by people to produce things for people. He thus described soil survey interpretation as "the organization and presentation of knowledge about characteristics, qualities, and behavior of soils as they are classified and outlined on maps" in order to identify alternatives or opportunities in the use and management of soils and to present this information to the people who must make choices from these alternatives. Because these choices are influenced by the preference and skills of these people, their income expectations, and their combined resources, soil survey interpretation must deal with both natural and economic aspects of production. Aandahl remarked that in the past a gap existed between natural and social scientists and that failure on the part of natural scientists to provide yield estimates had been the major reason. However, the gap was disappearing, and a slight overlap now occurred, related to soil productivity in the form of yield predictions for defined use, management system, and inputs. If the use is to be crop production, the definition of management should cover a combination of practices such as rotations, strip cropping, fertilization, and liming and also require an indication of the farmer's skill. The resulting yield estimates are building blocks for many kinds of economic analyses.

Aandahl's paper reflected the growing interest in economics and a tendency to relate soil productivity to the response to management practices rather than to the traditional concept of the soil's inherent ability to produce. Since then, multidisciplinarity and integration of knowledge have become keywords in land evaluation. Yet, in reality, the disciplinary gap was not easily bridged, partly because of loosely defined concepts and lack of quantification in the interpretation procedures and partly because soils and not people continued to serve as the starting point in soil survey interpretation.

Soil productivity is an example of a commonly used broad concept that is difficult to define and to quantify. In the U.S. Soil Survey, it was expressed in terms of yields of individual crops under alternative physically defined systems of management for each soil type or phase (Ableiter and Barnes, 1950). Aandahl (1960) defined productivity as the aggregate of the responses—quantity and quality of plants harvested—to all management systems. It is a collective term that can be expressed by yield predictions for alternative ways of using and managing soil. A full statement of the productivity of a soil would require a very large number of input–output combinations pertaining to alternative management systems at various levels of skill, from which, after conversion in monetary units, the most efficient combinations should be selected. Such yield predictions involved a qualitative procedure, even though the outcome was a number such as amount of produce or an index. Aandahl (1960) described it as a careful and complete analysis of all relevant information from many sources. Such an analysis was not based on the usual statistical approach but on a combination of judgment, knowledge of principles of plant growth and management, and data from census and field trials. It involved weighing the importance of each kind of information and continuously checking the consistency of the various yield predictions.

This method of analysis was the standard approach in most interpretive groupings of soils whether made for soil productivity estimates, conservation needs, irrigation suitability, or tax assessment. It comprised the assignment of classes, categories, or values to tracts of land for practical use.

2. USDA Land Capability Classification

The Land Capability Classification (LCC) of the USDA is the best-known example of interpretative groupings of soils and the one most widely used and adapted. At the Cincinnati meeting, Klingebiel (1958) gave an account of its structure, assumptions, procedures, and use. His paper forms the core of the official text on this classification system, published a few years later (Klingebiel and Montgomery, 1961). The LCC system was originally developed to assist farmers planning farm layouts, crop rotations, and conservation practices. The classification of specific tracts of farmland on the basis of their use capability is the result of an analysis and also a method of communication. The meaning of the soil groupings is widely propagated by extension workers, mainly on the basis of analogy with sample areas. Soil map units are grouped on the basis of responsiveness to management and similarities in hazards, limitations, or risks. The groupings of the capability classification are defined for arable farming as the most preferred land use. Capability refers to long-term proper use of soils for crop production without land degradation. Special consideration is given to the erosion hazard and soil conservation needs.

The classification is hierarchical: at the highest level, eight classes are distinguished on the basis of the range in alternative uses, with priority for arable cropping. At the second level, subclasses are defined on the basis of four kinds of conservation problems, i.e., (1) runoff and erosion, (2) wetness and drainage, (3) rooting and tillage limitations resulting from shallowness, drought risk, stoniness, or salinity, and (4) climatic limitations. At the third level capability units are formed by grouping soils similar in potential for plant growth, response to management, and conservation needs.

According to Klingebiel (1958) the grouping of soils in classes was not according to the most profitable use, nor was it a productivity rating, although a classification as suitable implied a favorable ratio of output to input. This ratio is based on long-term economic trends and a moderately high level of management—one that is practical and within the ability of the majority of farmers. The first four classes are for arable farming, and other uses are not considered. In general, the combined sustained economic yield of general farm crops decreases from class I to class VIII. Normally, land is classified on the basis of its actual soil and climatic limitations, but if soil improvements are economically feasible, the conditions after improvement are considered. Such improvements may include the removal of trees or stones, leveling, or the installation of irrigation and drainage systems. Differences in costs of reclamation do not influence the classification. Generalizations about crop choice can only be made at the third level of the classification, based on soil and climate factors. Economic land factors such as distance to market, size of parcels, and kinds of roads are not considered.

The judgment on which the grouping is based integrates all soil and some economic aspects of the land's use potential, but there is no formal weighing procedure to account for the separate effect of each soil factor, and economic factors are not explicitly used.

3. USBR Land Classification for Irrigated Land Use

The other well-known and still widely applied classification system of U.S. origin is the land classification for irrigated land use of the U.S. Bureau of Reclamation (USBR, 1953). This system is designed for selection of land for irrigation and evaluation of project areas. Five land classes are defined in terms of suitability for irrigated farming: three suitability classes for arable land, one special-use class, and one nonarable class. During surveys an additional class of unspecified suitability may be distinguished.

The criterion for the suitability classes is the payment capacity of the land, i.e., the money available for a farmer to pay water charges. This criterion applies to the financial situation of individual farmers and is a measure of overall land productivity. Inversely, the payment capacity can be used to fix water charges according to the productive value of the land. Its quantification in financial terms requires information on land development costs, operational and maintenance costs, and farm budget analyses.

The prediction of the payment capacity is based on selected soil, drainage, and topographic factors. However, there are no universal rating criteria, because each irrigation project is unique in its socioeconomic and climatic setting, which affects the payment capacities through farm structure, cost–price ratios, crop choice, presence of markets, skill, income level, and credit facilities.

For a specific irrigation project with defined farm types and cropping patterns, estimates are made of the effects of deficiencies in the soil and of topographic and drainage characteristics on construction work, farm water requirements, needs for soil improvement, salt leaching, land leveling, special irrigation practices, yield levels, and yield risk. Such estimates make use of experience from other irrigated areas and of local experimental data. That information is used to calculate the range in payment capacities in the project area. For the purpose of land classification, it is used to define meaningful class limits in terms of soil, topographic, and drainage factors.

Thus, with the assistance of agronomists, economists, and engineers, the original financial criterion is translated into a set of physical criteria. The classification always refers to one irrigation development option only, and the formulation of alternative development scenarios is not considered part of the land classification. Also some other important choices, for example, on the type of farming systems and on the amount and timing of supply of irrigation water, especially in relation with competitive water demands in multipurpose projects, are fixed boundary conditions during the classification and not taken into account as variables.

If the decision to develop an irrigation project has been taken on political or national economic grounds, the question whether it is worthwhile is redundant. The evaluation procedure can then start by grouping soils on the basis of physical criteria only.

4. Storie Index Rating

Another land evaluation system developed in the United States is the Storie Index Rating, of which a series of revisions exists (Storie, 1933, 1976). It is devised for the agricultural rating of soils for purposes of land tax assessment and land use zoning. The approach can be characterized as a multiplicative parametric method to express the joint influence of soil factors on suitability for and productivity of agricultural crops. It does not include management factors. The Storie Index is assessed by assigning percentage ratings to each factor (ideal soil condition is 100%) and then multiplying these ratings:

$$\text{Storie Index} = A \times B \times C \times X$$

where factor A is a soil profile factor determined by depth, permeability, and gravel content of the soil, B is the texture factor, C the slope factor, and X a miscellaneous factor. The index is normalized to a percentage rating. It is a simple quantified expression for soil productivity, but its reliability depends on the choice of factor determinants, their weighing, and the validity of the assumed multiplicative interaction between the factors. Assigning values to subratings requires subjective judgment, too.

The standard land evaluation systems developed in the United States have a long history, have been tested under a wide range of conditions, and have proven their usefulness through application. That is not always the case with systems that have found their way into the literature through presentations at workshops and congresses but without explicit examples of application. Therefore, developments outside the United States cannot be illustrated so closely with specific evaluation systems.

B. Conventional Land Evaluation in the International Context

Since 1950 many interpretive evaluation systems have evolved in different countries, most of them derived from the U.S. systems and supplemented with local expert knowledge in the form of analytical interpretation tables.

In addition, some other approaches to land evaluation have been developed in association with landscape geography, in which landscape units were identified on the basis of their visual appearance on air photos. An example is the Australian Land System approach devised for the inventory of large areas at reconnaissance scale (Christian, 1952; Christian and Stewart, 1968). The map units are recurrent patterns of distinct types of land, recognizable on air photos. The use of air photos leads to a holistic approach to land resources inventory whereby land is the integrated expression of land form, soils, and vegetation. The system uses hierarchical levels of generalization to distinguish landscape components: land systems, land units, sites. The basic assumption is that visible differences in complex map units reflect differences in land resources that can be translated into terms of use potential and management requirements. Similar mapping procedures have also been applied in more detailed studies combined with fieldwork to inventory attributes that cannot be interpreted on air photos. The map units can serve as main delineations of evaluation units, even if the soils are heterogeneous. The appraisal of land for a range of uses is based on interpretive procedures similar to those applied elsewhere in association with soil surveys. These procedures require information from each map unit for the assessment of parameters.

The landscape approach formed the basis for a type of land evaluation known as the integrated survey (Vink, 1966; Zonneveld, 1972), which has influenced the mainstream of land evaluation especially through the use of the land concept as basic areal evaluation unit instead of the narrower soil concept, and through the application of hierarchical categories for the aggregation of geographical units.

The common element in many of the evaluation systems adopted throughout the world was their aim at a final synthetic appraisal through a balanced integration of ecological and economic aspects. The result of the appraisal was often condensed into a suitability class rating. The formal evaluation procedures were established by people with field experience on the basis of reasoned intuition, and their application required informed judgment. This reliance on subjective interpretation leading to comprehensive appraisals is typical for the conventional approach in land evaluation.

The conventional systems differed from each other in the original purpose for which they were designed, in terminology, in the number and kind of soil properties they took into account, and in the logic of the procedures followed to arrive at a suitability rating. These differences hampered exchange of information between systems. Hence, in the international context a need was felt for standardization of terminology and methodology, in particular in support of land use planning in developing countries. In response to this need, the *Framework for Land Evaluation* (FAO, 1976) was developed jointly by FAO and a Dutch working group in consultation with a number of international experts.

To some extent the Framework philosophy can be considered as a merger of the genetic and the factor approaches, as suggested by Visser (1950a). It starts from the natural soil unit and then proceeds with the analytical breakdown of the soil into separate soil factors, as an intermediate step to the final synthesis. For its concepts the Framework has drawn heavily on the American experience in land classification and soil survey interpretation, in combination with the experience from integrated surveys. It differs from its American predecessors in its exclusive orientation toward comprehensive land use planning by national and local governments, although it is claimed that its principles and procedures have universal applicability. The fundamental principles of the Framework refer in particular to the necessity to compare specific alternative kinds of land use that should be economically viable and ecologically sustainable. The most typical feature of the Framework procedures is the comparison of present or future land conditions with the most preferred conditions through an iterative adjustment process called "matching." The Framework procedures are usually complex but lead to simple conclusions.

Since its publication, the Framework has been applied in many countries, although not in the United States, and considerable research effort has been invested in elaborating its procedures. However, both in the applications and in the research, the economic aspects have been treated stepmotherly.

The Framework has become the most quoted reference in land evaluation, and most authors agree on its importance for the development of land evaluation as a discipline. Beek (1980) describes the Framework as the climax of this quarter century of international methodological reassessment and as a milestone in the evolution of a realistic approach to land evaluation. Brinkman (1989) states that the Framework has become a de facto minimum standard in many countries. Verheye (1987) proposes to apply the Framework for a scientifically sound and objective assessment of the physical land potential for crop production in the European Community (EC) and for use as a basic tool for land use planning at the EC level.

Criticism of the Framework is limited mainly to footnote comments, and very few written accounts exist that expose its shortcomings (Van Diepen, 1983; Fox, 1987; Melitz, 1986; Radcliffe, 1983, 1988). It is clear that the Framework cannot be ignored when discussing land evaluation, but it should be realized that the Framework, as the climax of an evolutionary development, also forms the final stage of that development based on intuitive expert knowledge. As such, it marks the end of an era.

C. Development of Quantitative Land Evaluation

Another prominent landmark along the evolutionary path of land evaluation was the CSIRO Symposium in Canberra (Stewart, 1968) a few years before publication of the FAO Framework. Whereas the Framework still focused on the conventional type of land evaluation founded on empirical knowledge, the CSIRO Symposium addressed scientific research aspects of land evaluation including the choice of land parameters for suitability assessment, the use of remote sensing, automatic data processing and interpretation, the influence of spatial and temporal variability in soil parameters, and the use of mathematical, deterministic simulation models. This symposium has stimulated the emergence of alternatives for the intuitive inferences, correlative relationships, static representations, and qualitative class ratings, so typical for conventional land evaluation.

Mabbutt (1968), reviewing concepts of land classification, focused on the distinction between the landscape approach and the parametric approach. This parallels the distinction between the genetic soil profile type of approach and the factor approach made by Visser (1950a). Mabbutt defined the parametric approach as the division and classification of land on the basis of selected attribute values. It deals with the quantification of those land attributes that are considered relevant for the intended land use and was developed as an alternative for the landscape approach because of dissatisfaction with its qualitative and subjective results. The main difference between the approaches is the way the land units for evaluation are identified. In the landscape approach they are comprehensively conceived and fixed by delineation on a map; in the parametric approach they are the result of a choice of attributes and combinations of attribute values, not necessarily mappable but well suited for statistical analysis. Moreover, the parametric approach could take better advantage of automated scanning with improved sensors and electronic data handling. Mabbutt therefore predicted that the parametric approach would increasingly replace the landscape approach but concluded that both methods could be profitably combined to reinforce each other.

Nix (1968) was the first to propose the use of systems analysis and simulation in land evaluation for biological production. Once a working model of crop ecosystems was developed, the inventory of land resources could be directed toward the spatial and temporal variation in the model's key parameters. Nix foresaw the possibility of yield predictions for any crop under a defined system of management at any location by coupling such a model to an appropriate land inventory and using historical weather records.

In Canberra, the foundations were laid for approaches to land evaluation in the computer era, thus marking the beginning of a new era of quantitative land evaluation. There is no clear distinction between qualitative and quantitative, or quantified, land evaluation. Quantification may start as "the stiffening of qualitative systems through the infusion of parametric ingredients" (Mabbutt, 1968). Some authors consider the use of computers as a criterion for a quantitative approach, like, for example, automated qualitative expert systems (Rossiter, 1989). Others restrict quantitative land evaluation to the use of dynamic simulation models of land use systems (Driessen, 1986). A different view was held by Vink (1960), who considered quantitative as including a farm budget analysis. The original Framework (FAO, 1976) also related quantitative evaluation to the use of economic criteria in the evaluation procedure. Later (FAO, 1983), this concept has been revised, and quantitative now refers to the way of expressing the evaluation results, independent of the procedure, so that qualitative procedures may result in quantitative results, and vice versa.

In this review a method is considered quantitative when numerical parameters are used in calculation procedures to produce numerical results.

In the past two decades separate research efforts have been directed to the quantitative description of ecological processes and the integration of economic and ecological information. The purely ecological orientation, called quantitative physical land evaluation, deals with processes in the soil and related ecological compartments, with the aim of assessing limits to agroecological productivity, environmental tolerance, or technical feasibility (van Keulen et al., 1987).

For an orderly discussion on land evaluation a separation is made between conventional interpretive land evaluation, of which the Framework forms the climax, and quantitative physical land evaluation. This is followed by a discussion of the prospects for integrating ecological and economic considerations to land use decision scenarios through quantified optimization procedures.

IV. The FAO Framework as a Standard in Land Evaluation

A. Principles, Concepts, and Procedures

The *Framework for Land Evaluation* (FAO, 1976) sets out basic concepts, principles, and procedures for land evaluation that are "universally valid, applicable in any part of the world and at any level, from global to single farm." The universal validity refers to scale and precision of the applications and to its potential for application across disciplinary boundaries. A classic publication of the Framework school is Beek's thesis (1978). A concise introductory treatment is given by Dent and Young (1981). The Framework has been followed by a series of guidelines for rainfed agriculture (FAO, 1983), forestry (FAO, 1984), and irrigated agriculture (FAO, 1985). These guidelines provide an expansion of the basic concepts and details on the operational aspects of the procedures recommended in the Framework. Land evaluation for extensive grazing has been discussed at a workshop in Addis Ababa (Siderius, 1984).

The Framework as such does not constitute an evaluation system but is primarily designed to provide tools for the construction of local, national, or regional evaluation systems in support of rural land use planning. The ultimate goal is to contribute to the wise use of land resources by man. The two basic means to achieve this goal are land improvement and land management improvement.

The Framework defines land evaluation as "the process of assessment of land performance when used for specified purposes, involving the execution and interpretation of surveys and studies of landforms, soils, vegetation, climate, and other aspects of land in order to identify and make a comparison of promising kinds of land use in terms applicable to the objectives of the evaluation." It clearly places land resource inventory at the center of the land evaluation process and considers both its execution and interpretation as land evaluation. Remarkably, the definition does not include the study and survey of agronomic, technological, economic, and social aspects of land use as land evaluation activities.

1. Principles and Basic Concepts

The Framework defines six principles that are fundamental to land evaluation:

1. Land suitability is assessed and classified with respect to specified kinds of use.
2. The suitability classes are defined by economic criteria.
3. A multidisciplinary approach is required.
4. Evaluation should take into account the physical, economic, social, and political context of the area concerned.

5. Suitability refers to land use on a sustained basis.
6. Evaluation involves comparison of two or more alternative kinds of use.

The basic concepts of the Framework include land, land (mapping) unit, major kind of land use, land utilization type, land characteristic, land quality, diagnostic criterion or diagnostic factor, land use requirement, and land improvement. Land has replaced soil as the basic unit. Land includes soil, vegetation, hydrology, landform, and climate. As a consequence, the concept of soil characteristic was replaced by land characteristic, which continued to refer to measurable attributes. In contrast, however, the new concept of land quality was defined differently from the earlier concept of soil quality. A land quality is not necessarily a nonmeasurable, inferred attribute, as was soil quality, but an assessment factor to express the adequacy of soil conditions in relation to a specified land use. The concept "land quality" is discussed in Section IV.B.

2. Procedures

The procedures to be followed in land evaluation depend on the objectives, the level of detail of the study, and the degree of integration of economic information. The Framework deals mainly with the ecological aspects of procedures in which physical resource surveys precede economic and social analysis. Repeatedly, links and overlaps between the physical and economic analyses are acknowledged.

The Framework's evaluation procedures comprise the following activities:

1. Selection of relevant kinds of land use and their requirements.
2. Description of land units and assessment of land qualities.
3. Matching: comparison of land use requirements with land qualities for each land use on each mapped land unit.
4. Provisional suitability classification.
5. Economic and social analysis.
6. Final suitability classification.

It is important to note that there is a cyclic element in the procedures, leading to successive revisions of evaluation results. The focal point in the evaluation procedure is that at which the data from different sources are combined and compared, leading to the suitability classification. The complex action of synthesizing dissimilar information is called matching.

For the purpose of matching, the Framework recommends describing the land in terms of land qualities. This means that a transformation of basic land data is required, or, in the terminology of the Framework, a conversion of land characteristics into land qualities. Both are properties of the land, but the advantage of using land qualities would be that they have a distinct influence on land use, independently of other land qualities, and that they constitute the integrating expression of a large number of interacting land characteristics. The total number of land qualities is therefore smaller than the number of land characteristics. FAO (1983) lists 25 land qualities relevant for rain-fed agriculture (Table 1), compared to many hundreds of land characteristics. The disadvantage of land qualities is their greater complexity.

Land qualities are usually rated on a scale ranging from 1 (very good) to 5 (very poor). These ratings are compared with the requirements of a given land use. These requirements are expressed as suitability class limits per land quality. The Framework recommends four classes: S1, highly suitable; S2, moderately suitable; S3, marginally suitable; and N, not suitable.

It is possible, however, to use land characteristics directly in the matching procedure without their conversion into land qualities. The advantage of such a direct use is reduced complexity. But the disadvantages are that their influence on land use is not unique because of interactions between land characteristics and that their number is large.

Table 1. Land qualities for rain-fed agriculture*

LQ1	Radiation regime: total radiation, day length
LQ2	Temperature regime
LQ3	Moisture availability: total moisture, critical periods, drought hazards
LQ4	Oxygen availability to roots (drainage)
LQ5	Nutrient availability
LQ6	Nutrient retention capacity
LQ7	Rooting conditions
LQ8	Conditions affecting germination or establishment
LQ9	Air humidity as affecting growth
LQ10	Conditions for ripening
LQ11	Flood hazard
LQ12	Climatic hazards
LQ13	Excess of salts: salinity, sodicity
LQ14	Soil toxicities
LQ15	Pests and diseases
LQ16	Soil workability
LQ17	Potential for mechanization
LQ18	Land preparation and clearance requirements
LQ19	Conditions for storage and processing
LQ20	Conditions affecting timing of production
LQ21	Access within the production unit
LQ22	Size of potential management units
LQ23	Location: existing and potential accessibility
LQ24	Erosion hazard
LQ25	Soil degradation hazard

*From FAO (1983).

Table 2. Schematic procedure for the assessment of the suitability of land units for a given use on the basis of land qualities.

Land inventory: land units, LU1, LU2,. . .		A given land use or land utilization type (LUT)
↓		↓
Description of LUs in terms of land characteristics (LC)		Description of LUT in terms of land use requirements (LUR)
↓ conversion table LC → LQ		↓ rating table
Description of LUs in terms of rated → land qualities (LQ)	Matching	← Specification of suitability class in terms of minimum required LQ rating
	↓	
	Overall suitability class rating for the given LUT on each LU	

The matching of land qualities with land use requirements results in separate suitability class ratings for each land quality for a given land use. An overall suitability class is found by combining the various suitability ratings. The usual procedure here is to take the lowest single rating as class-determining for the comprehensive rating, in analogy with Liebig's "law of the minimum." This class-determining land quality is indicated with a symbol and thus retained as an evaluation result.

The essence of the matching procedure is shown schematically in Tables 2 and 3. The iterative element in this matching is not illustrated, but the Framework prescribes checking the results against experienced judgment and common sense and, if the result is not satisfactory, modifying either the definition of the land use or the various conversion and rating tables.

The sketchy outline presented here may suffice to illustrate some of the typical features of the Framework approach:

1. The synthesis of dissimilar information by the creation of class ratings.
2. The derivation of land qualities from land characteristics.
3. The use of class ratings rather than physically defined properties to describe soil conditions in terms of land qualities.
4. The analytical breakdown of land in land qualities.
5. The analytical breakdown of land use in land use requirements.
6. The matching procedure resulting in a suitability class rating for each land quality.
7. The combination of the various suitability class ratings to estimate the overall suitability class reflecting the combined effect of all land qualities.
8. The iteration in the procedures to adjust earlier findings and to refine retrospectively the rating and conversion tables.

Table 3. Application of the matching procedure to the suitability assessment of four land units for one given use on the basis of five land qualities*

	LQ rating per land unit				Suitability class rating per LQ and per land unit				Minimum required LQ rating per suitability class			
LQ	LU1	LU2	LU3	LU4	LU1	LU2	LU3	LU4	S1	S2	S3	N
LQa	2	1	1	1	s2	s1	s1	s1	1	2	4	
LQb	1	1	3	2	s1	s1	s3	s2	1	2	3	
LQc	1	1	1	2	s1	s1	s1	s2	1	2	4	
LQd	3	1	4	1	s3	s1	n	s1	1	2	3	
LQe	2	1	2	3	s1	s1	s1	s2	2	4	5	
*Overall class rating:					S3d	S1	N	S2bce				

B. About Land Quality

1. Original Land Quality Concept

No other term in land evaluation has created so much confusion as land quality. Outside this discipline land quality is usually understood as a measure of the degree that the land is free of pollution or that the land as a whole is suitable for its intended use, in analogy with the term water quality. In land evaluation, however, the analogon of water quality is land suitability. Unfortunately, the term land quality is not uniformly understood, as became evident again recently at the ISSS symposium on *Land Qualities in Space and Time* (Bouma and Bregt, 1989).

The land quality issue is not new. The term is related to the older concept "soil quality," which in the longstanding terminology of the U.S. Soil Survey (Soil Survey Staff, 1951) was an inferred soil property such as fertility, tilth, productivity, erodibility, irrigability, response to drainage, workability, and crop adaptability. Aandahl (1958) also mentioned permeability, erosion hazard, and moisture-holding capacity as soil qualities; Kellogg (1961) soil drainage, response to management, and suitability. Normally, such soil qualities cannot be measured directly, in contrast to soil characteristics, which are directly observable features such as color, structure, and texture. Accordingly, in the Framework land characteristic is defined as "an attribute of land that can be measured or estimated." However, the distinction between characteristic as measurable and quality as inferred property is not maintained in the Framework. It distinguishes qualities from characteristics on the basis of complexity and influence on a given land use and defines land quality as "a complex attribute of land which acts in a manner distinct from the actions of other land qualities in its influence on the suitability of land for a specified kind of use." The Framework adds that land qualities can sometimes be estimated or measured directly. However, specific definitions for individual land qualities have never been given. The Framework merely lists their names, such as moisture availability, conditions for germination, resistance to soil erosion, etc. (Table 1) and provides suggestions for the selection of land characteristics to be used for their assessment (FAO, 1976, 1983).

Beek (1978) explains that the land quality concept emanated from the wish to reduce the enormous amount of land data to manageable proportions without loss of information. The concept has been developed for the exclusive purpose of synthesizing measurable single properties of the land into assessment factors that have a specific influence on the "land use processes." Such land qualities are usually ranked on an ordinal scale, very high–high–medium–low-very low, using threshold values of component properties to distinguish different levels.

The functional description of land conditions and land use requirements is considered the foundation of successful land evaluation (Beek, 1978).

The main function of land in a land-use system is to be accommodating to the crop and to the farmer on a sustainable basis , i.e., without environmental degradation. Land qualities are directly linkable to land-use requirements such as crop requirements, management requirements, and conservation requirements. The crop requires favorable land conditions in terms of sunlight, temperature, water and nutrient supply, rooting space, no problems of toxicity, acidity, frost, etc.; the farmer requires the land to be workable and accessible, and floods, erosion, or pollution should be absent.

A land quality indicates the ability of the land to fulfill specific requirements such as supplying water or nutrients to a crop or an easy plowing job to the farmer. The values of land qualities are obtained by weighing the specific impact of selected land characteristics on a given land use. For instance, the land quality "moisture availability" can be determined by combining selected land characteristics such as seasonal rainfall, drought hazard, water-holding capacity of the soil, and depth of the groundwater table in spring. For each characteristic a diagnostic level must be indicated, taking due account of interactions between characteristics, and these subratings for the various relevant land characteristics must be converted into one final land quality rating. Such a land quality assessment is followed by a matching procedure to assess how well each land quality meets the requirement of the specific land use. Radcliffe (1983) remarked that, in fact, it is not the land quality that is matched with the land use requirement but land attributes capable of quantitative expression, which are taken as diagnostic of the quality (or of the requirement).

2. Sources of Confusion

In addition to the mix-up of the land quality and soil quality concepts there is also confusion about the elements of the land quality definition that refer to complexity, absence of interactions, and use specificity. This confusion hampers a consistent identification of land qualities.

The word "complex" in the definition of land quality referred to its formation by synthesis of characteristics. However, some land qualities can be assessed on the basis of one single characteristic, such as salt content to assess the excess of salts. Complexity in the assessment is also lacking when a single characteristic is used as a surrogate for the synthesis of a number of characteristics.

A land quality is selected for its distinct influence on land use. The main reason to use land qualities is to avoid the problem of interaction. This assumption, however, is a simplification. On the one hand, many interactions, complementarities, and overlaps exist between land qualities; on the other hand, some land qualities integrate the actions of other land qualities or have multiple influences on land use.

Well-known cases of interactions between land qualities are the influence of soil toxicity on moisture availability and of moisture availability on nutrient availability. complementarity of land qualities usually exists between oxygen availability and moisture availability. Examples of overlapping land qualities are physical soil degradation hazard and soil erosion hazard, workability and potential for mechanization (tractor plowing), accessibility and potential for mechanization. An example of multiple influences is the land quality "conditions affecting germination and crop establishment," which expresses a judgment on overall crop performance during a specific period by integrating the effects of other land qualities such as moisture availability, previous workability for seedbed preparation, rooting conditions, nutrient availability, temperature regime, toxicities, and damage to seedlings from erosion and runoff.

Another example of multiple influences is the land quality "temperature regime." Temperature interacts with radiation (another LQ) in the basic growth process of photosynthesis; it determines the phenological development of crops; it influences crop respiration losses and crop water use (a land use requirement); it influences, often in combination with air humidity (another LQ), yield formation and yield quality such as sugar content or color of fruits (LQ: conditions for ripening); it affects developments of pests and diseases (another LQ); it may cause damage during very hot weather; and it determines the risk of frost incidence (LQ: climatic hazards). These examples demonstrate that the application of land qualities in land evaluation cannot fully achieve its intended purpose of avoiding interactions between land characteristics.

The soil moisture regime is not in FAO's list of land qualities (Table 1), obviously because it does not act in a distinct manner independently of other land qualities. But here is an example of a single land characteristic with multiple influences on many land qualities such as availability of moisture, oxygen, and nutrients in the root zone, trafficability, soil temperature, occurrence of pests and diseases, resistance to erosion, and conditions for germination. This explains why sometimes the characterization of the soil moisture regime is regarded as the core activity in land evaluation (Bouma, 1989b), and many so-called land qualities are derived from it.

Another important function of the soil in relation to plant growth is the supply of plant nutrients, which is covered by two land qualities: nutrient availability and nutrient retention. Availability refers to the capacity of the soil to supply nutrients, and retention to its capacity to retain added nutrients. These two qualities are not entirely distinct and can be assessed either separately or jointly (FAO, 1983). If one accepts a combined rating, the land quality could be called "soil fertility," although that term is avoided in the Framework. Soil fertility assessment deals with the availability of all major and minor plant nutrients and their combined influence on growth and yield of a given crop or vegetation growing on a given soil.

Results of soil fertility assessment are often related to specific elements, yielding such remarks as "low phosphate levels are found in most soils, and there is copper deficiency in others." In an analysis of the productivity of Sahelian rangeland, Breman and de Wit (1983) concluded that plant productivity in regions with less than 150 mm annual rainfall was limited by moisture rather than by soil fertility, whereas in regions with higher rainfall phosphorus shortage resulted in a slow start of plant growth early in the rainy season, and shortage of nitrogen limited the production in the final stages. Clearly, such an analysis requires the study of the supply of each nutrient, its dynamics, and interactions with other nutrients and growth factors in a given land-use system. For soil fertility assessment, availability of individual nutrients could be considered separate land qualities (N availability, P availability, etc.), but FAO (1983) rejects this concept "first because of the similarity in the principles, by which each nutrient is treated, and secondly because through the operation of the law of the minimum it is frequently only one or two nutrients that are limiting to the crop." This illustrates again that the land quality concept is not unambiguously defined or understood.

Land quality assessment is the analytical disaggregation of land on the basis of cause–effect relationships between land and its use. Because land qualities can only be defined with reference to land use, Radcliffe (1983) saw no rational basis for their definition as "attributes of the land." However, land qualities are defined as attributes of the land and as such often rated irrespective of the land use. A rating for each land quality is assigned to each mapping unit. The adequacy of a land quality for a given use is then assessed by means of land-use-specific matching tables. In reconnaissance evaluations this may provide sufficient information on the overall potential of the land. But in detailed evaluations separate assessments of land qualities should be made for each specific land use, whereby for a given land unit a specific land quality could be assessed differently for alternative uses. Melitz (1986) demonstrated this with an example of the very different moisture requirements of oil palm *(Elaeis guineensis)* and groundnut *(Arachis hypogaea)*.

The Framework recommends the use of land qualities rather than land characteristics as the basis for land evaluation. However, the distinction between land characteristics and land qualities is in practice not very clear, because the two terms as defined in the Framework are not mutually exclusive.

A land characteristic is a physical parameter whose value can be indicated in quantitative units on a physical scale. The values of land characteristics are determined by measuring or estimating and relate to the land.

The term land quality is used in two senses. In the original Framework philosophy, a land quality is a judgment expressed as a dimensionless variable on an interpretive ordinal scale. The determination of this value is based on reasoning. Land qualities relate to both land and land use.

Radcliffe's (1983) statement that land qualities are not capable of direct quantitative expression referred to that land quality concept. Melitz (1986) has elaborated the land quality concept by distinguishing supply by the land with respect to a land quality and sufficiency of the supply for a given land use. He pointed out that, for example, good trafficability may be interpreted in two ways: (1) compared to other land, the land offers good trafficability (supply), or (2) the trafficability is sufficient for a given kind of ranching.

Over the years the meaning of land quality has been extended to include the description of interactions among land characteristics and, hence, a description of characteristics that were selected to serve as indicators for the assessment of a given land quality. Land qualities are then confused with land characteristics or with functions of several land characteristics. As a consequence, a land quality is sometimes expressed as a physical parameter or a set of physical parameters without immediate reference to a comprehensive graded value. A land quality can then be measured or estimated, and it would thus be a land characteristic as well.

Examples of land qualities expressed in physical units are the length of growing season in days, expressing high or low moisture availability, and the calculated annual soil loss in tons per hectare as a measure of soil erosion hazard. Formally, such measured or derived physical parameters are not land qualities. The true expression of a land quality is the ranked value resulting from the comparison of the physical value of the assessment parameter with critical threshold values.

In land evaluation the defined terms land qualities and land characteristics are often mixed with undefined terms such as land or soil attributes, properties, features, factors, aspects, traits, parameters, and data. In principle, these terms are free for defining some particular kind of land or soil information, but it is doubtful whether this would improve understanding of procedures and communication.

3. Its Use in Qualitative Procedures

In Framework land evaluations the results of extensive analyses and data manipulation are condensed in a few land quality ratings. It is evident that the claimed advantage of the small number of land qualities, in comparison with the number of land characteristics, is only valid for manual data manipulation in the final synthesis but not during the preceding assessment of the land qualities. In studies of comparable detail no reduction in calculus is achieved by using qualities. Sys (1980) concluded from a comparative study on land suitability for maize *(Zea mays)*, oil palm *(Elaeis guineensis)*, coffee *(Coffea arabica)*, and rubber *(Hevea brasiliensis)* that procedures based on land characteristics and on land qualities gave identical results. This is not surprising if one realizes that the same data base was used, specified in terms of land characteristics. Theoretically, there could be a difference in the selection of subsets from this data base, but in reality data selection in an evaluation on the basis of land characteristics is not random. In either procedure the reliability of the evaluation results depends entirely on the competence of the land evaluator.

Radcliffe (1983) demonstrated that exclusion of the land quality definition leads to a more logical, reduced sequence of operations, which is a distinct advantage, particularly in automated data analysis. The redundancy of the land quality concept as an intermediate step in the matching procedure is also reflected by the structure of the Automated Land Evaluation Systems (ALES: Rossiter, 1989), a computer program that allows land evaluators to build expert systems to evaluate land according to the FAO Framework. In ALES, land is described in terms of land characteristics, and each land utilization type in terms of land use requirements. Class limits for land use requirements are on the basis of land characteristics. Such land use requirements are identical to land qualities, and no provision is made to separate them.

In general, in land quality assessments insufficient attention is paid to land use specifications. For example, the land quality workability or ease of tillage is assessed according to the texture/structure/consistence relationships of the topsoil, corrected for the presence of stones (FAO, 1983). This ignores the fact that ease of tillage also depends on the depth of tillage, the type of implement, the available power, and the required tilth conditions in terms of clod or crumb structure. Another omission is the soil moisture condition at the time of the year that the soil should be tilled. That signals a more general problem of the land quality concept, namely, that it is not designed to handle temporal variability in site conditions. This applies in particular to land qualities that are influenced by temperature and moisture conditions on the supply side and by the cropping calendar on the demand side.

An easy solution is then to assess the land quality separately for specific periods, such as before, during, and after the growing season. This may require a large number of assessments, especially if alternative land uses are evaluated or if one land use refers to a crop rotation. For each land unit separate assessments are needed when significant differences exist in planting date and growth duration among crops. The assessment would still be based on rather static land characteristics. The separate assessments for different periods would still yield results pertaining to only one land quality per land-unit/land-use combination. The grade of the land quality would be determined by the most limiting, lowest grade of all separate assessments. A better solution is to model the land-quality/land-use requirement over the relevant period (FAO, 1983). This requires the application of quantified procedures and well-defined quantifiable concepts.

4. Its Use in Quantitative Procedures

In the Netherlands a change in the land quality concept coincided with the shift from broad qualitative to specific quantitative physical land evaluations. In the Dutch system of qualitative soil survey interpretation on the basis of semidetailed maps at scale 1 : 50,000 (Haans, 1980), a distinction was made between graded assessment (judgment) factors and the site conditions on which they were based. The assessment factors were defined in terms of supply, e.g., moisture supply capacity, and rated independently of the kind of land use. The suitability class was found by considering the relevance of the assessment factors for each use. Some assessment factors applied to only one form of land use, for instance, workability to arable farming, bearing capacity of the topsoil to pasture, and soil fertility to forestry. Other factors such as drainage status and moisture supply capacity were relevant for all three uses. The site conditions on which the assessment was based included soil-dependent processes, growth conditions, and soil behavior, the latter conceived as the response to management practices.

The distinction between judgment factors and physical parameters became diffuse when more emphasis was placed on the use of simulation models of the soil water balance for quantification of soil conditions. Under the influence of the Framework, the term land quality came into use, however, increasingly referring to physical concepts obtained by calculations. A typical example of this modified land quality concept is the assessment of environmental problems from soil surveys by Breeuwsma et al. (1986), who used the digitized soil map at 1 : 50,000 to derive interpretive maps showing "gradations of land qualities." The land qualities calculated from soil properties of the unsaturated zone were (1) travel times of water from the soil surface to the ground water table, (2) cation-exchange capacity (CEC), and (3) phosphate sorption capacity (PSC). These land qualities were expressed in terms of classes that covered the range of observed spatial variability.

A second example of modeling of land qualities relates to the temporal variability in trafficability. Wosten and Bouma (1985) used the probability that a threshold value for the pressure head in the topsoil was reached as a measure for adequate trafficability. Such threshold values are soil and land use specific and also depend on the depth for which the pressure head is calculated. The probability of occurrence of days with adequate trafficability was established for successive 10-day periods during the year, based on simulation of the soil water regime using 30 years of weather data.

Temporal variations in land quality values will usually be accompanied by variations in the spatial patterns of those values. Burrough (1989a) reviewed ways in which geographical information systems (GIS) and models of spatial and temporal processes can be linked for the purpose of land evaluation. With respect to land characteristics he distinguished spatial data such as point, line, and areal data and time series data. These basic data are stored in a GIS, where they can be recombined to estimate values of appropriate land qualities by using models.

The direct output of a model may be the growth pattern of a crop over time, a yield index, the crop's water use, the soil water deficit, or a time series of soil water potential values. Such output can be used to express land quality values or are themselves considered land qualities. In the operational sense, the GIS provides model input in the form of land characteristics, receives model output in the form of land qualities for a given location, and processes the output to obtain full geographical coverage.

The application of deterministic simulation models has led to new ways of expressing land qualities that do not agree with the Framework concept. The Framework terminology has not been developed to make meaningful distinctions in kinds of information needed in relation to modeling and manipulation of data on soil and other environmental aspects. Therefore, Bouma and van Lanen (1987) proposed to revise the current Framework concepts of land quality and land characteristic, introduced new concepts such as properties, and made further subdivisions on the basis of the nature of the data and the way they are obtained (Table 4). They redefine land qualities as dynamic attributes as opposed to characteristics and properties that are static. This implies a complete departure of the land quality concept as a complex judgment factor, and time series data would become qualities rather than characteristics. The land qualities are further subdivided into measurable single qualities such as soil moisture content and compound qualities that follow from calculations, such as soil moisture deficit. The land characteristics are directly available static data, as opposed to properties, which are not routinely collected during soil surveys and have to be derived from characteristics or can be measured. Characteristics can be subdivided into class characteristics (e.g.,

Table 4. Summary of features of land characteristics, properties, and qualities*

Term	Static	Dynamic	Obtained during soil survey	Obtained directly	To be calculated only
Soil/land characteristics	X		X	X	
Soil/land properties				X	
Single land qualities		X		X	
Compound land qualities		X			X

*From Bouma and van Lanen (1987).

soil texture class) and continuous characteristics (e.g., clay content). These proposals add to the semantic confusion by using old terms for new concepts, as was done in the Framework, but also form a clear plea for the need of a new terminology.

However, such a terminology should not be designed for its exclusive use in the context of land evaluations on the basis of soil survey data but should also be understandable for other disciplines. For deterministic process models, useful distinctions among data can be made on the basis of their function in the process description: state variables, rate variables, system constants, forcing variables, and boundary conditions. But a boundary condition in one model may be a rate or state variable in another model; e.g., fluctuations in groundwater depth with time may be calculated as an endogenous state variable, or they may be introduced as an exogenous forcing variable. On the basis of operational procedures a distinction among input, output, and model parameters may suffice. But in any case there should be no ambiguity on definitions or meaningless jargon. This holds especially for situations where basic and generated data are stored in data bases that are shared by people from various disciplines. Multidisciplinary teamwork requires well-defined concepts that serve well-defined purposes. Much confusion results from sloppy definitions and multiple interpretations of concepts. If new concepts must be defined, we recommend checking first to see if an equivalent exists already and, secondly, not coining words that already have a specific connotation in any of the branches of science that could possibly participate in land evaluation. Finally, we recommend serious consideration of an alternative for the term land quality.

C. Constraints Related to Application of the Framework

Complete application of the FAO Framework would require the observance of all its principles, the correct use of its concepts and terminology, and the adherence to its procedures. Such "model" applications, however, do not exist to our knowledge. Most applications ignore or violate the Framework recipes to a considerable extent. The minimum requirement for a land evaluation to be considered as an application of the Framework seems to be the use in some way of the terms land qualities and land characteristics. It should be noted that FAO (1983) considers partial application of the principles legitimate and a proof of the Framework's flexibility. However, some intriguing questions remain with respect to the application of the Framework:

Why do land evaluations fail to comply with the Framework, and what is the difference between Framework applications and other conventional land evaluations?

Why do frictions and misunderstandings develop in multidisciplinary land evaluation teams?

Why do land evaluations so often fail to fulfill their promise to serve as a tool for integrating knowledge from different sources and disciplines in support of land use planning?

Are these problems related to human imperfection rather than to the inadequacy of procedures? In the following an attempt is made to find answers to these questions with reference to the Framework. Starting point for this discussion is the realization that the Framework emerged at a moment that traditional soil survey was increasingly confronted with the limits of its applicability and a breakthrough toward integration with other disciplines seemed imperative. The difficulties encountered are to a large extent comparable to those in other multidisciplinary research efforts. For instance, the evolution of agronomy into farming systems research has much in common with the emergence of land evaluation from soil surveying.

1. Multidisciplinarity and Universality

The Framework addresses the issues of land and its best use with respect to productivity and sustainability. These issues are judged differently by ecologists and economists, and even within disciplines differences of opinion and working procedures exist. As a consequence, tensions may develop in a multidisciplinary team, especially with regard to the expectation of the contributions of the various disciplines. It is claimed that the Framework is a basic document for all disciplines involved in land evaluation, although in fact its postulates originate from soil survey interpretation rather than from the other contributing disciplines. The scope and procedures of the Framework are easily misunderstood by the co-workers in land evaluation, not only by those with a background in crop science or human science but also by colleagues dealing with land resources. The Framework can therefore act as a source of frictions in multidisciplinary teams.

A similar problem exists with regard to the proper understanding of land evaluation results by the users, whether farmer, planner, government official, or politician.

The fundamental question remains whether in such a broad multidisciplinary field as land evaluation any universally applicable set of principles, concepts, and procedures can be defined, or whether different objectives of evaluations call for adaptive approaches and a more flexible terminology, originating from different sources.

The need for a more specific terminology stems from differences in objectives, geographic scale, time horizon, nature of studied processes, level of generalization in process description, and kind of models used, from the use of computer technology, and from the specific mix of cooperating disciplines. The current Framework concepts fall short of describing the recently developed procedures, so its principles are reduced to a background philosophy not necessarily applicable to the real world. Yet, although any handbook on procedures or glossary of terms will be outdated soon, the formulation of some unifying concepts seems worthwhile, which requires updating of the Framework.

2. Concepts of Land and Land Use

Although it carries a multidisciplinary label, the Framework has been developed primarily for soil surveyors in developing countries who had to contribute to integrated rural development plans, often in the form of physical land use plans for watersheds. For that purpose the soil surveyors broadened their professional domain from soil survey to land evaluation. Soil was replaced by land, which comprises all "reasonably stable or predictably cyclic attributes of the biosphere." Land refers to more or less permanent features of the physical environment; i.e., it presents a static concept and does not take into account variable soil and weather conditions.

The Framework holds that land comprises the physical attributes "to the extent that they exert a significant influence on present and future uses of the land by man." In this part of the definition the concept of land is reduced from a general geographic concept to a set of properties used in a specific land evaluation study for the characterization of land. This is unfortunate and creates confusion. It implies that future land uses must be known before land can be fully described in terms of its relevant attributes. It also implies that a land attribute, say soil color or clay skins, is not part of the land unless its influence on any possible land use is known. The underlying assumption is that the specific influence of land attributes on various uses is known, which often is not the case and, in fact, is a reason for further research. Purely economic and social characteristics such as land tenure are not included in the concept of land; these are part of the land use.

In the Framework philosophy, land and land use are interdependent but separate entities. However, it is sometimes difficult to distinguish between these entities. A not purely physical attribute as location in relation to markets is part of the land, whereas other physical attributes of land, such as size and shape of parcels, degree of fragmentation of holdings, and standing crops are considered part of the land use. An overlap of attributes is unavoidable in the case of land use for forestry or nature reserve, where standing trees and vegetation are shared by the land and its use. A peculiar example illustrating the nearly compulsive separation of land and its use is the installation of drains or the construction of terraces, which according to the Framework are attributes of land use if carried out by the farmers themselves but are considered a change of the land if carried out with public funding.

3. Prime Concern Is for Land

The most striking feature of the Framework in comparison with preceding approaches is its emphasis on economic and social aspects, especially simultaneous evaluation for different levels of technology and the full specification of land utilization types as output of land evaluation. This gives the impression that the Framework is oriented toward people and social change.

However, this emphasis on socioeconomics obscures the notion that land evaluation is primarily concerned with land, conceived in terms of mapping units, areas demarcated on a map, usually a soil map. In Framework-type land evaluation, the land-mapping unit is the starting point, and the best use of each piece of land its goal. Production activities are limited to land use activities. People are considered to the extent that they participate in land use, and then not as actors but as management skill or labor. The role of actor is reserved for the land itself: the land performs, the land qualities act.

According to the Framework, land and land use are equally fundamental to land evaluation. But it also says that the description of land use need not comprise the full range of farm management practices. It suffices to include only those related to land management and improvement. When in detailed land evaluation studies nonsoil information is included in the description of the land utilization types, which are thus extended into farming systems, that information is not used in an operational way in the suitability assessment procedures.

4. Elaboration of Alternatives Is not Oriented to Decision Making

Land use planning becomes necessary when the land resource base can no longer support the rural population under the current use. Then, usually conflicts arise between groups of land users competing for the limited land. The solution of such problems requires the formulation of land use development objectives that can be used for the optimization of land use for competing demands. The Framework is not designed to contribute to decision making in such conflicting situations, especially not if choices must be made between bad and worse rather than between good and better. Instead, the Framework is oriented to the presentation of reasoned information on several promising land use alternatives that are selected under the tacit assumption that economic, social, and political priorities point harmoniously in the same direction.

5. Few Criteria for Suitability Assessment

The only normative criteria in the Framework to judge the relevance of the selected land use alternatives are productivity and conservation of land. These are valid criteria when sustained growth of agricultural production is pursued. But this is a narrow objective by most current evaluation standards for development projects, both in the economic and in the ecological field. Criteria to assess agroecosystem performance can be productivity, sustainability, stability, equitability, and autonomy (Marten, 1988). Related criteria are resilience and adaptability of systems. Criteria generally accepted by development economists for the evaluation of land development projects include growth of production, risks, food and income security, employment, income distribution, health, and economic independence of the country.

The criterion that the proposed land uses must "be sustained, that is, capable of being continued over an indefinite period of time" is not unambiguous. In the practice of land evaluation this is understood as absence of on-site land degradation, and even that in a very limited sense, hardly quantified or quantifiable. Off-site effects are mentioned in the Framework but not included in the suitability assessment procedure except in the forestry guidelines (FAO, 1984). The Framework does not discuss procedures for making comprehensive environmental impact assessments. Nonland factors that may influence sustainability such as the dependence on fossil energy or on imported inputs are also not considered.

6. Conflicting Criteria for Suitability Assessment

Sustainability is not necessarily compatible with suitability in the sense of profitability. The Framework definition of suitability classes as successive grades of profitability of a land use over land units cannot accommodate land uses that are environmentally or socially desirable but not profitable at all. The definitions of suitability classes should be modified to allow ranking on the basis of more assessment criteria than only profitability.

In the practice of land evaluation conservation is usually given priority over productivity. If soil conservation is required, it is implicitly assumed that the benefits always compare favorably with the required inputs or that the principle of economic viability does not apply anymore. For watershed land use planning, land evaluations often lead to recommendations for exclusive protective forest land use, judging the alternative of shifting cultivation inappropriate, even on land where it has been practiced for centuries. The consequence of forced migration of the watershed inhabitants is again part of the context, not of the evaluation procedure. Such land use recommendations do not provide a good starting point for the development of upland agriculture (Fox, 1987).

7. Incomplete Basis for Comparison between Alternative Land Uses

The function of land evaluation is to present comparisons of the most promising kinds of land use. The comparison of several alternatives is even proposed as a basic principle. In the Framework it is explained that a tract of land might be classified as highly suitable for rainfed agriculture, because the value of crops produced substantially exceeds the cost of farming, but only marginally suitable for forestry, on grounds that the value of timber only slightly exceeds the cost of obtaining it. However, the Framework states also that suitability classes for different uses cannot be compared, because suitability class limits are defined separately for each use. This means that if a given piece of land is rated highly suitable for rice cropping and marginally suitable for yams, the yam may yet give a higher net return and thus be the most preferred land use. So the "best" use of each land unit cannot be derived from suitability ratings. This holds in particular if within a land evaluation study the suitability classification is made for different levels of input and technology. Thus, land evaluation would allow land units to be ranked according to their suitability for a given use but not according to different use possibilities for a given piece of land. This has consequences for the correct interpretation of tabular legends to suitability maps, showing for each land unit a row of suitability ratings for different land uses (Table 5). Comparisons should only be made within columns per land use type and not within rows per land unit. This lack of the possibility of comparison implies that land evaluation inevitably betrays its very principle of comparability.

In fact, it is even disputable whether the suitability ranking of different land units for a given use (LUT) is theoretically possible, because a LUT is partly defined by the properties of the land. If the land differs, the LUT must differ too. For example, there are many different ways of growing yams (*Dioscorea* spp.) at a low input level, distinguished by differences in cultivar, field preparation, associated crops, planting date, propagation, soil, and climate, even if the socioeconomic context may be quite similar. It is only in a very generalized sense that land can be evaluated for a use such as "improved traditional yam cultivation," because at a detailed level each land use is adapted specifically to the land on which it is practiced.

Table 5. Tabular legend to land suitability map, giving suitability class ratings* for each land utilization type

	Land utilization types			
Land unit	LUT1	LUT2	LUT3	LUT4
LU1	S1	S1	S3	S1
LU2	S2	S1	S2	NS
LU3	S3	S2	S2	S1
LU4	NS	S1	S3	S1

*S1, highly; S2, moderately; S3, marginally suitable; NS, not suitable.

When land units are evaluated for a given use, the classes are often distinguished on the basis of increasing inputs to obtain the same standard yield level. This can lead to the nonrealistic conclusion that the highest expenditures are required on marginal lands. In actual farming practice it is commonly observed that the most capital-intensive form of a "given use" is practiced on the best land units and that the lowest expenditures are made on the least suitable lands (de Wit et al., 1987). The existence of a positive relationship between farm investment and quality of the land has been substantiated for large areas in Ontario, Canada (Dumanski et al., 1987).

8. Aim to Explain Leads to Complexity

According to the Framework the function of land evaluation is to bring about an understanding of the mutual relationships between land and the use to which it is put. This is in fact a research objective, and the question may be asked whether land evaluation is a science or a practical tool in land use planning (Radcliffe, 1988). Land evaluation can be considered an attempt to reconstruct "rational" decisions on land use through synthesis of information on land and land use in a given context. The methodology is usually an intuitive integration of many aspects by way of a complex matching procedure. To gain insight an analytical procedure is applied, consisting of splitting up both the land and the land use into abstract elements and subsequently recombining these elements to arrive at a suitability judgment on the basis of assumed relationships. The way of reasoning can be traced back by scanning a number of tables. The validity and consistency of such sets of tables are among the main concerns of a land evaluator and form the main reason for their frequent revision in the iterative matching process. The final check on consistency is through a field check, which, according to the Framework, "is essential to ensure that the suitability classes arrived at by the above procedures are in accord with experienced judgment." The tables can then be used for prediction of future conditions and for extrapolation to areas with less field data. However, experience has taught that the range of conditions to which interpretation results can be transferred is limited.

The Framework seems to acknowledge that a holistic approach to land evaluation as applied in a field check gives more consistent results than the analytical matching procedure and that the observations in the field check may overrule the analytical result. This may lead to the peculiar situation that the analytical procedure is calibrated first using the field observations, and in a later stage the field observations are presented as the prediction result of the analysis. Van Diepen (1983) discussed the danger of circular reasoning in the matching procedure. Melitz (1986) concluded that all the work of unraveling the complex relationships among land, its use, and its suitability in order to develop ratings or models that fit the facts from experience merely serves to explain these facts. He proposed new working procedures that circumvent the processing, analysis, and matching for all land qualities as prescribed by the Framework. These new procedures are essentially based on correlations of land units with land use experience. In fact, they elaborate the empirical approaches of the pre-Framework period.

It is important to distinguish between the methodological requirements of land evaluation research and the need for sound judgments and working speed in the practice of land evaluation. In combination, this calls for scientifically objective procedures that are complementary to the sound judgments from empirical appraisals rather than serving as a substitute for them.

V. Land Evaluation in Practice: Selected Case Studies

Procedures in land evaluation based on soil surveys are illustrated in this section with two examples of conventional land evaluation in support of land use plannning. The first example preceded the Framework and deals with soil survey application for the planning of polders in the Netherlands; the second one is a recent application of the Framework for polder development in Portugal.

A. Preframework Land Evaluation: The Case of the Dutch Polders

Soil survey information was used as a basis for land use planning in the Dutch polders, reclaimed from sea bottom land over the period 1925 to 1980. Four large polders comprising about 160,000 ha were successively brought under cultivation. The purpose of the soil survey in the polders was to obtain results for practical application such as agricultural land use planning and civil engineering (Smits and Wiggers, 1959).

1. Soil Survey as Basis for Land Evaluation

The soil survey was carried out in three stages at increasing intensity. The results of the first survey, carried out by under water sampling, were needed for the preliminary design of the future polder and included (1) a physical soil suitability map to design the outer boundaries of the polder by excluding zones with unfavorable soils and to make a provisional land use map, and (2) forecasts of the settling (shrinking) of the soil and of seepage after reclamation. Based on this information specifications for drainage and irrigation systems were developed, and estimates were made of the costs of reclamation and buildings.

The results of the second survey, executed directly after the polder emerged from the water, served to refine the results of the first survey. This included a final layout of the network of canals and roads, more detailed specifications for the digging of canals and spacing of field ditches, and additional details on soil suitability and land use.

The third and most detailed survey was executed during the last stage of reclamation, prior to renting the land to farmers, and consisted of describing soil profiles in freshly dug field ditches. The results of this survey were applied to specify the dimensions of the required drainage and irrigation systems and to design soil profile improvements such as deep plowing or mixing soil layers up to a depth of 2.0 m.

2. Land Evaluation for Land Use Planning

The broad types of land use for the new polders that were politically, economically, and socially acceptable were:

Arable land only.
Mixed farming with from one-sixth up to five-sixths of the area under grass.
Horticulture (vegetables and fruits).
Forestry.
Villages, roads.

Many soils were suitable for each of these uses, but the criteria were clear: extensive areas of horticulture were not wanted for fear of overproduction, and furthermore use as arable land was preferred because at that time it had a greater profit-earning capacity than cattle farming. With this in mind, the general pattern of land use was established on the basis of the prevailing soil conditions, which included:

Droughtiness.
Workability, which determines the timing of the field work, the required traction, and the ease of lifting potatoes and sugarbeets.
Susceptibility to surface sealing, interfering with the need for a fine seedbed.
Risk of wind erosion.
Nutrient status of the soils (K, P).

The size of the future farms was determined only after the general land use pattern was established, primarily on the basis of political, financial, and social considerations. These included a minimum farm size to provide adequate income and a desire to establish a harmonious structure of the rural population through variations in the size of the holdings. The size of the farms was only partly determined by soil conditions. Small farms were established only on soils of good quality, with a minimum of limitations: easy to till, allowing a great deal of freedom with respect to choice of crops, and with a low risk of crop failure. Soil information was also used for fixing the rent and, after settlement of farmers, for advising them on farm management aspects such as tillage requirements, crop rotation, and fertilizer application.

The number of soil characteristics used during the three stages of the soil survey to evaluate soil conditions with respect to their consequences for engineering and agricultural use was surprisingly small, namely two: the clay content (here the fraction of particles smaller than 16 μm) and the specific surface of the fractions exceeding 16 μm. The latter parameter, called the *U*-value, is a measure of the coarseness of all sand fractions combined. The soil survey was based on the description of textural profiles only, distinguishing 11 grades in clay content and six grades in *U*-value.

3. Comparison with the Framework Standard

It is interesting to compare this land evaluation for the Dutch polders with the standards and guidelines of FAO. The Framework advises against the use of land characteristics, such as soil texture, in suitability assessments, because a given land characteristic may influence suitability in many different ways. The guidelines (FAO, 1983) state "that soil texture has direct or indirect effects on some 14 land qualities, for example, moisture availability, nutrient retention, workability, and erosion hazard. It is thus impossible to say that any particular texture is 'good' or presents 'no limitations': sandy textures are favorable as regards workability, but adversely affect moisture availability or nutrient retention." The Dutch pedologists, however, found it most pragmatic to select soil texture as the basis for their land evaluation, because all relevant land qualities could be related to this single land characteristic. This view is supported by Sys (1985), who distinguishes between land characteristics and land qualities in accordance with the Framework but proposes to make one rating for all land qualities expressed by one characteristic and to avoid separate ratings for each single land quality. Consequently, Sys (1985) recommends that "the single rating of texture should be done with regard to the capacity to retain nutrients, water availability, permeability."

For the description of land utilization types in an evaluation for a future setting, modified by major land improvements, the FAO Framework recommends much more detail. In this evaluation there was no comparison of alternative kinds of land use, no subdivision in suitability classes, no transformation of soil characteristics into land qualities, no rating tables, no conversion tables, no iterative procedures; there was no integration of physical and economic analysis, no matching between land use requirements and land qualities, and no specification of a land utilization type as a result of the evaluation. Instead, the physical land evaluation was executed from the beginning within the boundaries set by the political decisions.

This is an example of a clearly decision-oriented approach, each step in the process leading logically to the next step and resulting in a reduction in the number of possible land use scenarios. In particular, the soil survey did not venture to formulate new alternative land uses for each soil unit or try to design future land use systems at the farm level.

B. Framework Land Evaluation: The Case of Leziria Grande

As an example of a Framework application, the land evaluation study for the Leziria Grande Project, a polder area in Portugal, is selected (Beek et al., 1980; Socio-Economic Working Group, 1984). The Leziria Grande is an island of 13,000 ha in the estuary of the Tagus River. The immediate aim of the study was to present development alternatives that would lead to land use intensification through improvement of the technical infrastructure. A second aim of the study was to prove the applicability of the Framework in project selection, planning, and evaluation and to explore the feasibility of a systems approach in land evaluation as described by Beek (1978). The data base for the study included a soil survey and crop yield projections by soil type under various land development stages. Because of the experimental character of the applied methodology, this data base was unusually extensive, and data analysis was detailed.

1. Soil Survey as Basis for Land Evaluation

The soil survey was carried out in two stages, the first at a semidetailed level (1 : 25,000), and the second at a detailed level (1 : 10,000). The final soil map distinguished more than 100 mapping units on the basis of differences in parent material, textural profile, calcareousness, presence of salts, drainage conditions, infiltration rate, subsoil permeability, depth of groundwater table and topsoil conditions. For practical reasons these mapping units were grouped, on the basis of the rating of 14 land qualities, into 15 land evaluation units that each could be considered as homogeneous in their need for land improvements for the various land utilization types and in terms of soil productivity. The relevant land qualities were oxygen availability in the root zone during winter, risk of waterlogging during winter, risk of waterlogging under irrigation, drainability, loss of cultivable area because of surface drainage, workability, ease of tillage, quality of seedbed, risk of slaking (crust formation), risk of compaction, availability of nutrients, soil salinity, risk of salinization, and prospects for desalinization. Drainability was selected as the most important quality because of its effect on land reclamation costs. Each land quality was assessed by a rating of its component characteristics. For instance, workability was related to the texture of the surface soil, microrelief, depth to groundwater table in winter, and number of rainless days in autumn.

2. Land Evaluation for Land Use Planning

Each land evaluation unit (major soil type) was evaluated for its suitability for a specific land utilization type. In the applied matching procedure, the required land reclamation measures were estimated in physical and financial terms, and the expected effects on output in terms of recurring inputs and yields, production costs and benefits. In this analysis each combination of crop/farm type/land unit/land development status is considered a separate option with a unique input–output relationship. Land development status refers here to the land quality ratings under a given land development alternative at a given target date. Four development alternatives were formulated, ranging from simple rehabilitation of the existing drainage system to total irrigation development. To this was added the "future without project" as a fifth alternative. The land quality situation for each development alternative was estimated for five reference years, 0, 1, 10, 20, and 30 years after implementation, to take into account the gradual improvement in land conditions and management, resulting in a gradual increase in productivity.

Eleven farm types or land use alternatives were selected, including medium- to large-scale rainfed and irrigated crop production, mixed farming, rice cropping, extensive grazing, and horticulture. A complicating factor for yield projection was soil salinity. The expected yield increases had to be attributed both to gradual desalinization of the soils and to improved cropping practices. The data from land evaluation included yield projections and related physical inputs, maximum admissible irrigation and cropping intensities, labor and machinery requirements, and crop calendars for about 50 eligible crops. They are differentiated according to soil type, reference year, and technical development alternative. The application of the systems approach implies that the yield predictions are based on an analysis of input–land quality–output relationships (Beek, 1978). Much effort was spent on land quality assessment for defined land use systems to provide a basis for yield projections through matching. However, in spite of this data analysis, yields were predicted largely against the background of a farm management survey in consultation with experienced agronomists, and taking account of analogous developments in similar areas.

3. Comparison with the Framework Standard

The close observation of the Framework rules in this study led to a very extensive data base, which required a first reduction for the purpose of land evaluation and subsequent drastic reductions for the purpose of the feasibility study. Beek et al. (1980) called it a paradox that, "while a grouping of land units is supposed to be based on estimates of inputs and outputs, the land units must be grouped at the beginning to avoid estimating too many inputs and outputs." Even after the reduction in number of land units, the evaluation report listed many thousands of possible combinations. These results could not be readily applied by economists in the feasibility study, using linear programming techniques for data processing and optimization. Finally, only 21 combinations were selected for project feasibility assessment.

In a separate study, a pragmatic procedure originating from consultancy practice (ILACO, 1981) was followed for the selection of a land development alternative. This alternative procedure gave the same result as the more time-consuming Framework approach. The Socio-Economic Working Group (1984) concluded that the much criticized pragmatic approach of land development consultants, taking a holistic view on the land and its use possibilities, does not necessarily produce inferior results than the more sophisticated Framework approach. The splitting up of the area into groups of soil types was considered worthwhile, but not the splitting of land into land qualities, because that information could not be satisfactorily synthesized and related to economic parameters. One of the reasons why greater precision was hard to achieve with such an analysis of physical and technical variables was the difficulty of assessing the human factor in terms of productivity.

VI. Yield Estimates and Productivity Ratings

Crop yield estimates differentiated by soil type and management system are one of the major concerns in any land evaluation for agricultural purposes. Yield estimates usually refer to long-term average yields for individual crops and are given either as an absolute or relative single value or as a yield class. Increasingly, variability in yield level is used as a measure of crop performance under defined conditions (Dumanski and Onofrei, 1989; Hazell 1986). For some applications, such as land tax assessment or land use zoning, a measure of overall performance of agricultural land use is needed. For such purposes an indicator crop is selected, or a composite index including various crops is developed. In the context of land evaluation, yield estimates serve as a basis for comparison of productivity of different kinds of land. Primarily biological productivity is considered, as defined by ecological and technological constraints. Economic aspects of land productivity are sometimes integrated in the biological assessment or treated separately in a second stage.

The method of assembling or generating information on yields varies among evaluation systems. The major approaches to yield estimation, often applied in combination with each other, are:

1. Intuitive estimates without a formal procedure, based on local experience.
2. Reasoned intuitive estimates obtained by application of hypothetical rating schemes such as matching (FAO, 1983), based on known or assumed relationships between crop performance and observed soil properties. This method is sometimes referred to as "systematized farmer's knowledge" and is the usual approach in conventional land evaluations.
3. Collecting yield data by soil type through interviews with farmers, from long-time farm records, or from regional yield statistics.
4. Use of statistical (empirical) models developed on the basis of regression analysis by relating observed yields to selected environmental and management factors.

5. Use of deterministic models in which growth-controlling factors and processes are described on the basis of knowledge from the natural sciences. The most advanced form of this approach is dynamic simulation modeling. For application in land evaluation, deterministic models are usually simplified (summary models).

In the first four approaches economic and management factors are included to varying degrees; the deterministic approach uses physical factors only.

A. Systematization of Observed Yield Levels

1. Estimation of Innate Soil Productivity

Soil survey reports usually contain estimates of local yields for various crops on different soils. The question is whether these data can be used directly to construct a reliable single-number index of soil productivity, which would be of great value in making land use decisions. For example, in the United States the concept of "innate soil productivity" has come into use in the context of preparing a national map of prime farmlands. Gersmehl and Brown (1986) have explored the validity of this concept by analyzing regional differences in yield relationships for the locally most important crops when grown on the major soils. County soil surveys were the source of soil and yield data. A single-number index of crop yield potential can be constructed by ranking soils in terms of productivity. This is easy when the yields for alternative crops on different soils are highly correlated but becomes difficult in the opposite case. Their analysis indicated striking regional disparities in the validity of such an index. A simple index appeared well applicable in the Midwest but much less in the southern states. This variation is attributed to differences in local crop mix, unique soil traits, climatic calendar, management techniques, and errors in the data base. More specific reasons for a low correlation are that crops with very contrasting soil requirements are grown or that specific soil traits favor selected crops or that climatic gradients affect one crop more than others or that crop calendars are different; for example, soybean is grown either as a main crop or as a second crop after wheat. It is further anticipated that a productivity index will change over time as a result of technological innovations, introduction of new crops and new varieties, changes in world market conditions, erosion, and possibly climatic change. These factors will alter the indices and modify their spatial pattern.

Gersmehl and Brown (1986) concluded that any single index of soil productivity cannot be applied anywhere, and where it works well, it should be adaptable to changing conditions, which may require a change in the logic of the assessment procedure. In addition, a map showing only index numbers is a poor data base, and for the purpose of understanding and reviewing such an index map, all basic data needed to assess the index should be stored in a geographical information system.

2. Soil Potential Rating

Soil productivity in relation to current and potential yield level also has an economic dimension, which may be easily overlooked if the soils are ranked on the basis of current yield levels. Recently the U.S. Soil Conservation Service (1983) developed a land evaluation procedure called soil potential ratings, which uses observed yield levels in combination with relative production costs as indicators for soil performance. This is in contrast to most traditional procedures that use soil properties as indicators (McCormack, 1987). The procedure starts with the definition of a performance standard, specified by a standard yield attained on one of the most productive soils in the area, under a standard set of management practices and a standard level of input costs. For each soil type the corrective measures needed to achieve the standard yield are specified. Performance and corrective measures are translated in terms of annual returns and costs, and the soil potential rating is calculated as the difference between management costs and returns, normalized to an index between 0 and 100. Thus, the index value is based on relative costs and prices. Extra reductions to the index value are applied to account for noncorrectable soil limitations or adverse effects.

The advantage of the soil potential rating in comparison with productivity indices based on yield level only is that the costs of overcoming soil limitations are considered. As a result, different soils producing similar yields may have very different soil potential ratings, for instance, because of differences in costs for drainage, tillage, or erosion control.

3. Matching

The matching concept has been described earlier as a typical Framework procedure (Section IV.A.2). It is usually based on the most limiting value of all diagnostic factors, refined by subjective combination of less severe limitations (Beek and Bennema, 1972; Veldkamp, 1979; Nyandat and Muchena, 1980; Bunting, 1981; Wielemaker and Boxem, 1982; FAO, 1983; Breimer et al., 1986). The matching procedure gives agroecological suitability class ratings, which are related to yield classes.

B. Statistical Analysis of Observed Yield Levels

Statistical analysis of yield data for the formulation of yield functions by soil type was increasingly applied during the 1950s. Initial attempts used graphic techniques such as polyfactor analysis developed by Visser (Visser, 1950b; Ferrari, 1950). This analysis allowed quantitative determination of the effect of each individual growth factor on yield on the basis of data from randomly distributed small plots in farmer's fields over a wide area. The following factors were used: pH, NPK values from soil analyses, clay content, and groundwater depth. Distance to farmstead was used as a measure of management level. The influence of each factor on yield is represented by a curve. These curves can be combined to describe the influence of several factors. Graphic analyses are very instructive but laborious tools and were replaced by computer-aided analyses in the form of linear and multiple regression analyses and covariance analysis. These latter methods were first applied by Odell (1958).

1. Linear and Multiple Regression Analysis

Odell (1958) determined the productivity of Illinois soils by multiple regression analyses of crop yields by soil type on the basis of farm records covering up to 25 years. The selected independent variables used to describe corn and soybean yields were summer rainfall, summer temperature, supply of N, P, and K, an index for inclusion of legumes in the rotation, and year. The estimated yield function is a summation of linear and quadratic terms of the form:

$$\text{yield} = (au^2 + bu) + (cv^2 + dv) + \ldots\ldots + r$$

where a, b, c, d are regression constants, u, v independent variables, and r the residual term. Odell (1958), using a yield function based on data from 150 observations (15 fields over 10 years), indicated that the accuracy of corn yield estimates was 25% for the yield on a single field in a particular year, 10% for the mean yield of a field over 10 years, and 2.5% for the mean yield of 15 fields over 10 years.

Many of such equations have been developed. The number of independent variables may vary from one to 10 or 20. They have the great disadvantage of having only local validity because they are crop, soil, and management specific. Many variables not explicitly considered in the equations are implicitly assumed constant. Very reliable predictions can be made only with models developed with a good data base and validated with another one.

The approach is still widely applied, and an arbitrary example is that by Olson and Olson (1986), who developed a linear equation for predicting average 20-year corn yields under high management for all soil map units in New York State. When tested at an independent experimental plot, the predicted yield was within 3% of the actual average corn yield over 19 years, which compares well with Odell's 2.5%. Initially, seven soil and climate variables were considered. Through an analysis of variance, parallel variables were detected and others redefined. Through multiple regression analysis an equation for predicting corn yields was found that contained only four parameters:

$$Y_{\text{corn}} = INTERCEPT + B1 \cdot RAINSTOR + B2 \cdot TEMP + B3 \cdot BASES + B4 \cdot OCARB$$

in which $RAINSTOR$ = rainfall − soil storage, an index for water stress, to be calculated separately for droughty and ponding conditions from soil water storage capacity, evaporation, and rainfall; *TEMP* = growing degree days; *BASES* = sum of bases in milliequivalent per pedon; and *OCARB* = organic carbon in grams per pedon.

The equation should not be used to predict corn yields for individual years. The aim of the equation is to rank soils on the basis of their long-term ability to produce corn. Note that the choice of parameters and the shape of the yield function are quite different from Odell's. It is important to eliminate parallel variables and to account for interactions between variables. Statistical techniques to detect such relationships include analysis of variance, multivariate analysis, principal component analysis, and cluster analysis.

2. Parametric Approaches: Statistical and Empirical Indices

In eastern Europe research on interactions of multiple growth factors continued in the tradition of Mitscherlich (1909, 1955) and aimed at development of mathematical descriptions based on fundamental relationships among several growth factors that were universally valid. Such equations appear very systematic and convey the nature of the assumed interactions: additive, multiplicative, logarithmic, logmultiplicative, or limitational (Kaltofen, 1975). Such complex statistical parametric models are intended to be explanatory. Most so-called parametric systems are empirical and simple, usually based on a small number of factors (1 to 10) of which the ratings are combined by addition or multiplication. An extensive review of the numerous systems is given by McRae and Burnham (1981).

Additive systems have been applied, especially in Germany since the 1920s and in eastern Europe, mainly for land taxation purposes. They are still applied but are becoming outdated (Koreleski, 1988). Multiplicative systems, on the other hand, are becoming increasingly popular because their results appear realistic and their logic of multiplicative interactions plausible.

The working group that developed the FAO Framework objected to the use of parametric methods because the underlying principles were not compatible with the systems analysis approach (Brinkman and Smyth, 1973). Disadvantages of the parametric methods are misleading accuracy, arbitrariness in the choice of factors, and a too-great flexibility (McRae and Burnham, 1981). The objections and disadvantages refer especially to situations in which parametric equations are formulated with no other verification than expert judgment, which is the rule in the land evaluation practice (FAO, 1983).

a. The Land Index

The Storie index is the classical example of a multiplicative parametric system. Many versions have been developed worldwide. One of the mainstreams of the approach is the school of Sys at the State University of

Ghent, Belgium. In the procedures proposed by Sys (1978, 1985) a land index (LI), which may be a soil index, a climate index, or a combined soil–climate index, is calculated as

$$LI = \prod_{i=1}^{q} R_i \times 100^{1-q}$$

in which R_i is the rating index of the ith factor, and q is the number of factors.

Factors used in calculating the land index are land characteristics or possibly land qualities. The kind of factors chosen varies with circumstances (purpose, climate, soil, crop, management level, available data). The number of factors should be kept at a minimum to avoid too low land index values. Each factor is rated on a scale from 20 to 100 for important factors and from 80 to 100 for less important factors; i.e., the different factors are weighed a priori. Soil factors may be adjusted with depth correction indices. The basic equation of Sys (1978; 1985) includes 10 soil factors to describe the influence of topography (slope), wetness (drainage class and flooding), physical soil characteristics (texture, stoniness, depth), fertility characteristics (CEC, base saturation, organic matter), and alkalinity or salinity. When applied for crop suitability assessment in a region, the number of factors discriminating index values between soil units is much less than 10, as not all factors are limiting. The rating criteria are crop specific.

A large amount of data on optimum and marginal soil conditions have been compiled in the form of crop requirement tables for 30 major annual and perennial crops of the (sub)tropics (Sys and Riquier, 1979; Sys, 1985, 1986). They constitute one of the most extensive data bases of its kind for land evaluation purposes and yet are only indicative, as the diagnostic criteria are in the form of threshold values for class ratings, and there are few classes, resulting in a wide range of values within each class. The applicability is further limited by the paucity of references to agroecological zones, locations, crop varieties, and management conditions for which the criteria were originally elaborated. More detailed information with a higher interpretive value in the form of factor response curves is not available.

Land indices are often the final result of land evaluations, but in principle they could be used for yield prediction. For wheat (*Triticum aestivum* L.) in Belgium, Sys (1978) found a high correlation between a land index value and yield on farms when grouped by farm type. The farm types differed in size and mechanization level. The correlation was much lower when all farm types were combined. The relationship could be described by a simple regression equation of the form

$$YIELD = A \cdot LANDINDEX + B$$

Embrechts et al. (1988) found a similar relationship between a land index and yield of oil palm *(Elaeis guineensis)* in Indonesia.

b. The Productivity Index Model

Recently, the productivity index (PI: Pierce et al., 1983) was developed to assess the relative yield potential of eroded soils on the basis of a minimum set of soil variables. This PI model was chosen to quantify the effect of erosion on crop production in an international context (Rijsberman and Wolman, 1984). The PI uses a multiplicative equation:

$$PI = \sum_{i=1}^{r} (A_i \times C_i \times D_i \times WF_i)$$

where A_i, C_i and D_i express the sufficiency in soil layer i of available water-holding capacity, bulk density, and pH; r is the number of soil layers; WF is a weighing factor varying with depth, based on an idealized root distribution in an ideal soil. The sufficiency of soil factors is derived from factor–response curves. The factors are soil characteristics that change under progressive loss of topsoil and at the same time influence the rooting pattern and hence yield, the main attention being given to the change in moisture availability to the plant. Originally, the PI value was correlated to the yield of corn in the corn belt of the United States. Later, PI was used directly as a measure for production potential for field crops in general. The PI model has been applied to evaluate the effects of various long-term soil erosion scenarios. Application in the tropics warrants adaptation of the model. Although the PI model was designed for use with a small data set, it appeared that very few data were available to verify the relationships described by the PI model. Thus, the model served to identify basic research needs.

Despite their apparent quantitative approach, the parametric methods are qualitative assessments. Refinements are often attempted by adding new factors, applying more complicated mathematical functions, or replacing single factors by a mix of subfactors. Each refinement requires recalibration of the equation, and there are limits to the number of factors that can be statistically analyzed and correlated in a sensible way. In particular, statistical procedures are not suited for dealing with positive and negative feedbacks between factors of dynamic nature.

C. Deterministic Modeling

Yield estimates can also be made with deterministic models. Conceptually, such models calculate the yield response to growth-controlling environmental factors on the basis of knowledge of the fundamental relationships between crop performance and soil, weather, and water, as manipulated by the farmer. Simple forms of this approach are balance-sheet models describing the in- and outflow and retention of substances over time steps usually varying from 1 day to 1 year. Such balances may deal with water, soil mass, biomass, plant nutrients, etc. The most advanced form is dynamic simulation of processes in time steps usually varying from minutes to hours, depending on the actual rates of change. The relationships used in deterministic models are based on the results of research on basic processes, such as transport processes in the soil, crop transpiration, CO_2 assimilation, respiration, phenological development, nutrient uptake, etc. The major advantage of this approach would be the possibility of predicting yield of any crop at any location (Nix, 1968).

1. Soil Water Balance Models

Deterministic models in land evaluation have long been restricted to soil water balance models, ranging from very simple to very detailed, that were applied to estimate soil water availability. The various approaches have been described elsewhere (Addiscott and Wagenet, 1985; Feddes, 1985; Feyen, 1987). Bouma (1989a) has discussed their merits for application in land evaluation. Modeling of yield effects is more advanced for situations with water deficits than for those with excess water.

The basis for yield predictions from soil water balance models is crop water use or transpiration, as influenced by soil moisture status. Early investigations in crop water relationships had shown that dry matter production was proportional to crop water use (Briggs and Shantz, 1913). De Wit (1958) demonstrated that the proportionality factor is species dependent and governed by weather conditions. Schematically, this can be explained by the fact that transpiration is a linearly increasing function of energy availability, whereas CO_2 assimilation shows a saturation-type behavior. In addition, transpiration is directly influenced by air humidity and wind speed, but assimilation is not. These differences in response to varying weather conditions make it necessary to recalibrate models relating crop growth directly to water use if they are applied to other regions or to years with exceptional weather. The influence of water stress on final yield also depends on its timing within the growth cycle. In all models, yield reductions from drought stress are derived from the soil moisture deficit, but large differences exist in the way transpiration is calculated and in the applied relationship between transpiration deficit and yield reduction. The simplest way is to apply a water use efficiency expressing harvested crop yield per unit of water evapotranspired (Doorenbos and Kassam, 1979). More complex procedures are followed in physiologically based crop models that use daily values of relative transpiration as a reduction factor for photosynthesis, reflecting the reduced exchange of carbon dioxide as a result of partially closed leaf stomata (van Keulen and Wolf, 1986). An intermediate approach is use of response functions that describe the effect of reduced transpiration on growth (Feddes, 1985).

2. Crop Growth Models

Over the last two decades the systems-analytical approach to crop ecology has led to the development of many crop growth simulation models for quantitative estimates of the growth and production of the main agricultural crops under a wide range of weather and soil conditions (de Wit and van Keulen, 1987). Strongly simplified versions of such models have been implemented in land evaluation studies on the scale of continents for the Agroecological Zones Project (FAO, 1978–1981) and for Canada (Stewart, 1983). With increasing computing capacity, the application of more detailed models in land evaluation becomes feasible, e.g., the CERES models used by the International Benchmark Sites Network for Agrotechnology Transfer (IBSNAT) project (Jones and O'Toole, 1987) and the WOFOST model of the Centre for World Food Studies (van Keulen and Wolf, 1986; van Diepen et al., 1989).

WOFOST simulates crop growth during one growing season in daily increments, using a state variable approach. Major crop processes described are CO_2 assimilation, respiration, and phenological development, and the major exchange processes with the surroundings are transpiration and water and nutrient uptake. The rates of these processes are determined by the state of the crop and controlling environmental conditions.

The effects of the principal growth constraints are evaluated by making separate calculations of (1) the potential yield, reflecting the production ceiling determined by the crop's genetic potential under the ambient radiation and temperature regime, (2) the water-limited yield, relecting the influence of limited water supply and possibly of excess water, and (3) the nutrient-limited yield.

Many authors point to the data requirements of deterministic models as one of the major constraints to their wide-scale application. Burrough (1989a) characterizes this as "the parameter crisis." This suggests that the collection of high-quality data could improve the situation. Recently, attention has also been given to the unavoidable stochastic nature of input data and their influence on model performance (Burrough, 1989b).

An associated problem on the output side is that a deterministic model by definition produces results pertaining to a single event. Because of the spatial and temporal variability of the environmental factors, there is a degree of stochastic uncertainty in these results. Research is continuing to combine deterministic and stochastic approaches in a single stochastic process model (Burrough, 1989a), but stochastic models have hardly been applied in crop production analysis (Ritchie and Crum, 1989).

A dependable description of the growth processes requires a large number of parameters on crop and environment. For instance, WOFOST uses 20 crop parameters and five major crop state variables for the calculation of daily growth, and the values of most parameters and variables change over time. Each parameter and function in the model has its own inaccuracy. These errors accumulate in the simulated final crop yield (Spitters, 1990). For this reason, deterministic models usually need calibration or "fine tuning" when applied to new situations in spite of their original promise of universal applicability. It is, however, impossible to calibrate the complete parameter set by comparing overall model performance with measured growth data, as many different parameter sets may lead to practically identical results. Sensitivity analysis can be applied to study the influence of parameter values but can not improve their accuracy.

A model with few parameters is easier to calibrate and, when properly calibrated, can give more accurate predictions than a multiparameter model, although its explanatory value is lower. There is a trade-off between explanatory and predictive capability of simulation models; hence, it is important to find the optimum balance between detailedness for explanatory purposes and conciseness for predictive purposes. If a deterministic model is used as an analytical tool, greater accuracy is required in predicting relative than absolute yield levels. The calculated yields may be used for evaluating relative differences between regions, long-term yield variability, or the relative importance of the growth factors water and nutrients as a basis for establishing development priorities.

3. Nutrient Models

Application of deterministic nutrient models is a recent phenomenon in land evaluation. Traditionally, the influence of soil fertility on crop yield was described as low, moderate, or high on the basis of rating tables converting soil chemical data into nutrient status classes. More specific statements on soil fertility are possible with application of the fertility capability classification (FCC) system (Sanchez et al., 1982), which is essentially a threshold model to indicate whether or not there is a certain fertility problem.

Quantitative yield estimates can be derived from the amount of nutrients taken up by the crop and the concentration of nutrients in the various plant organs (van Keulen and van Heemst, 1982). It is then necessary to know the net supply of plant nutrient from the soil and from the application of fertilizers and the possible range of nutrient concentrations in the plant organs. The QUEFTS system (Janssen et al., 1986, 1989) provides a procedure to calculate nutrient-limited yield from such information. It represents the highest yield that can be obtained if no other growth factors are yield-limiting.

The QUEFTS system was developed for the quantitative evaluation of the native fertility of tropical soils using maize as an indicator crop. The procedure comprises a number of successive steps. First, the amounts of nitrogen, phosphorus, and potassium that are potentially available for uptake by a maize crop from an unfertilized soil are estimated using empirical relationships between chemical soil properties and nutrient uptake. These amounts of nutrients can also be derived from results of field trials. The actual nutrient uptake is calculated from the potential supply and an empirical relationship between actual and potential uptake. From the actual uptake of nitrogen, phosphorus, and potassium, the attainable yield is calculated taking into account the minimum and maximum concentration of each nutrient in the various plant parts and the proportional uptake of elements under well-balanced supply. The QUEFTS system also allows calculation of the yield response to fertilizer application for given recovery fractions (ratio between uptake and application) for each applied nutrient. These recovery fractions must be obtained from fertilizer trials.

The QUEFTS system is based on the seasonal nutrient balance for annual crops. Dynamic nutrient balance models that take into account the main processes such as transport of nutrients by mass flow and diffusion/dispersion, exchange with the soil complex, chemical transformations, etc. are difficult to use in practical situations and are still largely a research option (de Willigen and Neeteson, 1985). The application of such models is hampered by their lack of universality because of insufficient knowledge of the basic processes, their extended data requirements, and problems with validation and verification.

D. The Economic Dimension of Yield Assessment

So far mainly physical yield estimates have been considered. It is one step further to include economic factors also in the analysis in order to derive production functions that can be used to determine economically optimum combinations of inputs for a farm with given land resources and production goals. This line of research was stimulated by farm economists such as Heady (1952a), who used linear programming techniques to determine the most profitable production situation for farms. The application of such optimization techniques requires quantification of technical input–output parameters for each production activity on a farm (Heady, 1952b; Heady and Shaw, 1954). These are in part soil specific. Until recently, however, the possibilities of applying such linear programming techniques have been largely ignored within the context of land evaluation, save a few isolated applications (Diltz, 1980).

VII. Use of Models in Regional Land Evaluations

Land evaluation studies for very large areas have been conducted by FAO's Agroecological Zones (AEZ) Project covering the Third-World countries (FAO, 1978–1981; Higgins et al., 1987) and by Dumanski and Stewart (1981) for the national territory of Canada. Such studies aim at providing information on the adequacy of land resources to feed present and future populations and on the location of zones with comparative advantages or limitations for agricultural production. Such information should be useful for national and international policy development.

Both studies present their results in terms of suitability classes for the major crops when grown under purely rain-fed conditions. The suitability classes for each crop on each soil in each climatic zone are assessed on the basis of calculated crop yield potentials, expressed as the ratio of an estimated yield attainable under average management and a theoretical maximum attainable under perfect management.

In both studies long-term mean climatic data were used; hence, the vagaries of the weather within a growing season and the variations over the years are smoothed out. The time interval for which the growing conditions are evaluated is one growing season. For example, the concept "length of growing period" is used for the assessment of the land quality "moisture availability." The growing period is defined as the period in which water availability and temperature permit crop growth and is calculated from climatic data. It is assumed that crops can grow without moisture stress for the duration of that period and that locations with the same length of growing period have identical moisture regimes. These assumptions are not entirely valid. First, rainfall in one location may be more dependable than in another because of differences in rainfall distribution, even if the duration of the growing period is the same, Secondly, if mean monthly rainfall data are used as input, water-balance models may predict sufficiency of crop water supply even for drought-prone regions. However, years with average rainfall do not exist, and it is precisely the deviations from the average that interfere with farming practice: first rains too late, dry spells too long, too heavy rains, end of the rains too early. The AEZ study has dealt with this problem qualitatively by reducing the expected yield to 50% to 75% percent because of moisture stress if the length of the crop growth cycle equals the length of the growing period. No reduction is applied when the growing period exceeds the crop growth cycle by 30 days or more.

At the Centre for World Food Studies (CWFS), a different approach has been followed to deal with temporal and spatial variability. The leading principle is that because crops do not grow under average conditions, crop production should be calculated preferably with observed site data rather than with averaged data. For studies at a regional scale the calculations are repeated for a large number of selected sites and for a large number of years. If desired, the results of the calculations can be averaged. In other words, calculate first and average later. The use of averaged input data is valid only if linear relationships exist between input data and the calculated results. If these relationships are nonlinear, the results of calculations with averaged input data can be erroneous (de Wit and van Keulen, 1987).

The consequence is that even for broad studies the assessment should be based on calculations for points. Therefore, it is important that these points be representative. This approach is a departure from common procedures in agroecological macrostudies, which use generalized stereotypes in assessment procedures as a proxy for the sum of the constituent parts. In the CWFS approach the stereotypes ideally serve for the selection of sites only. An example of this approach is the handling of weather variability (Wolf et al., 1989). First, climatic zones are delineated on a map. There are no a priori criteria for defining the zones, because the only purpose of the zonation is to identify representative weather stations. On the basis of observed daily weather data, the effect of moisture stress on crop performance and yield is evaluated by calculating a soil water balance for each day of the growth cycle. If no daily weather data are available, a statistical weather generator is applied to generate at least daily rainfall data on the basis of monthly means. In a crop growth simulation model, these daily effects are integrated, and the yield reduction from moisture stress is derived. These calculations are then repeated for many years. As yield indicators, the average calculated water-limited yields, the coefficient of variation in yield, and the ratio of water-limited yield to maximum potential yield can be used.

A similar procedure is proposed for the handling of soil information, but here the practical problem is that the reported soil data often have passed through several stages of generalization. For example, the results of many hundreds of granulometric analyses for many soil profiles and soil horizons are often condensed into one statement that the soils are medium to fine textured. As a result, comprehensive soil data sets for single profiles are scarce.

VIII. Application of Land Evaluation in Land Use Planning

Land evaluation for rural purposes deals with the assessment of potentials and constraints of land for agricultural use. Traditionally, assessment of productivity and in particular of crop yield has taken priority over all other aspects of land use performance. Presently, the effects of land use on quality of the environment and environmental sustainability of agricultural production systems have become major issues. These problems range from pollution with nitrate, phosphate, and biocides associated with intensive agriculture in situations of structural overproduction in some industrialized countries to declining soil fertility and erosion in association with low-input farming against a background of overexploitation of the natural resource base and scarcity of external inputs in the least-endowed countries.

Land evaluation may contribute to the solution of these problems by presenting feasible land use alternatives that meet the objectives of planners or policymakers. It is necessary for the land evaluation to be focused on the planning objectives. A common criticism is that land evaluations are carried out in isolation, especially when "standard" evaluation systems are applied, based on few, poorly selected, or unstated criteria (Ive et al., 1985). Failure of land evaluations to influence land use decisions is often also related to deficiencies in the institutional context such as a lack of clearly defined objectives, nonrelevance of proposed land use alternatives, a lack of follow-up during the planning and implementation stages, possibly because of the nonexistence of a planning authority or a lack of an effective strategy for reaching the land users.

Land use planning in its simplest form is the allocation of land to various categories of use according to criteria formulated during the land evaluation process. Land use planning becomes more complex when there are land use problems. Such problems are region specific and may have agronomic, economic, social, and political dimensions. Then, land use planning deals with multipurpose use of land, trade-offs between different functions of the land, and conflicting interests between the different categories of land users and between collective and individual goals and needs, so that land evaluation criteria are not obvious.

The approaches to solve these land use problems range from applying pragmatic empirical rating systems in support of local land use planning to applying analytical mathematical models to determine optimum solutions. An example of the empirical approach forms the land evaluation and site assessment (LESA) procedure for guiding the conversion of farmland to urban uses (Wright et al., 1983). As examples of the analytical approach are taken the application of multiple-goal programming (van Keulen, 1990; de Wit et al., 1988) for assisting in solving the problem of conflicting goals in regional land use planning and the related concept of integral land evaluation (Smit et al., 1984).

A. A Pragmatic Approach to Local Land Use Planning: The LESA System

LESA is a system designed by the USDA Soil Conservation Service as a tool for helping to make sound land use decisions, especially in the context of urban encroachment on agricultural lands. The aim is to preserve the best agricultural land (Wright et al., 1983; Dunford et al., 1983). The LESA procedure consists of two parts: land evaluation and site assessment. For the evaluation of land for crop production, three existing kinds of land evaluation are combined: land capability classification, current category of land use (e.g., prime farmland), and soil potential rating or, if the latter is not available, soil productivity rating. The soils are ranked according to capability and then rearranged into about 10 agricultural groups, each containing between 6% and 15% of the total farmland. For each group a relative land evaluation value is determined on a scale ranging from 100 for group 1 (the best soils) to 0 for group 10 on the basis of weighted average yield.

The site assessment part of LESA takes into account the effects of location of the evaluated tract of farm land, based on such factors as distance to urban area and services, adjacent land uses, zoning regulations, size of site, compatibility with adopted plans, and uniqueness of the site. Rating criteria are developed for each factor. For each site all factors are rated, and the total of all factor ratings is the site assessment value, which has a maximum of 200 points. High values indicate preference for agricultural use. The relative land evaluation value is then added to the site assessment value to yield the final LESA rating. Sites with the lowest LESA ratings are preferred for nonagricultural use. This procedure will usually protect the most valuable farmland. It is empirical, adaptable to local conditions and priorities, and therefore leads to consistent and acceptable propositions in local land use planning.

B. Mathematical Programming Techniques for Land Use Planning

1. Application of Multiple-Goal Programming

Multiple-goal programming is a mathematical procedure derived from linear programming that can be applied in the case of conflicting goals in regional land use planning (van Keulen, 1990; de Wit et al., 1988). The application of the interactive multiple-goal programming method to explore different land use planning options requires an input–output model, a set of goal variables, and an interactive multiple-criteria decision method (Spronk and Veeneklaas, 1983).

The contribution of land evaluation to the procedure consists of the specification of an input–output model, for which many of the existing productivity assessment procedures developed for land evaluation purposes can be usefully applied. The integrating role of the matching procedure, as usually applied in the suitability assessment, is replaced by an interactive multiple-criteria decision method. Compared to land evaluation, the focus of the procedure has shifted from identifying the best land use alternative under a given land use policy toward identifying the best land use policy under a given set of socioeconomic goals and technical constraints.

The input–output model contains the technical coefficients that describe the range of techniques available for agricultural development and associated activities in a rural environment. Each activity is defined by its relevant input and output coefficients, which are derived from a well-defined way of producing a given product. Cropping activities require coefficients that define the yield of both marketable product and crop residues, the material inputs required to realize that yield such as fertilizer, irrigation water, and biocides, and the labor requirements, both in quantity and time. The coefficients should also include quantitative estimates of the environmental effects (such as pollution, soil erosion, and soil fertility coefficients) associated with each activity. The production techniques should include a range of activities that are either currently practiced or have potential applicability in the region. All alternative techniques can be included in the activity set, irrespective of a judgment on their relevance, because the results of the analysis will indicate their appropriateness according to the specified rural development scenario. For each production technique, coefficients may be derived from statistical data, from practical experience, or from empirical or theoretical models. Data on new production techniques and on polluting effects of production techniques are scarce and often controversial and therefore require attention from research. An additional problem is the incorporation of the requirement of sustainability in situations in which currently practiced techniques exhaust the land. It is unrealistic to exclude such activities, and a solution might be to consider rate of degradation as an additional output and to define minimization of that output as the goal variable "sustainability."

Goal definition forms a crucial point in the multiple-goal programming method for regional rural development planning. Preferably a large number of development goals should be defined in physical rather than monetary terms. The goals should cover all the major interests in the region so as to ensure that all options for technically feasible development pathways are kept open. Major development goals that may be chosen at the regional level are target volumes of food and cash crops produced, total income generated, an equitable income distribution, a reasonable level of employment, and nonagricultural goals including nature and recreation and, at the farm level, maximum income and minimum risks. Sustainability should also be treated as a major goal and needs to be translated into physical terms such as minimum loss of nitrate or minimum use of biocides.

In the interactive procedure a feasible, widely acceptable solution is attained by successive optimization of the various goal variables. The solution is a mix of activities arranged in such a way so as to define an optimum development pathway over a planning horizon. In the first cycle of the procedure, the lower bounds of all goals considered are set at their minimum values to ascertain that a feasible solution is obtained that satisfies all these minimum requirements concurrently. Next, each of the goals is optimized on its own, with the lower bounds of the other goal variables as minimum goal restrictions. The combined result of this first iteration cycle indicates for each of the goal variables the most favorable value that can be obtained and the worst value that has to be accepted if realization of one of the other goals would be given absolute priority. However, the ideal solution in which all the goal variables reach their optimum values simultaneously does not exist, and an acceptable compromise has to be found. This is done in a stepwise procedure whereby the solution space of the first round is reduced by tightening the minimum restrictions on some goals and repeating the optimization for the other goal variables. In due course a solution is obtained, where none of the goals can be improved without sacrificing on one of the others. The result is that the costs of realizing one goal more completely can be explicitly expressed in terms of sacrifices on the other goals. The planner can thus quantify the opportunities for exchange between the various goals within the feasible solution space. The selection of a specific combination of goals is clearly a political choice.

Under the condition that the technical coefficients of the input–output model can be correctly quantified and that the goals are clearly formulated, the application of the interactive multiple-goal programming technique makes it possible to identify alternative technically feasible development pathways for a region and to evaluate the costs of achieving one goal in terms of sacrifices on other goals. The method is a tool for the analysis of rural development possibilities and gives an indication of the set of production techniques required for achieving any of the selected combinations of goal values. These technical requirements for each development option provide a starting point for further economic and social analysis and, finally, for the formulation of a plan of action.

2. Integral Land Evaluation

A similar approach using mathematical programming techniques has been developed by the Land Evaluation Group at the University of Guelph in Canada under the name "integral land evaluation" (Smit et al., 1984; Land Evaluation Group, 1983). Integral land evaluation deals with the determination of the critical importance of land for specific uses in order to meet basic societal goals such as economically acceptable production levels and needs of goods, services, and amenities. The mathematical model provides quantitative information on the flexibility of land use and the feasibility of land use options given the available land resources and socioeconomic objectives. To demonstrate the capability of the approach, the Land Evaluation Group has developed a prototype land evaluation model for Ontario, known as LEM2. The model was run for three scenarios characterized by increasing food production targets. The model results indicate that, first, the best soils become critically important for the most demanding crops such as fruits, vegetables, and tobacco, and that finally even the poorest soils are fully needed for forage production. The model quantifies this trend through a decreasing flexibility in the choice of land use under the subsequent scenarios and an increasing critical importance of the available agricultural land. Therefore, larger proportions of the land should be assigned to those uses with the greatest comparative advantage.

With different information and analytical tools, the procedure can be applied at other geographical scales to address any kind of "what if" questions, such as the evaluation of alternative erosion control policies. Such a study was carried out by assessing the effects of soil erosion on grain corn yields grown on various soil types in Huron County over a 25-year period under alternative scenarios of erosion control (Land Evaluation Group, 1983).

C. Computer Packages for Application of Land Evaluation in Land Use Planning

The main components of an integrated land evaluation system are:

1. A geographic information system with information on climate, land (including soil), land use, crop yields, use limitations, population, administrative subdivisions, etc.
2. A data base management system.
3. Analytical tools (models) to assess physical land use performance under given technological conditions.
4. Analytical tools for evaluating the adequacy of alternative land use options for given socioeconomic or environmental goals.

Integrated land evaluation systems are available as software packages that can be used on microcomputers, such as the LUPLAN package developed at CSIRO, Australia (Ive et al., 1985), and the Comprehensive Resource Inventory and Evaluation System (CRIES) of Michigan State University (Schultink, 1987).

The scope of LUPLAN is related to the LESA method but is more comprehensive, because it is not exclusively oriented to farmland protection. It is a computerized aid to land use planning. Its major components are a geographic data base, a land evaluation module, and a land use allocation module. The land evaluation module can calculate suitability ratings according to the USDA land capability system, the Storie index, or any other rating system, modified as desired. The ratings of several systems may be combined to reflect policy guidelines. LUPLAN calculates a suitability score (attractiveness score) for each land use on each mapping unit. Initially, the land use with the highest score is identified as the most preferred land use. The resulting preferred land use plan is reviewed by tallying the extent to which each policy guideline has been satisfied. If the plan as a whole is not acceptable, the relative importance of the guidelines can be adjusted and an alternative land allocation plan generated.

The CRIES system is designed for application in developing countries. The method focuses on evaluation of alternative land use options and policy scenarios in terms of derived public and private benefits. The evaluation procedures can be applied to farming systems and to regional or national aggregates. The assessment of the physical resource potential is carried out by single grid cell area or larger aggregates and serves to identify the (unrealized) production potential of the resource base. The major components of CRIES are a geographical information system based on grid cells and an agroeconomic information system. It includes separate modules for water balance calculations, yield predictions, calculation of erosion rates, statistical analysis, and linear programming. For the assessment of comparative production advantages at the single-farm level, it uses the MULBUD model (Etherington and Matthews, 1985), which is particularly suited to evaluate land use options for multiple enterprises characterized by mixed and diverse activities, typical for small farms in the tropics.

We expect that many more systems with similar capabilities will become available in the near future.

IX. Conclusions

The FAO Framework for land evaluation is still one of the basic documents in land evaluation and the most widely quoted reference. The present review shows how it is rooted in earlier approaches to conventional land evaluation, of which at its conception it reflected the state of the art. As such it represented a standard and provided incentives to reconsider existing procedures worldwide. On the other hand, its terminology became a source of confusion; because it was based on redefinition of existing concepts, it was liable to multiple interpretations; and it did not fit into the conceptual frameworks of other disciplines. A comprehensive, critical analysis of its principles, concepts, and procedures reveals inconsistencies and vagueness on essential points. In particular, the operational aspects of data analysis and data synthesis in land evaluation procedures, the integration of data from the biophysical, technological, and socioeconomic disciplines, and the links with land use planning are weakly articulated in the FAO Framework.

During the last decade, land evaluation research has focused increasingly on the application of quantitative procedures. In the biophysical domain, the temporal variability of land data can be accounted for by application of dynamic simulation models, and the advance of geographic information systems allows handling of massive amounts of spatially distributed data. In the domains of social sciences and planning, working procedures are being developed that can absorb land evaluation data for use in analytical models and planning scenarios.

In the present situation the Framework is becoming an outdated document from an operational point of view, but with a function as background philosophy. Yet, if land evaluation is to develop further as a discipline, it will also need a theoretical basis, marking its position among other disciplines. The recent development of new approaches calls for a thorough revision of the FAO Framework, including an adaptation of the standard glossary of land evaluation. The latter is being attempted with the Framework terminology as a starting point, but it is doubtful whether this will lead to a coherent vocabulary for use among disciplines.

The unique feature of land evaluation is that it deals with the full complex of land resources and their use potential, based on functional relationships between land and its uses. Realization of applicable results in that domain requires integration of information from various disciplines. In current land evaluation practice, such integration still relies heavily on subjective judgment, but operational tools are becoming available to replace part of the subjective elements, for instance, by the use of models of land use systems and/or mathematical programming. The often advocated combination of qualitative and quantitative approaches (Visser, 1950b; Vink, 1960; Mabbutt, 1968; Bouma, 1989b) remains valid, but this requires methodological research, too. The complete integration of all existing information in an operational way could benefit from the application of operations research. The use of information theory in land evaluation has great promise but appears to have received very little attention so far.

Changes in procedures also call for the use of other kinds of data than hitherto collected. With respect to soil data, the common procedure of averaging and categorizing observed data is less meaningful and less necessary than it used to be. For the purpose of model calculations, it is preferable to use multiple sets of observed point data per map unit rather than one idealized prototype point. Automated data storage makes this approach feasible. Numerical models need numerical data, but in many cases these have to be derived from descriptive information. The development of pedotransfer functions to generate the required numerical parameters from readily available soil survey data is therefore very important. But the need for in situ measurements of model parameters also must be stressed.

No doubt, increasingly detailed and extensive data banks and pedotransfer functions will become available. The next problem then is how to deal with the stochastic nature of each single observation and how to avoid undue propagation of errors. There is always a point where the use of more detailed data does not produce more accurate results because of scale-specific variability in natural processes and also because of our limited understanding of the functioning of biological systems of which the soil forms a basic part. Thus, careful selection of the appropriate degree of detail in model specification is required, even where computer capacity is not limiting.

References

Aandahl, A.R. 1958. Soil survey interpretation—theory and purpose. *Soil Sci. Soc. Am. Proc.* 22:152–154.

Aandahl, A.R. 1960. Soil productivity—concept and predictions. Trans. 7th Intern. Congress of Soil Science (Madison, Wisconsin, USA). 4:365–370.

Ableiter, J.K., and C.P. Barnes. 1950. Soil productivity ratings. *Trans. 4th Int. Cong. Soil Science (Amsterdam)* 1:360–364.

Addiscott, T.M., and R.J. Wagenet. 1985. Concepts of solute leaching in soils: A review of modelling approaches. *J. Soil Sci.* 36:411–424.

Beek, K.J. 1978. *Land evaluation for agricultural development*. ILRI Publication 23, Int. Inst. Land Reclam. Improvem, Wageningen.
Beek, K.J. 1980. From soil survey interpretation to land evaluation. 27, pp. 29–42. In: *Land reclamation and water management,* ILRI publication. Int. Inst. Land Reclamation and Improvement, Wageningen.
Beek, K.J., and J. Bennema. 1972. *Land evaluation for agricultural land use planning, an ecological methodology*. Agricultural University, Wageningen.
Beek, K.J., L. Reis, and R. Thiadens. 1980. Data analysis in land evaluation. pp. 63–80. *In: Land evaluation guidelines for rainfed agriculture. World soil resources report 52*. Food and Agriculture Organization of the United Nations, Rome.
Bouma, J. 1989a. Using soil survey data for quantitative land evaluation. pp. 177–213. In: B.A. Stewart (ed.), *Advances in soil science,* Vol. 9. Springer-Verlag, New York.
Bouma, J. 1989b. Land qualities in space and time. pp. 3–13. In: J. Bouma and A.K. Bregt (eds.), *Land qualities in space and time. Proc. ISSS Symp. Wageningen*. Pudoc, Wageningen.
Bouma, J., and A.K. Bregt (eds.). 1989. *Land qualities in space and time. Proc. ISSS symposium, Wageningen*. Pudoc, Wageningen.
Bouma, J., and H.A.J. van Lanen. 1987. Transfer functions and threshold values: From soil characteristics to land qualities. pp. 106–110. In: K.J. Beek, P.A. Burrough, and D.E. McCormack (eds.), *Quantified land evaluation procedures. Proc. Int. Workshop Washington*. ITC Publication 6, Enschede.
Breeuwsma, A., J.H.M. Wösten, J.J. Vleeshouwer, A.M. van Slobbe, and J. Bouma. 1986. Derivation of land qualities to assess environmental problems from soil surveys. *Soil Sci. Soc. Am. J.* 50:186–190.
Breimer, R. F., A.J. van Kekem, and H. van Reuler. 1986. *Guidelines for soil survey and land evaluation in ecological research. MAB technical notes 17*. UNESCO, Paris.
Breman, H., and C.T. de Wit. 1983. Rangeland productivity and exploitation in the Sahel. *Science* 221:1341–1347.
Briggs, L.J., and H.L. Shantz. 1913. The water requirements of plants. I. Investigations in the Great Plains in 1910 and 1911. United States Department of Agriculture, Bureau Plant Industry, Bulletin 284.
Brinkman, R. 1989. Land qualities and their variability—practical implications and some of FAO's responses. pp. 15–24. In: J. Bouma and A.K. Bregt (eds.), *Land qualities in space and time, Proc. ISSS Symp. Wageningen*. Pudoc, Wageningen.
Brinkman, R., and A.J. Smyth (eds.). 1973. *Land evaluation for rural purposes. Publication 17*. Int. Inst. for Land Reclamation and Improvement, Wageningen.
Bunting, E.S. 1981. *Assessments of the effects on yield of variations in climate and soil characteristics for twenty crop species. Technical Note 12*. Centre for Soil Research, Bogor, Indonesia.
Buol, S.W., P.A. Sanchez, R.B. Cate, and M.A. Granger. 1975. Soil fertility capability classification: 1 Technical soil classification for fertility management. pp. 126–141. In: E. Bornemisza and A. Alvarado (eds.), *Soil management in tropical America*. North Carolina State University, Raleigh.
Burrough, P.A. 1989a. Modelling land qualities in space and time: the role of geographical information systems. pp. 45–60. In: J. Bouma and A.K. Bregt

(eds.), *Land Qualities in Space and Time. Proc. ISSS Symp. Wageningen.* Pudoc, Wageningen.
Burrough, P.A. 1989b. Matching spatial databases and quantitative models in land resource assessment. *Soil Use Manage.* 5:3–8.
Christian, C.S. 1952. Regional land surveys. *J. Aust. Inst. Agric. Sci.* 18:140–146.
Christian, C.S., and G.A. Stewart. 1968. Methodology of integrated surveys. pp. 233–280. In: P. Rey (ed.), *Aerial surveys and integrated studies, Proc. Toulouse Conf. 1964:* UNESCO, Paris.
Dent, D., and A. Young. 1981. *Soil survey and land evaluation.* Allen & Unwin, London.
de Willigen, P., and J.J. Neeteson. 1985. Comparison of six simulation models for the nitrogen cycle in the soil. *Fertil Res.* 8:157–171.
de Wit, C.T. 1958. Transpiration and crop yields. *Versl. Landbouwk. Onderz. [Agr. Res. Rep.]* 64.6.
de Wit, C.T., H. Huisman, and R. Rabbinge, 1987. Agriculture and its environment: Are there other ways? *Agric. Syst.* 23:211–236.
de Wit, C.T., and H. van Keulen. 1987. Modelling production of field crops and its requirements. *Geoderma* 40:253–265.
de Wit, C.T., H. van Keulen, N.G. Seligman, and I. Spharim, 1988. Application of interactive multiple goal programming techniques for analysis and planning of regional agricultural development. *Agric. Syst.* 26:211–230.
Diltz, D. 1980. *Methodology for qualitative economic land suitability evaluation for agriculture. Land Resources Survey Sierra Leone. Techn. Report 9.* FAO/UNDP, Freetown.
Doorenbos, J., and A.H. Kassam. 1979. *Yield response to water. FAO Irrigation and Drainage Paper 33.* Food and Agriculture Organization of the United Nations, Rome.
Driessen, P.M. 1986. Quantitified land evaluation (QLE) procedures, a new tool for land-use planning. *Neth. J. Agric. Sci.* 34:295–300.
Dumanski, J., and C. Onofrei. 1989. Crop yield models for agricultural land evaluation. *Soil Use Manage.* 5:9–15.
Dumanski, J., M. Phipps, and E. Huffman. 1987. A study of relationships between soil survey data and agricultural land use using information theory. *Can. J. Soil Sci.* 67:95–102.
Dumanski, J., and R.B. Stewart. 1981. *Crop production potentials for land evaluation in Canada.* Research Branch, Agriculture Canada, Ottawa.
Dunford, R.W., R.D. Roe, F. R. Steiner, W.R. Wagner, and L.E. Wright. 1983. Implementing LESA in Whitman County Washington. *J. Soil Water Conserv.* 38:87–89.
Embrechts, J., P. Zulkarnain, and K. Sys. 1988. Physical land evaluation using a parametric method. Application to oil palm plantation in North-Sumatra Indonesia. *Soil Surv. Land Eval.* 8:23–29.
Etherington, D., and P. Matthews. 1985. *MULBUD user's manual.* National Centre for Development Studies, Australian National University, Canberra.
FAO. 1976. *A framework for land evaluation. Soils Bulletin 32.* Food and Agriculture Organization of the United Nations, Rome.
FAO. 1978–1981. *Report on the agro-ecological zones project. World soil resources report 48(1)—Africa, 48(2)—Southwest Asia, 48(3)—South and Central*

America, 48(4)—Southeast Asia. Food and Agriculture Organization of the United Nations, Rome.

FAO. 1983. *Guidelines: Land evaluation for rainfed agriculture. FAO soils bulletin no. 52.* Food and Agriculture Organization of the United Nations, Rome.

FAO. 1984. *Land evaluation for forestry. Forestry paper no. 48.* Food and Agriculture Organization of the United Nations, Rome.

FAO. 1985. *Guidelines: Land evaluation for irrigated agriculture. FAO soils bulletin no. 55.* Food and Agriculture Organization of the United Nations, Rome.

Feddes, R.A. 1985. Crop water use and dry matter production: State of the art. pp. 221–234. In: A. Perrier and C. Riou (eds.), *Les besoins en eau des cultures (conf. int. Paris).* INRA, Paris.

Ferrari, T.J. 1950. Growth factors and soil productivity. *Trans. 4th Int. Congress of Soil Science (Amsterdam)* 1:348–352.

Feyen, J. (ed.). 1987. *Simulation models for cropping systems in relation to water management. Agriculture series. EUR 10869.* Commission of the European Communities, Luxembourg.

Fox, J. 1987. Why land evaluations for agriculture go awry. pp. 127–129. In: K.J. Beek, P.A. Burrough, and D.E. McCormack (eds.), *Quantified Land Evaluation Procedures. Proc. Intern. Workshop Washington.* ITC Publication 6, Enschede.

Gersmehl, P.J., and D.A. Brown. 1986. Regional differences in the validity of the concept of innate soil productivity. *Ann. Assoc. Am. Geogr.* 76(4):480–492.

Glinka, K.D. 1914. *The great soil groups of the world and their development.* Translated from German by C.F. Marbut, 1927. Edwards Bros., Ann Arbor, MI.

Haans, J.C.F.M. 1980. Soil resource inventories and their application in the Netherlands. pp. 21–33. In: J. Lee and L. van der Plas (eds.), *Land resource evaluation (Wexford, Ireland). Agriculture Series EUR 6875.* Commission of the European Communities, Luxembourg.

Hazell, P.B.R. 1986. An overview of workshop issues. In: pp 15–46 P.B.R. Hazell (ed.), *Summary proceedings of a workshop on cereal yield variability.* IFPRI, Washington, DC.

Heady, E.O. 1952a. *Economics of agricultural production and resource use.* Prentice Hall, New York.

Heady, E.O. 1952b. Use and estimation on input–output relationships or productivity coefficients. *J. Farm Econ. Proc.* 34:775–786.

Heady, E.O., and R. Shaw. 1954. Resource returns and productivity coefficients in selected farming areas. *J. Farm. Econ.* 36:243–257.

Higgins, G.M., A.H. Kassam, H.T. Veldhuizen, and M.F. Purnell. 1987. Methods used by FAO to estimate environmental resources, potential outputs of crops, and population supporting capacities in the developing nations. pp. 171–184. In: A.H. Bunting (ed.), *Agricultural environments.* CAB International, Wallingford, UK.

ILACO. 1981. *Agricultural compendium for rural development in the tropics and subtropics.* Elsevier, Amsterdam.

Ive, J.R., J.R. Davis, and K.D. Cocks. 1985. LUPLAN: A computer package to support inventory, evaluation and allocation of land resources. *Soil Surv. Land Eval.* 5(a):77–87.

Janssen, B.H., F.C.T. Guiking, D. van der Eijk, E.M.A. Smaling, and H. van

Reuler. 1986. A new approach to evaluate the chemical fertility of tropical soils. *Trans. 13th ISSS Congress Hamburg* III:791–792.

Janssen, B.H., F.C.T. Guiking, D. van der Eijk, E.M.A. Smaling, J. Wolf, and H. van Reuler. 1989. A system for quantitative evaluation of soil fertility and the response to fertilizers. pp. 185–188. In: J. Bouma and A.K. Bregt (eds.), *Land qualities in space and time. Proc. ISSS Symp.* Pudoc, Wageningen.

Jones, C.A., and J.C. O'Toole. 1987. Application of crop production models in agro-ecological characterization: Simulation models for specific crops. pp. 199–209. In: A.H. Bunting (ed.), *Agricultural environments.* CAB International, Oxon, UK.

Kaltofen, H. 1975. Quantitative aspects of the coaction of several growth factors. pp. 71–112. Tagungsbericht 139 (in German). Akademie der Landwirtschaftwissenschaften der DDR, Berlin, GDR.

Kellogg, C.E. 1951. Soil and land classification. *J. Farm Econ.* 33:499–513.

Kellogg, C.E. 1961. *Soil interpretation in the soil survey.* USDA Soil Conservation Service, Washington, DC.

Klingebiel, A.A. 1958. Soil survey interpretation—capability groupings. *Soil Sci. Soc. Am. Proc.* 22:160–163.

Klingebiel, A.A., and P.H. Montgomery. 1961. *Land-capability classification. USDA Agric. Handbook 210.* Soil Conservation Service, USDA, Washington, DC.

Koreleski, K. 1988. Adaptations of the Storie index for land evaluation in Poland. *Soil Surv. Land Eval.* 8:23–29.

Krantz, B.A. 1958. Soil survey interpretation—interpretation of soil characteristics important in soil management. *Soil Sci. Soc. Am. Proc.* 22:155–156.

Land Evaluation Group. 1983. *Land evaluation group report '83. Publ. LEG-15.* Univ. School of Rural Planning and Development. Univ. of Guelph.

Mabbutt, J.A. 1968. Review of concepts of land classification. pp. 11–28. In: G.A. Stewart (ed.), *Land evaluation, Papers of a CSIRO symposium (Canberra).* Macmillan of Australia, Melbourne.

Marbut, C.F. 1928. A scheme of soil classification. *First Int. Congress Soil Science (USA, 1927).* 4:1–31.

Marten, G.G. 1988. Productivity, stability, sustainability, equitability and autonomy as properties for agroecosystem assessment. *Agric. Syst.* 26:291–316.

McCormack, D.E. 1987. Soil potential ralings—a special case of land evaluation. pp. 81–84. In: K.J. Beek, P.A. Burrough, and D.E. McCormack (eds.), *Quantified Land Evaluation Procedures. Proc. Intern. Workshop.* ITC Publ. 6, Washington.

McRae, S.G., and C.P. Burnham. 1981. *Land evaluation. Monographs on Soil Survey.* Clarendon Press, Oxford.

Melitz, P.J. 1986. The sufficiency concept in land evaluation. *Soil Surv. Land Eval.* 6:9–19.

Mitscherlich, E.A. 1909. Das Gesetz des Minimums und das Gesetz des abnehmenden Bodenertrages. *Landwirsch. Jahrb.* 38:537–552.

Mitscherlich, E.A. 1955. Zum Wirkungsgesetz der Wachstumfaktoren. *Z. Acker. Pflanzenb.* 99:261–266.

Nix, H.A. 1968. The assessment of biological productivity. pp. 77–87. In: G.A. Stewart (ed.), *Land Evaluation, Papers of a CSIRO Symposium (Canberra).* MacMillan of Australia, Melbourne.

Nyandat M.N., and F.N. Muchena. 1980. Land evaluation in Kenya with particular emphasis on the criteria used. pp. 26–48. In: *Third meeting of the Eastern African Sub-Committee for Soil Correlation and Land Evaluation. World Soil Resources Reports 51*. FAO. Rome.

Odell, R.T. 1958. Soil survey interpretation—yield prediction. *Soil Sci. Soc. Am. Proc.* 22:157–160.

Olson, K.R., and G.W. Olson. 1986. Use of multiple regression analysis to estimate average corn yields using selected soils and climatic data. *Agric. Syst.* 20:105–120.

Pierce, F.J., W.E. Larson, R.H. Dowdy, and W.A.P. Graham. 1983. Productivity of soils: Assessing long-term changes due to erosion. *J. Soil Water Conserv.* 38:39–44.

Pierre, W.H. 1958. Relationship of soil classification to other branches of soil science. *Soil Sci. Soc. Am. Proc.* 22:167–170.

Radcliffe, D. 1983. Land qualities—an aid to land evaluation? *Land Water* 17:18–20.

Radcliffe, D. 1988. Land evaluation: Science or planning tool? *Land Water* 31:16.

Rijsberman, F.R., and M.G. Wolman. 1984. *Quantification of the effect of erosion on soil productivity in an international context*. Delft Hydraulics Laboratory, Delft.

Ritchie, J.T. and J. Crum. 1989. Converting soil survey characterization data into IBSNAT crop model input. pp. 155–168. In: J. Bouma and A.K. Bregt (eds.), *Land qualities in space and time. Proc. ISSS Symp. Wageningen*. Pudoc, Wageningen.

Rossiter, D. 1989. ALES: A microcomputer program to assist in land evaluation. pp. 113–116. In: J. Bouma and A.K. Bregt (eds.), *Land qualities in space and time, Proc. ISSS Symp. Wageningen*. Pudoc, Wageningen.

Russell, E.W. 1973. *Soil conditions and plant growth*. 10th ed. Longmans, London.

Sanchez, P.A. 1976. *Properties and management of soils in the tropics*. John Wiley & Sons, New York.

Sanchez, P.A., W. Couto, and S.W. Buol. 1982. The fertility capability soil classification system: Interpretation, applicability and modification. *Geoderma* 27:283–309.

Schultink, G. 1987. The CRIES resource information system: Computer-aided land resource evaluation for development planning and policy analysis. *Soil Surv. Land Eval.* 7:47–62.

Siderius, W. (ed.). 1984. *Proceedings of the workshop on land evaluation for extensive grazing (LEEG), Addis Ababa. Publication 36*. Intern. Inst. for land Reclamation and Improvement, Wageningen.

Simonson, R.W. 1985. Soil classification in the past–roots and philosophies. pp. 6–18. In: *ISRIC Annual Report 1984*. Intern. Soil Reference and Information Centre, Wageningen.

Smit, B., M. Brklacich, J. Dumanski, K.B. MacDonald, and M.H. Miller. 1984. Integral land evaluation and its application to policy. *Can. J. Soil. Sci.* 64(4): 467–479.

Smits, H., and A.J. Wiggers. 1959. *Soil survey and land classification as applied to reclamation of sea bottom land in the Netherlands. Publication 4*. Int. Inst. for Land Reclamation and Land Improvement, Wageningen.

Socio-Economic Working Group. 1984. *Leziria Grande de VFX, Rural Development project feasibility study. Main report.* DGHEA/ILRI, Lisbon/Wageningen.

Soil Conservation Service. 1983. Soil potential ratings. Part 603.09. pp. 603:142–175. In: *National Soils Handbook.* USDA Soil Conservation Service, Washington.

Soil Survey Staff. 1951. *Soil Survey Manual. U.S. Dept. Agriculture Handbook 18.* Soil Conservation Service, USDA.

Spitters, C.J.T. 1990. On descriptive and mechanistic models for inter-plant competition, with particular reference to crop-weed interaction. pp. 217–236. In: *Theoretical production ecology: Reflections and prospects. Simulation Monographs.* Pudoc, Wageningen.

Spronk, J., and F.R. Veeneklaas. 1983. A feasibility study of economic and environmental scenarios by means of interactive multiple goal programming. *Reg. Sci. Urban Econ.* 13:141–160.

Stewart, G.A. 1968. Land evaluation. pp. 1–10. In: G.A. Stewart (ed.), *Land Evaluation. Papers of a CSIRO Symposium (CSIRO/UNESCO, Canberra).* Macmillan of Australia, Melbourne.

Stewart, R.B. 1983. *Modeling methodology for assessing crop production potentials in Canada.* Research Branch, Agriculture Canada, Ottawa.

Stockstad, O. 1958. Soil survey interpretation for engineering purposes. *Soil Sci. Soc. Am. Proc.* 22:164–166.

Storie, R.E. 1933. *An index for rating the agricultural value of soils (revised 1937).* Univ. of Calif. Agr. Exp. Sta. Bull. 556. Berkley, CA.

Storie, R.E. 1976. *Storie index soil rating (revised 1978).* Spec. Publ. Div. Agric. Sci. no. 3203. Univ. of California, Berkeley.

Sys, C. 1978. The outlook for the practical application of land evaluation in developed countries. pp. 97–111. In: *Land evaluation standards for rainfed agriculture. World soil resources report 49.* FAO, Rome.

Sys, C. 1980. *Land characteristics and qualities and methods of rating them. World Soil Resources Reports 52.* pp. 23–49. FAO, Rome.

Sys, C. 1985. *Land evaluation Parts I, II, III. ITC-State Univ. of Ghent. Publ. Agricole 7.* Administration Générale de la Coopération au Développement, Brussels.

Sys, C. 1986. Approach to soil survey interpretation for rice cultivation. *Pedologie* 36:199–217.

Sys, C., and J. Riquier. 1979. *Ratings of FAO/Unesco soil units for specific crop production. Report second FAO/UNPRA Expert Consultation Land Resource for Populations of the Future.* pp. 55–90. FAO, Rome.

USBR. 1953. *Bureau of reclamation manual. Vol V. Irrigated land use, Part. 2 Land classification.* US Dept. Interior, Washington DC.

van Diepen, C.A. 1983. Evaluating land evaluation. pp. 13–29. In: *ISM Annual Report 1982.* Int. Soil Museum, Wageningen.

van Diepen, C.A., J. Wolf, H. van Keulen, and C. Rappoldt. 1989. WOFOST: A simulation model of crop production. *Soil Use Manage.* 5:16–24.

van Keulen, H. 1990. A multiple goal programming base for analysis of agricultural research and development. pp. 265–276. In: *Theoretical production ecology: Reflections and prospects. Simulation monographs.* Pudoc, Wageningen.

van Keulen, H., J.A.A. Berkhout, C.A. van Diepen, H.D.J. van Heemst, B.H.

Janssen, C. Rappoldt, and J. Wolf. 1987. Quantitative land evaluation for agro-ecological characterization. pp. 185–197. In: A.H. Bunting (ed.), *Agricultural environments*. CAB International, Wallingford, UK.
van Keulen, H., and H.D.J van Heemst. 1982. *Crop response to the supply of macronutrients. Agric. Res. Rep. 916*. Pudoc, Wageningin.
van Keulen, H., and J. Wolf (eds.). 1986. *Modelling of agricultural production: Weather, soils and crops. Simulation Monographs*. Pudoc, Wageningen.
Veldkamp, W.J. 1979. Land evaluation of valleys in a tropical rain area. Ph.D. thesis, Agric Univ. Wageningen.
Verheye, W.H. 1987. Quantified land evaluation as a basis for alternative land use planning. pp. 144–146. In: K.J. Beek, P.A. Burrough, and D.E. McCormack (eds.), *Quantified Land Evaluation Procedures. Proc. int. workshop Washington*. ITC Publication 6, Enschede.
Vink, A.P.A. 1960. Quantitative aspects of landclassification. *Transactions 7th Intern. Congr. of Soil Science (Madison, Wisconsin, USA)* 4:371–378.
Vink, A.P.A. 1966. *Integrated surveys and land classification. Publ. S12*. ITC-Unesco Centre for Integrated Surveys, Delft.
Visser, W.C. 1950a. The trend of the development of land evaluation in the future. *Trans. 4th Int. Congr. Soil Sci. (Amsterdam)* 1:334–336.
Visser, W.C. 1950b. The quantitative basis of the evaluation of soil producibility. *Trans. 4th Int. Congr. Soil Sci. (Amsterdam)* 1:373–377.
Wielemaker, W.G., and H.W. Boxem (eds.). 1982. *Soils of the Kisii area, Kenya. Agric. Res. Rep. 922*. Pudoc, Wageningen.
Wolf, J., J.A.A. Berkhout, C.A. van Diepen, and C.H. van Immerzeel. 1989. A study on the limitations to maize production in Zambia using simulation models and a geographic information system. pp. 209–212. In: J. Bouma and A.K. Bregt (eds.), *Land qualities in space and time. Proc. ISSS Symp. Wageningen*. Pudoc, Wageningen.
Wösten, J.H.M., and J. Bouma. 1985. Using simulation to define moisture availability and trafficability for a heavy clay soil in the Netherlands. *Geoderma* 35:187–196.
Wright, L.E., W. Zitzmann, K. Young, and R. Googins. 1983. LESA—agricultural land evaluation and site assessment. *J. Soil Water Conserv* 38:82–86.
Zonneveld, I.S. 1972. *Land evaluation and land(scape) science. ITC textbook of photo-interpretation. Vol. VII*. Int. Inst. for Aerial Survey and Earth Sciences, Enschede.

Crop Residue Management

Rajendra Prasad and J.F. Power

Contents

I. Introduction

In less affluent countries such as those in South and Southeast Asia, grain is directly used for human consumption, and crop residues are the main source of fodder for animals. In addition, crop residues have several other uses (Table 1). Crop residues are certainly an asset in these countries and

1991 Springer-Verlag New York Inc.
Advances in Soil Science, Volume 15

seldom are left in the field. In advanced countries (Europe and America), draft animals have been replaced by internal combustion engines for farm power, and farmyard manure has been replaced by chemical fertilizer as a major source of nutrients. The development of synthetic fiber and a variety of inexpensive synthetic materials has replaced cereal straw for basketing, packaging, millinery, mat making, and similar uses (Staniforth, 1979). Consequently, with widespread use of combine harvesters, crop residues largely remain in the field and must be managed to provide the greatest advantage possible, especially for water conservation, erosion control, and maintenance of soil organic matter. In 1985, 918.58 million tons of cereal grain was produced in the developed world (Table 2), indicating that cereal residue production probably exceeded 1000 million tons per annum.

In parts of some countries (United Kingdom, Canada, and Australia) wheat straw is often burned, whereas in other countries (West Germany) there are strict laws against burning (Staniforth, 1979). In the United States and Canada, cereal stubble and straw are widely used to reduce wind and water erosion and to conserve soil water. Crop residues are frequently managed by stubble mulching wherein stubble is undercut to control weeds (McCalla and Army, 1961) or by no-tillage (or zero-tillage) in which herbicides are substituted for tillage to control weeds (Unger and McCalla, 1980).

A number of symposia have been organized, and reviews have appeared concerning crop residue management systems. Some major ones were: American Society of Agronomy (1978, 1983); Soil Conservation Society of America (1973, 1977, 1979); Great Plains Agricultural Council (1976, 1984); Ohio State University (1972); Plant Protection Limited (1973, 1975); United States Department of Agriculture (1977); Interna-

Table 1. Crop residues and their uses in less affluent countries

Crop	Residue	Use
Cereals	Straw	Cattle feed, thatching of dwelling huts and cattle sheds
Pigeon pea (*Cajanus cajans* L. Millsp)	Stem	Fuel for cooking food, thatching of dwelling huts and cattle sheds
Cotton (*Gossypium* sp.)	Stem	Fuel for cooking food, thatching of dwelling huts and cattle sheds
Linseed or flax (*Linum usitatissium*)	Stem	Fiber
Sugarcane (*Saccharum officinarum* L.)	Tops	Green fodder for cattle
	Bagasse	Fuel in sugar factory, raw material for making hard board or paper.
Peanuts (*Arachis hypogea* L.)	Green tops	Cattle feed

Table 2. Production of cereal grain in developed and developing countries*

Crop	Developed countries	Developing countries
Wheat	301,773	203,593
Rice	26,339	446,348
Barley	153,223	23,351
Maize	315,992	172,508
Rye	31,704	1,583
Oats	47,554	2,076
Millets and sorghum	33,871	70,721
Others	8,126	4,772
Total	918,582	924,953

*Source: Food and Agriculture Organization of the United Nations (1987). Data expressed in thousands of megagrams.

tional Soil Tillage Organization (1986); Knipscheer et al. (1983); National Agricultural Centre (1984); and Australian Society of Agronomy (1986). Considering the importance and relevance of managing crop residues in situ, two reviews were published in *Advances in Agronomy*. The first one was by McCalla and Army (1961), and the second by Unger and McCalla (1980). Considerable new research was conducted in the last decade relating residue management to soil chemical, physical, and biological properties and consequent fertilizer management practices needed for successful crop production. Of particular interest are data emanating from ^{15}N fertilizer experiments. Because of these advances, it is appropriate to incorporate this new knowledge into the framework of previous literature and to summarize the recommendations and conclusions. We have not attempted to review all available literature but only pertinent data that illustrate the main findings of recent research.

Because of variation in the nomenclature used with crop residue management practices, it was necessary to define specific terms for use in this review. The following abbreviations are used in this chapter to describe different tillage systems: NT, no-tillage, no-till, zero-tillage, conservation tillage, direct drilled; MT, minimum tillage, stubble mulch, light disking or chiseling; CT, conventional tillage, tillage involving moldboard plowing, or heavy disking or chiseling; RB, residue burned; and RI, residue incorporated.

II. Grain Yield

Leaving a residue cover on the soil surface as practiced in conservation tillage or NT may have positive, negative, or no effect on grain yield, depending on soil–climate conditions. For example, Kitur et al. (1984)

from Kentucky, Al-Darby and Lowery (1986) from Wisconsin, Christian and Miller (1986) from England, and Maurya (1986) from Nigeria reported no significant differences in grain yields of crops as a result of different residue management practices. Gerik and Morrison (1984) and Locke and Hons (1988) likewise reported from Texas no difference in grain yield of sorghum [*Sorghum bicolor* (L.) Moench] under different residue management practices. However, Musick et al. (1977), Unger and Wiese (1979), Unger (1984), and Baumhardt et al. (1985) reported greater sorghum yields with NT than with CT under both irrigated and dryland conditions. Greater grain yields with NT than with CT were reported from Georgia by Langdale et al. (1984) and Hargrove (1985); from Nebraska by Lamb et al. (1985), Power et al. (1986), and Follett and Peterson (1988); and from Nigeria by Osuji (1984). Nelson et al. (1977) observed that yields were greater with NT because of the better infiltration and less evaporation.

In contrast to these results, lower crop yields with NT were reported by Griffith et al. (1977), Papendick and Miller (1977), Costamagna et al. (1982), Touchton and Johnson (1982), Hargrove and Hardcastle (1984), Gallaher (1984), and Thurlow et al. (1984). On five vertisols in Queensland during 1979 to 1983, and with four residue management systems, Freebairn et al. (1986) found that residue incorporation and stubble mulch treatments gave greater wheat yield than did bare fallow (CT) and NT. Reduced yield for bare fallow and NT was caused by crown root *(Fusarium germanium)* or yellow spot *(Pyrenophera tritici-repentis)*. Graham et al. (1986) from England reported greatest wheat (*Triticum aestivum* L.) yields when straw was burnt and wheat was direct drilled (NT). Other workers from England reported similar results (Ellis and Lynch, 1977; Oliphant, 1982). From a 5-year study on a chernozem soil in Yugoslavia, Videnovic et al. (1986) also reported less corn (*Zea mays* L.) yield with MT than with CT.

The variability exemplified by the citations given above indicates that no one residue management system is superior under all conditions. Rather, a number of factors regulate plant growth and crop yield response to residue management systems. Factors identified as possible causes of yield reduction when large amounts of residues are on the soil surface include nitrogen immobilization, problems with seed placement, variable plant population, increased number of pests, i.e., slugs in humid climates (Oliphant, 1982), incidence of plant pathogens (Freebairn et al., 1986), production of phytotoxins such as acetic acid (Lynch, 1977, 1981, 1984; Lynch et al., 1980, 1981; Harper and Lynch, 1981), and potential for a soil environment too cool and/or wet for optimum biological activity (Doran and Smith, 1987).

Counteracting these potentially adverse effects of surface residues are beneficial effects, which include water conservation, soil erosion control, reduced winter kill, and maintenance or enhancement of soil organic matter with its attendant improvements in the soil environment. The specific influences of these factors are discussed in more detail later, but collectively they often have a favorable effect on crop growth and grain yield. In some instances, these effects may require several years before their positive influence on crop yield becomes evident. Meisinger et al. (1985) found that corn yields for MT improved with time when compared to those for CT (Figure 1). Nuttall et al. (1986), working with wheat in Saskatchewan, made similar observations. Where crop residues are removed or burned, it may take a number of years before adverse effects on soil organic matter content and yield are evident. However, if residues are removed or burned from an erosive soil, this process is accelerated.

In general, the following conclusions can be drawn: (1) Grain yields are seldom greatly affected by residue management practices under conditions of favorable precipitation, adequate soil water, good drainage, and adequate available N. (2) Increased grain yields with residue cover, as obtained with no-tillage (NT) or minimum tillage (MT) practices, are most likely in areas having limited precipitation and soil water and with adequate weed control. These effects may require several years to become apparent and may require additional chemical fertilizers. (3) Reduced grain yields with the presence of residue cover are most likely in areas receiving adequate to excessive precipitation, low temperatures, poor drainage, poor weed control, or low fertility levels.

III. Soil Chemical Properties

A. Organic Matter and Nitrogen

Crop residues in reduced tillage systems decompose slowly because mixing with the soil is often inadequate for complete microbial decomposition (Dick, 1983; El-Haris et al., 1983). Martin and Stott (1981) pointed out

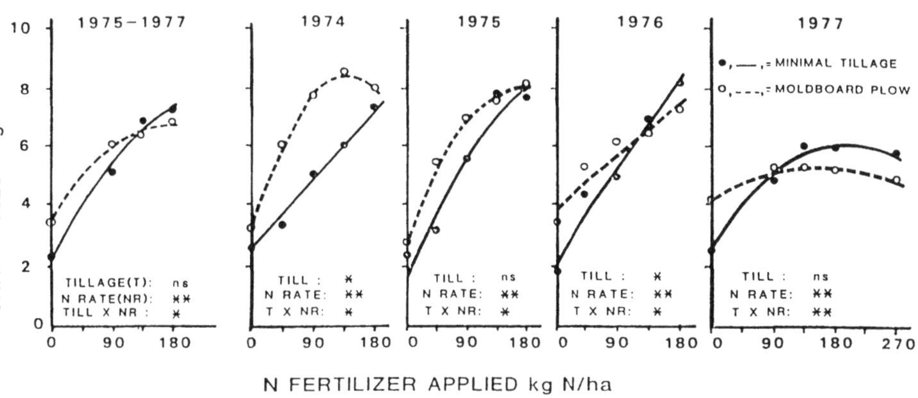

Figure 1. Corn grain yield in relation to N fertilizer applied to moldboard plow tillage (open circles) and minimal tillage (solid circles) plots on a Mattapex silt loam. (From Meisinger et al. 1985).

that humus decomposes at the rate of 2% to 5% per year and that new (active) humus decomposes much more rapidly than old humus. They indicated that 99% of the humus turned over in 25 to 60 years. When soil organic matter (OM) content is increased, the bulk of change presumably would be associated with active new humus, which could beneficially affect soil structure and nutrient retention (Elliott and Papendick, 1986).

Effects of residue management practices on OM content of cultivated soils have been studied in comparison to that for native grass meadow. After 18 years of continuous corn at Wooster, Ohio, soil organic carbon (OC) was 11%, 23%, and 25% less with NT, MT, and CT, respectively, than soils with grass meadow (Dick, 1983). With a corn–oats–meadow rotation at Hoytsville, Ohio, after 18 years soil OC had not changed in the NT soils, although it decreased 12% and 14% in MT and CT soils. Lamb et al. (1985) and Follett and Peterson (1988), both working in Nebraska, reported that with NT wheat-fallow, soil OC was only slightly less than that for native prairie but decreased significantly with CT. Inclusion of a sod crop in the rotation likewise increased soil OM (Stevenson, 1982). Biederbeck et al. (1980) in the Canadian Prairies found that after 20 years, residue burning reduced soil organic C and total N by about 15% to 20% and 4% to 10%, respectively, as compared to a chopped straw treatment. Returning crop residues to the soil, in contrast to their removal from the field, increased organic matter content as has been reported by a number of workers (Shipley and Regier, 1977; Rasmussen et al., 1980; Hooker et al., 1982).

Bauer and Black (1981) reported higher OC and total N content in soil to a depth of 45.7 cm after 25 years under stubble mulch than under CT. In a 4-year study with sorghum-wheat double cropped on a Cecil sandy loam, a kaolinitic, thermic Typic Hapludult, in Georgia, OC in NT was significantly greater than that in MT and CT soils: 18.6%, 13.7%, and 11.8% for NT, MT, and CT, respectively (Langdale et al., 1984). Havlin et al (1990) obtained similar results with corn and sorghum rotations in Kansas. In a 10-year study on Mattapex and Matapeake silt loams in Maryland, Weil et al. (1988) concluded that OC in the surface 0 to 8 cm was greater for NT than for CT. There was no difference at deeper depths. In comparison, total N was greater in the 8- to 28-cm depth in CT than in NT soils. Consequently, the C:N ratio was the same for NT and CT soils. In contrast, Langdale et al. (1984) found no change in total N content of soil with depth, so the C : N ratio changed with depth in accord with OC changes. Maurya in Nigeria (1986) and Lamb et al. (1985) in Nebraska also reported higher OC in the surface 10 cm of soil with residue cover (NT) as compared to residue incorporation. However, Carter and Rennie (1982), working in the Canadian Prairies, found no change in OC and total N between NT and CT after 16 years of cropping. In a later study using ^{15}N, (Carter and Rennie, 1987), they found that microbial biomass organic ^{15}N was greater in NT than in CT soils (Table 3), indicat-

Table 3. Deposition of residual ^{15}N fertilizer in 0 to 5 cm soil depth with NT and CT*

Tillage practices	Percentage recovery of ^{15}N			
	Mineral	Biomass	Organic	Total
NT	0.4	5.9	14.9†	21.2
CT	3.4	9.9*	7.6	20.9
	Distribution of fertilizer N (% residual N)			
NT	2.1†	27.6†	70.3†	100
CT	16.3	47.3	36.4	100

*From Carter and Rennie (1987).

†Indicates significant difference between tillage treatments ($P = 0.05$).

ing greater N immobilization with NT. Microbially induced differences in mineral N pool size and recovery of residual N have been observed in several studies (Frederickson et al., 1982; Leitch and Vaidyanathan, 1983; Carter and Rennie, 1984a,b). Dowdell et al. (1983) from Great Britain also reported that the supply of mineral N was less in NT than in CT plots. Powlson and Jenkinson (1981) concluded that changing from traditional cultivation methods to reduced and no tillage had little effect on the total quantity of soil OM but did alter its distribution in the profile.

Rasmussen et al. (1980) found that burning crop residues had relatively little effect on soil OC content after 45 years when comparing RB to RI treatments. However, soil N levels were reduced by residue burning. This may have resulted from the fact that up to 35% of the carbon in charred residues remained on the soil surface after burning, with most of the N in these residues being volitalized during the burn. In England, Moss and Cotterill (1985) found greater soil OC in the surface soil after burning. On the other hand, Pikul and Allmaras (1986) generally found less OC in soils with RB compared to RI (Figure 2). Likewise, in Saskatchewan (Canada), Biederbeck et al. (1980) also found less OC with RB than with RI after 20 years of treatment. Nuttall et al. (1986) also reported that in the Canadian Prairies OC and total N in the surface 0 to 15 cm remained unaffected by residue management practices, which included spring burning and spreading of chopped straw on the soil surface. There are several possible reasons for conflicting results regarding the effects of burning. These include the degree to which crop residues are burned, time of burning, depth of sampling, effects of burning on soil bulk density, and tillage employed.

Moss (1979) reported that soil-reacting herbicides applied for controlling *Alopercurus myosuroides Huds* (quack grass) in winter cereals were often less effective on NT (after burning) than on CT soil. This was attributed to the presence of burnt straw residues on the surface soil, which

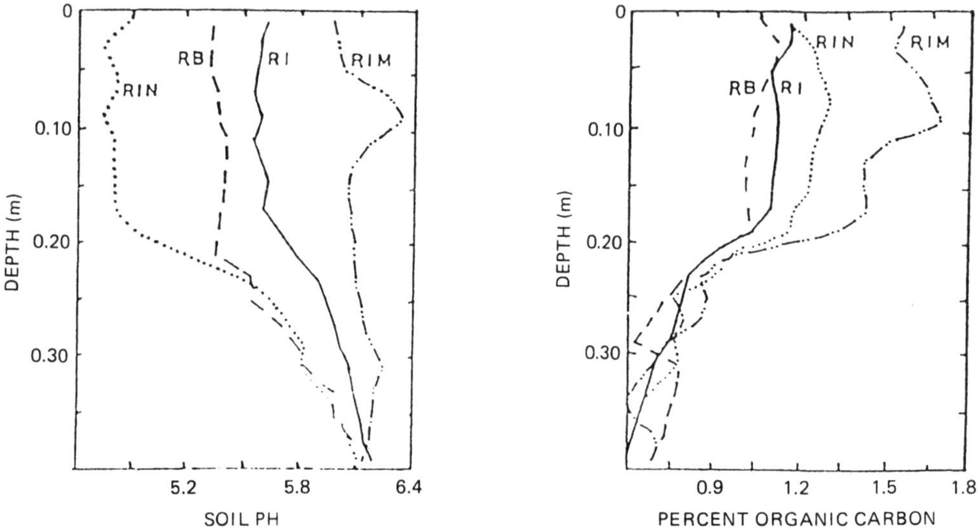

Figure 2. Effect of residue management practices on soil pH and organic C. RB, residue burned in fall after harvest; RI, residue incorporated into soil; RIM, residue plus 22 Mg ha^{-1} of strawy manure incorporated into soil; RIN, residue incorporated plus 90 kg N ha^{-1}. (From Pikul and Allmaras, 1986)

were highly adsorptive of herbicides (Embling et al., 1983) and capable of reducing herbicide activity (Toth et al., 1981; Addala, 1982). Because herbicide adsorption is positively correlated with soil organic matter content (Langdale et al., 1984; Maurya, 1986; Weil et al., 1988), organic matter content has been used as a measure of herbicide adsorption. Cussans et al. (1987) showed that adsorption measurements, determined as K_d values for chlorotoluron, were better for predicting problems with herbicide performance than was organic matter content. Moss and Cotterill (1985) reported that K_d values for surface 0 to 2.5 cm of ash-covered soils (where crop residues were burned) were two to four times of those for nonash soils (residue removed).

Available information on tillage effects on soil organic matter and N availability leads to the following conclusions: (1) tillage of all kinds leads to a decrease in soil OC and N as compared to native sod; (2) incorporation of crop residues (RI) as compared to their removal usually increases organic matter content of soil; (3) leaving crop residue cover on the soil surface (NT) leads to an accumulation of organic matter and total N in the surface soil; (4) burning of crop residues often produces a variable effect on soil organic matter content, depending on soil depth, tillage practices, degree of burning, time, and other factors.

B. Phosphrous

Most workers (Cruse et al., 1983; Langdale et al., 1984; Hargrove, 1985; Follett and Peterson, 1988; Weil et al., 1988) have reported that P accumulates near the soil surface in NT systems, resulting in increased soil test P levels. Data of Weil et al. (1988) and Hargrove (1985) are shown in Figure 3. Kunishi et al. (1982) reported that most of the P remained as unreacted Ca–P compounds in the 0- to 3.75-cm layer of a Matapeake silt loam soil. Stratification of P resulting from MT and NT is believed to result in improved P availability because there is less soil contact of organic P in crop residues and hence less P fixation (Blevins et al., 1983).

Weil et al. (1988), working on a Delanco silt loam, a fine-loomy, mixed, mesic Aquic Hapludult (near Clarksville, Maryland), found that although total and dilute-acid-extracable P were stratified more in the 0- to 2-cm layer under NT than CT, organic P did not show as much stratification. Nevertheless, organic P in subsoil (24 to 28 cm) layers was lower than that in the Ap horizon. A tendency for surface stratification of total and dilute-acid-extracable P to increase with phosphate fertilization was

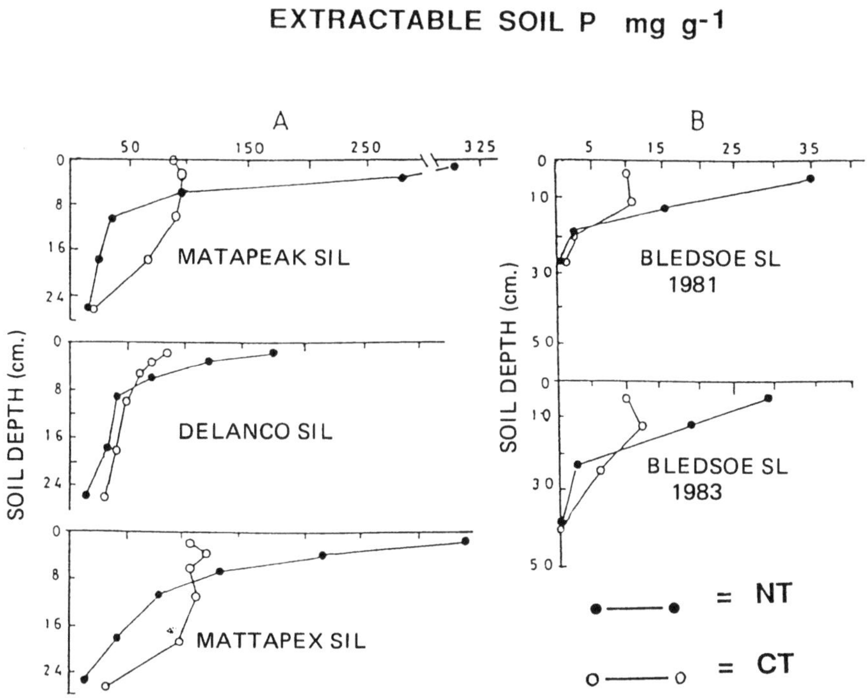

Figure 3. Dilute-acid-extractable soil P as influenced by tillage practices. A: Dilute acid extract data of Weil et al. (1988). B: Double acid extract data of Hargrove (1985).

noted. They further observed that inorganic P conversion into organic forms was less efficient than for N because P is more subject to absorption reactions with the soil than occurs with N. Follett and Peterson (1988), working on a site plowed from native sod in 1970, observed that $NaHCO_3$ extractable P levels (inorganic and organic) were 50% to 75% greater than those observed at a nearby site originally plowed in 1930. This was related to greater organic matter content in the more recently plowed site (5.3% compared to 2.7% in the 0- to 5-cm layer of the earlier site). Tillage treatment had little effect on total extracable P at the later site, but at the other site total extractable P decreased significantly as degree of soil disturbance resulting from tillage increased, that is, in decreasing order to native sod, NT, MT, and finally CT. At the early plowed site, a decline in total extractable P was associated with N fertilization as well as with degree of soil disturbance; NT had greatest P values for the 0- to 5-cm soil layer. Total extractable P and organic P in the 0- to 5-cm layer increased, whereas that in the 5- to 10-cm layer decreased as a result of N fertilization.

In England, burning cereal residues also resulted in higher extractable P content in the surface 0- to 2.5-cm soil layer: 55 and 24 ppm in soils with burned and unburned residues, respectively. Also in England, when straw was spread and burned after harvest, NT resulted in more $NaHCO_3$-extractable P in the 0- to 10-cm layer than CT, both at sowing and after harvest (Drew and Saker, 1978, 1980). Similarly Nuttall et al. (1986) in Canada found that spring burning of wheat straw resulted in increased bicarbonate-extractable P (Table 4).

Thus, leaving cereal residue on the surface (NT) or burning it results in a stratified accumulation of P in the soil, with greater concentrations of sodium bicarbonate or dilute-acid-extractable P in the surface of such soils compared to soils where residues were incorporated.

Table 4. Effect of tillage on sodium bicarbonate extractable P (mg P g^{-1} soil) in 0 to 0.15 m depth*

Year	Plow	HD cult	One-disc	Chop straw	Double disc	Spring† burn
1978	8.3	12.6	11.0	8.7	12.4	14.0
1979	5.8	7.9	8.1	7.1	7.8	8.0
1980	7.6	8.3	9.4	8.3	19.7	14.7
1981	7.9	10.7	16.3	7.5	9.4	15.6
1982	11.7	16.5	21.6	14.1	13.8	22.2
1983	10.8	15.4	16.2	12.1	15.1	27.4
SE (Tillage × year) 1.3						

*Source: Nutall et al. (1986). (Reproduced from Canadian Journal Soil Science 66:159–171).

†All other treatments applied in fall. The experiment was initiated in 1959.

C. Other Nutrients

Culley et al. (1987a) and Follett and Peterson (1988) reported higher exchangeable K in the 0- to 5-cm soil depth under NT than under CT. However, Langdale et al. (1984), after a 4-year study on a Cecil sandy loam in Georgia, reported that double acid-extractable K (Olsen and Dean, 1965) at the 0- to 1-cm soil depth was 146, 175, and 216 mg kg_{-1} soil for NT, MT, and CT, respectively. There was no significant difference between tillage treatments when the 0- to 75-cm soil depth was taken into consideration. Burning of crop residues in England resulted in increased K content in the surface 0- to 2.5-cm soil depth; 205 to 1110 ppm with burning compared to 125 to 406 ppm with no burning (straw removed or incorporated) (Moss and Cotterill, 1985).

Langdale et al. (1984) also reported that exchangeable Ca, Zn, and Mg for the 0- to 1-cm soil depth was greatest in NT soil, significantly greater than in MT and CT. Soil Mn content was not different. Hargrove et al. (1982) also reported a surface accumulation of Zn as a result of OM accumulation and suggested a possible association with soil organic fractions. From a study made on soil samples collected after 16 years of wheat-fallow cropping system in western Nebraska, Follett and Peterson (1988) found more extractable Zn, Fe, Mn, and Cu in the surface 5 cm soil under NT than under CT, with concentrations of most of these nutrients decreasing with soil depth. Accumulation of OM and higher soil water content near the soil surface in NT may reduce redox potentials in the soil and thus may increase Fe and Mn solubility in the soil.

Most studies have shown an increased accumulation of plant nutrients in surface soil layers under NT and MT. Hargrove (1985) observed that this stratification of nutrients under NT was not a detriment to plant growth, and, in fact, root activity and water and nutrient uptake from deep in the soil profile were greater for NT than for CT.

D. Soil pH

Culley et al. (1987a) reported a pH of 4.8 in NT soil as compared to 5.0 for CT soil at for the 0- to 10-cm soil depth. On an Alliance silt loam, a fine-loamy, mixed, mesic Aridic Argiustoll in Nebraska, NT and MT had lower soil pH than CT at the 0- to 5-cm soil depth (Follett and Peterson, 1988). Pikul and Allmaras (1986), from a 50-year study on a Walla Walla silt loam, a coarse-silty, mixed, mesic Typic Haploxeroll in Oregon, concluded that RB gave a lower soil pH than RI. Addition of 90 kg N ha^{-1} of an ammonium-containing fertilizer with RI generally reduced soil pH, whereas addition of organic manure with RI increased soil pH (Figure 2). In a study in England, pH of the 0- to 2.5-cm layer of soils with RB was less than that for soils where straw was removed or incorporated 10 to 15 cm deep. Average values were 6.6 and 7.2 for soils with burned and unburned residues, respectively (Moss and Cotterill, 1985).

In conclusion, leaving crop residue on the surface (NT) as well as RB leads to reduced pH in the surface 5-cm soil depth and often at even deeper depths.

IV. Soil Physical Properties

Residue management practices affect soil physical properties such as water content, temperature, aggregate formation, bulk density, porosity, and hydraulic conductivity. Although many of these properties are interrelated, these properties are discussed independently.

A. Soil Water

Bond and Willis (1969), in controlled environment studies, showed that increasing amounts of crop residues on the soil surface reduced evaporation rate and increased duration of first-stage drying. However, given sufficient time, the water content to which soils dried was not affected by the presence of residue on the soil surface. Thus, residue-covered soils tend to have greater water content than bare soil except after extended drought. This phenomenon has been verified in the field by several investigators (Aase and Siddoway, 1980; Cannell et al., 1980; Tanaka, 1985). More frequently, however, intermittent showers rewet the soil surface before initiation of third-stage drying in residue-covered soils. For example, Power et al. (1986) showed that soil without surface cover dried faster, and essentially all available water was removed by early August, whereas residue-covered soil still retained 50 to 70 mm available water at that time. Precipitation in mid-August increased water storage for all soils, but water content in no-residue soil still remained appreciably below that of residue-covered soil.

Reducing the evaporation rate of soil water is one of the major advantages of leaving crop residues on soil surface and has made NT and MT popular in midwestern United States, the Canadian Prairies, and in several other subhumid and semiarid parts of the world. Studies in the southern Great Plains (U.S.A.) have shown that precipitation storage during fallow was greater where surface residues were maintained by conservation tillage methods that where residues were incorporated (Unger and Wiese, 1979; Unger, 1984). In a Texas study, plant available soil water to 1.8 m depth at planting ranged from 203 to 217 mm with NT and from 152 to 158 mm with CT. The Pullman clay loam used in this study had a plant available water storage capacity of about 230 mm to a 1.8-m depth. Use of NT eliminated need for a preplant or early season irrigation for sorghum or corn. In another study at Bushland, Texas, soil water content to the 1.8-mm depth at corn planting was 298 mm in NT and 280 mm in CT soils (Unger, 1986).

In the Canadian Prairies, where crop production is limited mainly by the available water in soil at seeding, soil water content was 62 mm (1.2-m depth) under NT compared to 17 mm under CT in fields where the soil was left bare from autumn to spring seeding (Grevers et al., 1986). During the snow melt to seeding period (last week of March to first week of May), there was considerable reduction in soil water content to 40 cm with no noticeable changes below. Soil water loss during this period for CT was substantially greater than that for NT. During the period following snow melt (up to 2 months), this region receives strong, dry winds, resulting in great soil water loss from evaporation unless a vegetative barrier such as crop residues is provided. Recharge of the soil profile from snow melt was greater with NT. Reduced evaporation with NT was also reflected in water use efficiency, with values of 49.7 kg ha^{-1} cm^{-1} for CT, compared to NT values of 53.7 to 186 kg ha^{-1} cm^{-1}.

In Queensland, Australia (for details see Section II), residue mulch treatments retained more water (Figure 4). that did RB or RI (Freebairn et al., 1986). Osuji (1984) reported from Nigeria that water stored to the 30-cm depth was greater with NT than with CT during the early (March to July) as well as during the late (August to November) corn growing season. Negi et al. (1981) also reported substantially greater plant available water with NT.

Amount of residue cover is also important in determining water storage in soil under NT. For example, Baumhardt et al. (1985) showed that significantly more precipitation was stored during fallow with NT than with

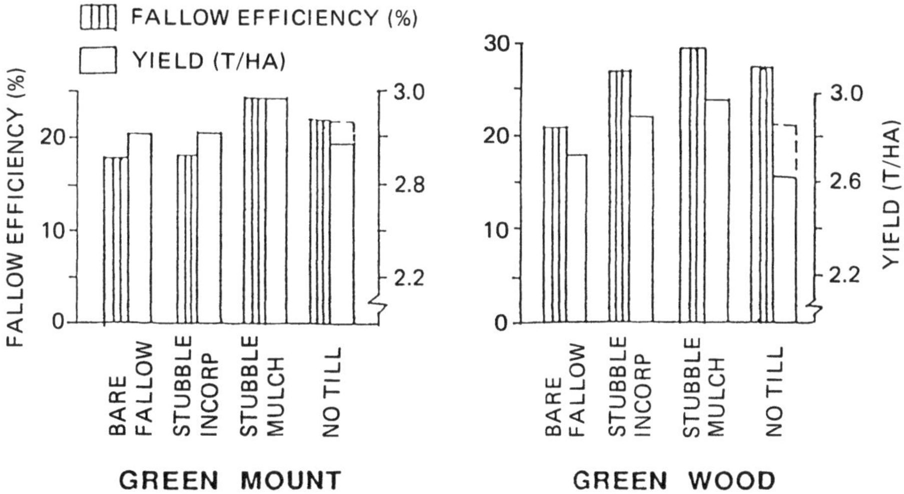

Figure 4. Mean fallow efficiency (percentage rainfall stored) and wheat yield for four residue management systems at two locations in Queensland. Dotted lines indicate mean yields if yield loss attributed to *Fursarium graminearum* is added. (From Freebairn et al., 1986. Reproduced by permission of Elsevier Science Publishers.)

CT when 1.1 kg m^{-2} residue was present but not when only 0.2 kg m^{-2} residue was present. Power et al. (1986) reported that on a Crete–Butler silty clay loam in Nebraska, soil water storage was about 30% greater when quantity of residue cover was increased from zero to 100% to 150% of that produced by the previous crop (Figure 5). In the same study in Nebraska, Wilhelm et al. (1986) developed the following regression equation relating corn residue cover to soil water storage at planting:

$$Y = 173 + 5X \; (r^2 = 0.84)$$

where Y is the soil water present at planting (mm) and X is the quantity of surface residues applied after harvest of the previous crop (Mg ha^{-1}). Thus, each megagram per hectare of corn residue left on soil surface increased water storage at planting by 5 mm.

In some studies, however, workers failed to obtain a significant difference in soil water content between NT and CT (Hill et al., 1985; Nuttall et al., 1986). As indicated earlier, such results may occur because of extended dry periods or because residue rates were too low to be of significance.

Although the effect of residue cover on soil water storage and retention depends on amount of residue cover, infiltration rate, water storage capacity of the soil, and evaporation potential of the climate all play major roles in determining quantity of soil water stored. Mielke et al. (1984) found in western Nebraska that cumulative infiltration for NT was double that for CT (Figure 6). Rainfall return frequencies indicated little likelihood of runoff from NT. Cumulative infiltration for mature sod was double that for NT. These results probably occur because of greater macroporosity, especially with NT and sod.

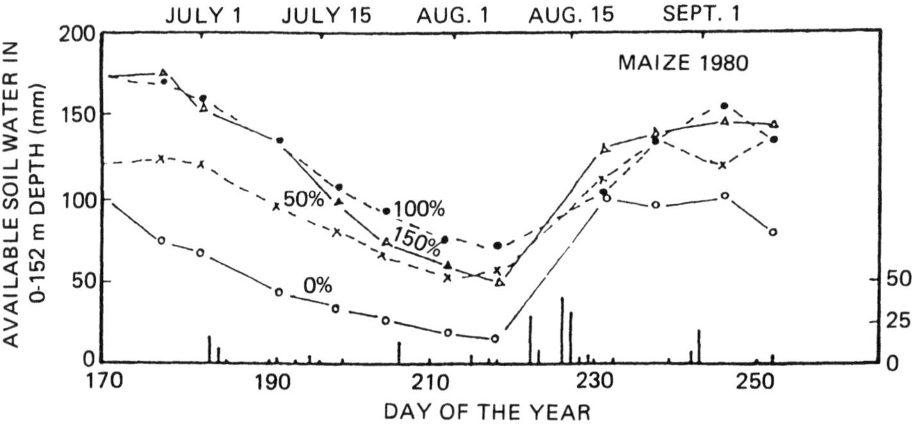

Figure 5. Available soil water as affected by crop residue rate (0, 50, 100, and 150% of the residues produced by previous crop). (From Power et al., 1986. Reproduced by permission of Elsevier Science Publishers.)

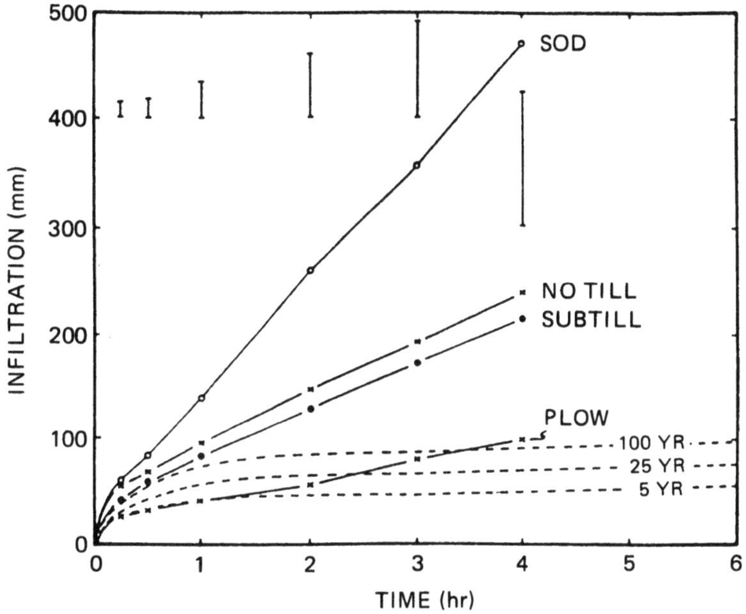

Figure 6. Cumulative infiltration for fallow tillage methods for Duroc loam near Sidney, Nebraska, compared to precipitation amounts for 5-, 25-, and 100-year return frequency. Bars are proportional to $LSD_{(0.10)}$ values. (From Mielke et al. 1984. From Transactions of the ASAE, copyright 1984 by the American Society of Agricultural Engineers.)

B. Soil Temperature

Residue management can significantly influence soil temperature. The main mechanisms involved are (1) change in radiant energy balance and (2) insulation (Unger and McCalla, 1980).

The radiation balance is influenced by heating of air and soil, evaporation of soil water, and reflection of incoming radiation by surface residue (Van Doren and Allmaras, 1978). Residue characteristics involved in the reflectance of incoming radiation include residue age, color, geometry (whether standing or matted on the surface), distribution, and amount (Unger and McCalla, 1980). Because reflectance is greatest with light-colored residue, the temperature difference between NT and CT is likely to be greatest with bright-colored residue and may decrease as residue ages.

The insulation effect of crop residues is controlled by the amount and associated thickness of residue cover. Unger (1978) reported that at Bushland, Texas, soil temperatures (0 to 75 mm) were 3° to 10°C less than air temperatures during a hot period (21 through 25 August 1973) and 2° to 13°C greater than air temperatures during a cold period (1 through 7 January 1974), depending on the amount of wheat straw spread on the soil surface. The difference in mean temperature between 0 (bare soil) and 12 Mg ha^{-1} of straw were much narrower, 1° to 3°C. Gerik and Morrison (1984), working on Austin silty clay, a carbonatic, thermic Typic Haplustoll (near Temple, Texas), found no significant difference in mean soil temperature under NT and CT. Effects of crop residues on soil temperature diminish as canopy closure of a growing crop approaches.

Lowering of soil temperature by residue cover can create a problem for seedling emergence and early growth of spring-seeded crops in cooler parts of the world. Seed-depth soil temperature is one of the most important factors controlling corn emergence and seedling growth up to 4 to 6 weeks after emergence (Alessi and Power, 1971; Watts, 1973). Cummins and Parks (1961) concluded that the minimum soil temperature required for corn germination is 10°C. Conservation tillage systems leave most or part of the previous crop residues on the soil surface, often resulting in colder seed-zone soil temperature in the spring at planting. Johnson and Lowery (1985) showed that for till-plant (TP), chisel (CH), and no-till (NT), respectively, the minimum soil temperature required for corn germination occurred 3, 4, and 10 days later than for conventional moldboard plow (CT).

Schneider and Gupta (1985) and Gupta (1985) suggested use of mean seed-zone temperature or seed-zone growing degree days (GDD) as an index of corn emergence under several conservation tillage systems. The GDD was calculated using the following expression:

$$\mathrm{GDD} = \left(\sum_{i=1}^{24} \frac{T_i}{24}\right) - 10, \text{ where } 10 \leqslant T_i \leqslant 30$$

where T_i is the soil temperature (°C) at 5-cm depth at the ith hour. Cumulative GDD for a certain period was calculated by summing up daily GDD for that period.

Based on a 3-year (1982 to 1984) study at Arlington, Wisconsin, Al-Darby and Lowery (1987) ranked different tillage methods in respect to seedling emergence in the following order: CT > TP > CH > NT, where TP and CH refer to till-plant and chisel, respectively. Final (100%) emergence for NT was delayed by 8, 2 to 3, and 2 days compared to CT in 1982, 1983, and 1984, respectively. Mean daily soil temperature at 5 cm under CT and NT differed as much as 4°C during the emergence period of 1983. Seedling emergence in all treatments started when accumulated GDD equaled 40 to 55 and reached maximum at 70 to 80 GDD.

Prevailing soil temperatures are also important in determining seedling growth in corn. Optimum temperatures for corn seedling growth range from 24°C to 29°C (Willis et al., 1957; Blacklow, 1972). Mock and Erbach (1977) recorded less plant growth and grain yield with conservation tillage systems than with CT; they measured a 30% reduction in grain yield with early planting for conservation tillage systems. Al-Darby and Lowery (1986) observed that during the first 5 weeks after emergence, corn growth with NT was consistently and sometimes significantly slower than with CT. However, this poor early growth did not result in reduced corn yield. Al-Darby et al. (1987) concluded that the depressed early growth was compensated by more soil water conserved with NT than with CT.

Hay (1977) also suggested that soil temperature was a major factor affecting temperature of the wheat seedling apex and hence leaf extension rate. He showed that stubble mulch insulated the surface soil layers from large temperature fluctuations, resulting in lower frequency of freezing but fewer hours above 5°C. In a study in Australia (at Murrumbateman, N.S.W.), soil temperatures were measured at different depths under a wheat crop sown by three different methods in a dry (1982) and a wet (1983) season (Aston and Fischer, 1986). Generally soil temperatures at any particular depth under CT were warmer during the day and cooler during the night than under NT. These differences persisted throughout both growing seasons but were least during the wet year. Wheat plants showed early vigor with CT and had a larger shoot dry weight per plant at the $4\frac{1}{2}$-leaf stage. Results of 3- to 6-year tillage experiments at five sites ranging westward from the high-rainfall zone near Canberra (Australia) to the drier fringe of the wheat belt near Yanco indicated that average yield of NT wheat was within 10% of that of CT, although yield reductions greater than 10% were recorded on occasion (Fischer, 1982).

Unger and Stewart (1976) pointed out that in warmer regions such as the southern United States, no temperature-induced delay of planting for NT is expected. In a study at a higher elevation, increasing straw mulch rate did delay time required for the soil to reach a favorable temperature for sorghum germination and growth, but soil temperature was near optimum before normal sorghum planting dates for the region. Additional straw mulch resulted in greater yield because of storage of greater amounts of soil water (Unger, 1988).

It may thus be concluded that although leaving residue cover on soil lowers soil temperatures and delays emergence of spring-seeded crops, additional water conservation in mulch-covered soil often offsets any adverse early season disadvantages. Reduced soil temperatures may be beneficial in warmer climates.

C. Aggregates

Soil aggregation refers to the cementing or binding together of several primary soil particles into secondary units. The binding substances include (1) oxides and hydroxides of iron, (2) organic substances directly from plants, (3) decomposition products of plant residues, (4) microbial cells, and (5) excretory products of microorganisms and gelatinous substances secreted by earthworms. Although oxides and hydroxides of iron are of importance in tropical soils, crop residue management practices affect most other binding substances.

Earthworms secrete gelatinous substances that coat and stabilize soil aggregates while feeding on organic materials (Edwards and Lofty, 1977). In most cultivated soils, however, microbial cells and their excreted products serve major roles in aggregate stabilization, depending on the environment, chemical composition of residues, and residue management (Elliott and Papendick, 1986; Gupta and Germida, 1988).

Decomposition of crop residues is controlled by the chemical composition of the residues, soil temperature, soil water, and associated macro- and microflora. Knapp et al. (1983a,b) indicated that the quantity of readily available C and N in wheat straw determined microbial biomass size and rate of straw decomposition. Reinertsen et al. (1984) postulated two readily available pools (one cold water soluble and one insoluble) that determined the size of the microbial biomass and resultant rate of decomposition. Elliott and Lynch (1984) aerobically degraded three wheat straws containing 1.09%, 0.5%, and 0.25% N in the absence of added N and found that straw containing 0.25% N resulted in most aggregate formation in the soils tested. From the work of Knapp et al. (1983a,b), Reinertsen et al. (1984), and Elliott and Lynch (1984), Elliott and Papendick (1986) postulated that extracellular materials such as polysaccharides might dominate the aggregation resulting from decomposing straw if the wheat straw contained low N and if external N was unavailable. Parr and Papendick (1978) had stated earlier that although the N content or C : N ratio of a crop residue can be useful in predicting decomposition rates, these values should be used with some caution since the C : N ratio says nothing about the availability of the C or N to microorganisms. Capriel et al (1990) showed close relationships between the aliphatic component of soil organic matter and both microbial biomass and aggregration, suggesting that aggregration is improved by enhanced aliphatic C content resulting from greater microbial biomass.

During the decomposition of a high-N-containing crop residue such as legume, binding of the soil by the hyphae of fungal cells themselves might be more important than binding by excretory products because N would not normally be limiting (Elliott and Lynch, 1984). The role of mycorrhizal fungi in regard to aggregate formation and stability is essentially unknown.

Skidmore et al. (1986), after a 14-year study at Garden City Experiment Station, Kansas, reported that although residue management practices (which included residue removal, RI, twice the amount of RI, and RB) did not significantly affect aggregation, the crops grown did. Aggregates > 0.84 mm, geometric mean diameter (GMD), and dry aggregate stability were significantly greater after wheat than after sorghum (Table 5). Soil aggregates from sorghum plots had greater organic matter content, were

Table 5. Soil physical properties as influenced by growing sorghum or wheat on Richfield silty clay loam*

Crop	Aggregates 00.85 mm (%)	Dry aggregate stability (%)	Wet aggregate stability (%)	Geometric mean diameter (GMD, mm)	BD ($Mg\,m^{-3}$)	Saturated hydraulic conductivity $\mu m\ s^{-1}$
Wheat	70.1	88.9	39.6	2.86	1.23	6.7
Sorghum	47.1	83.7	52.5	1.18	1.05	49.4
LSD (0.01)	2.6	1.6	7.1	0.23	0.07	14.4

*Source: Skidmore et al. (1986).

smaller, more fragile, less dense, less stable when dry, and were more stable when wet than were aggregates from soil with winter wheat. Armbrust et al. (1982) also found that aggregates from winter wheat plots were less erodible and more stable when dry than were the aggregates from soil producing sorghum. The work of Elliott and Lynch (1984) and Elliott and Papendick suggests that these results need to be investigated from the viewpoint of chemical composition of wheat and sorghum straw.

Studying the effects of tillage practices over a period of 6 years in the Canadian Prairies, Nuttall et al. (1986) found that aggregates >0.84 mm and <12.7 mm were most abundant with straw chopped and spread and were least abundant with fall plowing. Values for spring burning of straw were intermediate. Gupta and Germida (1988), also working in the Canadian Prairies, found that compared to native grassland, cultivation affected especially aggregrates in the 0.25 to 1.00 mm range.

Information on the effect of residue management on soil aggregation is limited. We do know, however, that aggregation depends on the amount and chemical composition of the residues and on prevailing soil water and temperature regimes. In general, crop residues on the soil surface usually increase soil aggregation.

D. Bulk Density, Cone Index, Porosity

Soil bulk density (BD) and cone index (CI) are directly related; therefore, residue management practices that affect one will also affect the other.

Blevins et al. (1977) in Kentucky, Bauder et al. (1981) in Minnesota, Mielke et al. (1984) and Follett and Peterson (1988) in Nebraska soils, Hargrove (1985) from Georgia, and Skidmore et al. (1986) in Kansas reported no significant differences between different residue management practices in respect to soil BD. In over 50 years of study on a Walla Walla silt loam (near Pendleton, Oregon), RB resulted in a well-defined tillage pan at a depth of 19 to 29 cm (Pikul and Allmaras, 1986; Rasmussen and Smiley, 1989). Maximum BD of this pan was 1.31 Mg m^{-3} at 23-cm depth. Residue incorporation (RI) gave maximum BD of 1.30 Mg m^{-3} at 19-cm depth. Addition of manure (RIM) or inorganic fertilizer N (RIN) reduced BD and maximum BD to 1.21 to 1.23 Mg m^{-3}, respectively, which was located at 23-cm depth. In the upper 19 cm, there was no significant difference between RI and RB or between RIM and RIN. Changes in soil compaction since 1931 did not extend below the upper 30 cm of soil. The BD of the upper 30 cm of soil in 1931 was 1.16 Mg m^{-3}, whereas that in 1984 was 1.24, 1.17, 1.26, and 1.19 Mg m^{-3} for RI, RIM, RB, and RIN treatments, respectively. The percentage of large transmission pores (>50 μm as suggested by Greenland 1977) decreased as BD increased.

Hill and Cruse (1985) reported that in a 7-year study on a Mollisol in Minnesota, BD tended to be greater with NT compared to CT at a depth of 10 to 12 cm. Soil strength measured with a falling cone penetrometer was also significantly greater in the 5- to 13-cm soil depth in NT plots. Also in Minnesota, Culley et al. (1987a), working on a poorly drained Webster loam, found that moist BD (dry weight of soil per unit volume of moist soil) was essentially constant with NT (near 1.4 Mg m^{-3}), whereas with CT it was between 1.2 and 1.3 Mg m^{-3} in the surface 5 cm and increased to 1.4 Mg m^{-3} at about 25-cm depth. Dry BD values were 0.2 to 0.4 Mg m^{-3} more than moist BD values, and for both NT and CT, BD increased below 25-cm depth. They further observed that cone index (CI) was more sensitive to wheel tracking than was BD. Radcliffe et al. (1988), using a tractor-mounted hydraulically driven cone penetrometer (Clark and Reid, 1984) in Georgia, found that CI in NT exceeded 4 MPa at a depth of 10 to 20 cm, indicating the presence of a compacted zone. Values of CI of 2 to 3 MPa are reported to prevent root growth (Taylor and Gardener, 1963; Taylor and Burnett, 1964).

Pidegon and Soane (1977) in Scotland observed greater moist BD and CI for NT than for CT. Equilibrium BD values under NT were attained within 3 years, but CI of the subsoil apparently continued to increase with time. They (Soane et al., 1982) also suggested that NT soils eventually acquired sufficient strength to carry traffic without more compaction.

Gantzer and Blake (1978) observed lower total porosity (increased moist BD) to a depth of 30 cm beneath NT rows throughout the fifth year of a Minnesota study, in comparison to CT. On a Cecil sandy clay loam in Georgia (Radcliffe et al., 1988), percentage of pores >29 μm was greater under CT than under NT. However, in a sandy loam soil in Nigeria, the total porosity at 5-cm depth was more in NT plots than in RI plot (Maurya, 1986).

Skidmore et al. (1986) from Kansas reported that residue management treatments did not affect BD significantly, but the crops grown did. Average total porosity for the soils in sorghum and wheat plots was 0.63 and 0.5 m^3 m^{-3}. About 29% of the soil pores for soil under sorghum were greater than 30 μm in diameter (drained at -10 J kg^{-1}) compared to 15% for soil under wheat. This resulted in about a twofold greater hydraulic conductivity for soil under sorghum compared to that for wheat. Culley et al. (1987b) reported that NT soil cores from 0- to 10-cm depth became unsaturated on application of the initial potential (-0.5 kPa), indicating the presence of macropores (as defined by Greenland, 1977). Macropores accounted for about 5% of the total porosity of this layer, well within the range of macroporosity observed by Germann and Beven (1981). Surface soil cores from CT exhibited no such macroporosity.

E. Hydraulic Conductivity and Infiltration

Gantzer and Blake (1978) found less saturated hydraulic conductivity (K_s) near the soil surface under NT than under CT soon after planting, but differences were not significant at harvest. Culley et al. (1987b), using a permeater (Reynolds et al., 1983), found that K_s under NT was 1.9 times that under CT. Zuzel et al (1990) attributed almost all the enhancement of Ks at the surface to effects of crop residue cover.

Biederbeck et al. (1980) found that K_s of the top 15 cm of soil from straw-burned plots at Melfort, Saskatchewan, was significantly less (P = 0.10) than that of soil from chopped straw plots (6.9 and 2.8 $\mu m\ s^{-1}$, respectively). Rasmussen and Smiley (1989) reported similar results in Oregon. However, at another site (Indian Head), differences between the two treatments were not significant; the average K_s was 3.2 $\mu m\ s^{-1}$.

Skidmore et al. (1986) reported that although most soil physical properties were not significantly affected by tillage treatments, K_s in soil cores taken from sorghum plots was much greater than that in soil cores from wheat plots. A comparison of water release characteristics also showed a distinct difference in soil structure between sorghum and wheat plots.

Pikul and Allmaras (1986) studied the effects of over 50 years of residue management treatments on K_s of Walla Walla silt loam (near Pendleton, Oregon). Hydraulic conductivity was least in a tillage pan (Table 6). At depth of maximum BD, K_s was three times greater in soil with residue incorporated and manure added (RIM treatment) than in soil for other treatments. High levels of organic matter accumulation seemed to be the factor responsible for enhanced K_s. While measuring K_s at depth of maximum BD, earthworms were always uncovered in RIM treatment but not in RIN (residue incorporated with additional inorganic N fertilizer). Edwards and Lofty (1982) found that all forms of N fertilizers increased earthworm population, but organic N manures increased populations more than inorganic N fertilizers. However, inorganic N fertilizers lowered soil pH below 4.5 (Figure 2), which retards earthworm activity (Barley, 1961). Lower soil pH was also found in commercial fertilized fields

Table 6. Saturated hydraulic conductivity (K_s, $\mu m/s$) of selected long-term residue management treatments*

Treatment	Bare surface	Stubble covered surface	Tillage pan	Subsoil
Control	1.08	1.97	0.34	1.74
Residue incorporated + 20 Mg ha^{-1} manure	2.24	3.61	0.85	1.80
Residue burned	1.52	3.15	0.29	1.76
Residue incorporated + fertilizer N	2.36	4.15	0.23	1.89
LSD (0.05)	0.89		0.25	

*Source: Pikul and Allmaras (1986).

in the Pacific Northwest, and lime additions would dramatically enhance the value of large quantities of residues returned for improved infiltration and decreased runoff (Kraft and Allmaras, 1985).

Radcliffe et al. (1988) did not find a significant difference between NT and CT in respect to K_s as determined by constant-head method (Klute and Dirksen, 1986). They reported that infiltration rate was higher under NT than under CT. Using a sprinkler infiltrometer similar to that described by Peterson and Bubenzer (1986), and with a sprinkling rate of 42 mm hr^{-1}, mean final infiltration rates were 37 mm hr^{-1} for NT and 16 mm hr^{-1} for CT (Radcliffe et al., 1988).

It can thus be concluded that leaving organic residues on the soil surface is likely to increase K_s and infiltration in soils. In suitable environments, earthworms may play an important role in this enhancement. However earthworms are often completely absent in soils in drier environments.

V. Soil Biological Properties

A. Macroorganisms

Leaving crop residues on the soil surface as in NT management provides a favorable environment for soil- and surface-residue-dwelling organisms because of reduced water loss, amelioration of temperature extremes and fluctuations, and presence of a relatively continuous substrate for decomposers (House and All, 1981; House and Stinner, 1983; Crossley et al., 1984). Predatory and saprophagous soil arthropods as well as crop-damaging herbivores such as ground beetles *(Coleoptera carabidae)* (House and All, 1981), spiders (Blumberg and Crossley, 1983), and decomposer fauna such as earthworms (Edwards, 1975; Barnes and Ellis, 1979) are reported to occur in larger numbers in NT than in CT soils. House and Parmele (1985) reported 2202 and 637 earthworms per square meter in NT and CT plots, respectively, under a sorghum–rye cropping system on a Hiwassee loam, a Kaolinitic, thermic Typic Rhodudult (near Athens, Georgia); the corresponding NT and CT values for sorghum–clover cropping system were 1210 and 191, respectively. Ground beetles, spiders, and other macroarthropods as well as microarthropods (*Acarina, Collembola,* insects) were also more frequent under NT than under CT in both cropping systems. However, Enchytraeid populations were greater under CT than NT. These results suggest that earthworms and microarthropods may assume a more dominant role in organic matter decomposition and nutrient flux patterns in NT agroecosystem.

House and Parmele (1985) pointed out that tillage substitutes mechanical energy for the biotic activity of the soil fauna (i.e., earthworm tunneling and redistribution of residues). When tillage is not practiced, as in

NT cropping, organic matter decomposition will depend mainly on biotic activity. Nitrogen mineralized as a result of activity by earthworms has been estimated to be as high as 100 kg N ha^{-1} (Edwards and Lofty, 1977).

Wallwork (1983) reported that many microarthropods such as oribatid mites and *Collembola* have well-developed mouth parts capable of fragmenting organic matter while feeding on bacteria and fungi adhering to plant residues. The fragmentation of plant material increases its surface area and thus accelerates microbial activity, which in turn enhances organic matter breakdown and mineralization (Seastedt and Crossley, 1980; Seastedt, 1984). Although soil fauna of natural terrestrial ecosystems influence organic matter decomposition and mineralization processes (Crossley, 1977; Peterson and Luxton, 1982; Seastedt, 1984), this catalytic role has not been demonstrated for soil fauna in agricultural systems. More research is needed to understand fully the role of earthworms and other soil fauna in the decomposition and nutrient release from crop residues.

B. Microorganisms

McCalla (1958) reported that bacteria, actinomycetes, fungi, earthworm, and nematode populations were greater under stubble mulch than under CT. Doran (1980) reported that the surface layer (0 to 7.5 cm) of most reduced-tillage soils had higher microbial populations, higher phosphatase and dehydrogenase enzyme activity, and higher levels of total N and potentially mineralizable N than conventionally tilled soils. Aerobic microorganism counts were 10% to 80% greater, and anaerobic bacteria (including denitrifiers) were 60% to 300% greater in the surface of reduced than in conventionally tilled plots. This might be expected because soil beneath residue cover is generally cooler, wetter, and less aerated than that under conventional tillage. Gupta and Germida (1988) made similar observations, and they also found fungi to be particularily sensitive to tillage effects.

The foliar disease yellow spot *(Pyrenophera tritici repentos)* considerably reduced wheat yields in Queensland. Where wheat residue was burned immediately prior to seeding, the yield of wheat was increased by 30% to 70% (Ward and Rees, 1985; Rees and Platz, 1986). Crown rot *(Fusarium graminearum)* may also reduce yields of no-till wheat (Wildermuth et al., 1986). In Australia populations of root-lesion nematodes *(Pratylenchus thornei)* were higher under no-till than tilled wheat and caused grain yield losses as great as 0.6 ton ha^{-1} (Thompson et al., 1982).

Cotton (*Gossypium hirsutum* L.) planted non-till into winter legumes (*Trifolium incarnatum* L. or *Vicia villosa* Roth) had 10% to 18% greater infection with the fungus *Rhizoctonia solani* (Rickerl et al., 1988). In contrast, legume-mulched plots also had higher populations of Collembola (microarthropod insects that feed on the mycelia and spores of fungi in the rhizosphere), which decreased disease inoculum density and seedling disease severity (Wiggins and Curl, 1979).

VI. Fertilizer Needs of Crops

A. Nitrogen

Nitrogen, being the most limiting nutrient for crop production, has attracted most attention. Many reports are available addressing the availability of N in crop residues to succeeding crops and the impact of N in crop residues on nitrogen fertilizer needs. Legume residues contribute a substantial amount of N to the succeeding crops.

1. Cereal and Other Nonlegume Residues

Langdale et al. (1984) conducted a 4-year study in which 100 kg fertilizer N ha^{-1} was applied to each crop for sorghum–wheat double cropped on a Cecil sandy loam in Georgia. With NT, about 56% of the fertilizer N was returned to the soil annually through the stover of both crops. Mineralized soil N contributed substantially to the N needs of crops. Total soil N increased by 75 kg N ha^{-1} yr^{-1}. About 21% of the fertilizer N applied to the wheat was lost to the atmosphere (Sharpe et al., 1988). Hargrove (1985) reported enhanced N uptake by NT corn: 227, 212, and 190 kg N ha^{-1} under NT-coulter, NT with 30 to 45 cm deep in-row chiseling, and CT, respectively.

Using ^{15}N-labeled fertilizer, Dowdell and Crees (1980) working on a calcareous clay soil at Buckland, U.K., found no difference in N uptake or dry matter production by winter wheat seeded into no-till and plowed soil. Only a fraction of the residual fertilizer N was recovered by subsequent wheat crops. About 30% of the labeled N remained unaccounted for in either treatment. Similar results were reported by Webster et al. (1985). Power et al. (1986), using ^{15}N-depleted fertilizer, showed that with NT, little if any of the N contained in corn stover (as high as 120 kg N ha^{-1}) was utilized by the following corn crop, and only about 5% of the residual immobilized fertilizer N was utilized. Increasing rates of residue cover increased N uptake from applied N by 4 to 11 kg N ha^{-1}. The major effect of surface residues on N availability to the following corn crop resulted from greatly enhanced mineralization and uptake of indigenous soil N (increasing from 73 to 124 kg ha^{-1} as residue rates increased). These data suggest that microorganisms in residue-covered soil mineralized about 50 kg N ha^{-1} more than in bare soil, probably because of more favorable water regimes. From their studies on a Reiff loam, a mixed, nonacid, thermic Typic Xerofluvent (near Davis, CA), Abshahi et al.

(1984), using ^{15}N, concluded that only 4% of the original 157 kg N ha^{-1} applied to sugarbeet was recovered by the subsequent wheat crop, and 30% remained in the soil (same as reported by Dowdell and Crees, 1980), largely as organic N in the 0- to 30-cm soil zone. They further observed that wheat grain yield with sugarbeet tops removed was optimized at 124 kg N ha^{-1}. When sugarbeet tops were returned (incorporated with a rototiller) to the soil, wheat grain yield was optimized at 62 kg N ha^{-1}. Thus, utilization of sugarbeet tops as a source of N for a succeeding wheat crop greatly increased apparent fertilizer N use efficiency.

Opinions differ on the relative availability of fertilizer N under NT, MT, and CT. Smith and Howard (1980), working with a loam overlying sandy clay loam subsoil at Edinburgh, reported that wheat yield and N uptake from surface-applied ^{15}N-labeled fertilizer was more with CT than with NT. However, Frederickson et al. (1982), working on a Palouse silt loam, a mixed, mesic Pachic Ultic Haploxeroll (near Pullman, Washington), found that when ^{15}N-labeled fertilizer was surface applied to spring wheat at 168 kg N ha^{-1} in May 1980, 31% of applied N was taken up by wheat under CT and 39% for NT plots. For fertilizer applied at 112 kg N ha^{-1}, respective values were 25% and 34% for CT and NT. More residual ^{15}N was left in CT soil, indicating more immobilization of applied N with CT. In the succeeding winter wheat crop, contribution from the immobilized fertilizer N was 2.4% to 8.4%, while that from straw residue was 7.8% to 11.4%. Locke and Hons (1988), working on a Weswood silt loam (near Burleson, Texas), reported that fertilizer N uptake and total N accumulation patterns for sorghum were similar for NT and CT.

Studies by Kitur et al. (1984) at Lexington, Kentucky, and by Meisinger et al. (1985) at College Park, Maryland, more adequately explain the relative availability of applied ^{15}N under NT and CT. In the study by Kitur et al. (1984) on a Maury fine silt, corn was NT planted in mid-May in standing rye (*Secale cereale* L.) residue, whereas rye residue in CT plots was incorporated before seeding corn (CT-RI). There was no difference between NT and CT-RI in respect to N uptake by corn or recovery of fertilizer N applied at 168 kg N ha^{-1} (40% to 45% recovery). At 84 kg fertilizer N ha^{-1}, recovery was 36% to 37% in NT and 62% in CT-RI plots. Although differences were not significant, there was a trend toward reduced corn grain yield at reduced N rates. From their study on a Mattapex silt loam, a mixed, mesic Aquic Hapludult, Meisinger et al. (1985) found that by dough stage corn had recovered 62% of fertilizer N at 135 kg N ha^{-1} in CT plots and 53% in MT plots. Similar recoveries of 53% occurred with CT and MT at 180 kg fertilizer N ha^{-1}. Corn was planted in MT plots directly into the killed rye mulch. No rye was fall seeded in CT plots. They further observed that N sufficiency level in corn was about 1.1% to 1.2% of total dry matter at the dough stage, as suggested by Stanford (1973). These values were often reached with 135 to 180 kg fertilizer N ha^{-1} on CT plots, but additional N was needed on MT plots to reach the same plant N concentration. Average fertilizer N needs of corn under MT and CT as calculated by Meisinger et al. (1985) are given in Table 7, which shows that corn grown under MT required 68 kg N ha^{-1} more fertilizer N than that grown under CT.

Myers (1979) showed that banding ^{15}N for sorghum grown under NT was efficient in enhancing ^{15}N uptake relative to broadcasting. Placement of fertilizer N in bands below the surface may reduce both ammonia volatilization and nutrient immobilization (Mengel et al., 1982; Reinertsen et al., 1984; Fox et al., 1986).

On the Canadian Prairies, wheat yields under NT showed greater response (6.1 kg grain per kg N) to available N (soil NO_3 N plus fertilizer N) than under CT (2.9 kg grain per kg N) (Grevers et al., 1986).

In summary, the impact of cereal residues on N fertilization of a crop and their residual effects on succeeding crop provides the following conclusions:

1. At low N rates, recovery of applied N is usually greater for CT than for NT or MT.
2. At higher N rates, recovery of applied N is often nearly equal for CT and NT.
3. Differences between NT and CT are more evident in early stages of crop growth.
4. Higher rates of N are often needed for maximum yield for NT than for CT. However, often maximum yield is also greater for NT.
5. Only small amounts (about 5%) of N contained in cereal residues left on the soil surface (NT) are utilized by the succeeding crop.
6. Residual fertilizer N, immobilized in organic forms in the soil, is generally greater for CT, especially when residues are incorporated.
7. In a number of studies about 30% of applied N could not be accounted for. This N was probably lost by leaching (McMahon and Thomas, 1976; Thomas et al., 1973; Tyler and Thomas, 1977), denitrification (Aulakh et al., 1984; Rice and Smith, 1982), and/or ammonium volatilization.

Table 7. Average fertilizer N needs of corn grown under MT and CT and receiving a surface application of N at planting on a Mattapex silt loam*

Tillage	Crop N requirement (NC) kg N ha^{-1}	Total dry matter production (DM) Mg ha^{-1}	Soil N supply (NS) kg N ha^{-1}	N fertilizer efficiency (EF) (%)	Fertilizer N need (NF) kg N ha^{-1}
MT	184	16.0	59	53.1	235
CT	164	14.5	76	54.4	167

*Meisinger et al. (1985). Abbreviations: NC, total N uptake at dough corresponding to a N concentration of 1.15%; DM, total dry matter at dough stage; NS, soil N uptake at dough stage; EF, fertilizer uptake efficiency at dough stage.

2. Legume Residues

This review refers only to recent information on the contribution of legume residues to following crops, with particular reference to NT and MT effects. Data from a few experiments on green manuring, especially those using ^{15}N, are available elsewhere (Watanabe, 1984; Abrol and Palaniappan, 1988; Rosegrant and Roumasset, 1988; Vander Heide, 1988; Westcott and Mikkelsen, 1988; Nagarajah, 1988; Bouldin, 1988).

Legumes are credited with supplying substantial amounts of N to the succeeding nonlegume crops grown in rotation (Kurtz et al., 1984; Greenland, 1985). Estimates of the fertilizer N value of alfalfa (*Medicago sativa* L.) to the following corn crop have been as high as 180 kg N ha^{-1} (Voss and Shrader, 1984; Baldock and Musgrave, 1980). Fox and Piekielek (1983) reported that four successive crops of corn removed 298 kg ha^{-1} residual N from plowing in alfalfa. In the subtropical and tropical regions of the world, where two or more crops can be grown in a year, growing of legumes contributed about 30 to 60 kg N ha^{-1} as judged by increased yields of cereals (Xiao, 1984; Hazarika and Baruah, 1981; Jakhro and Faroque, 1982; Singh et al., 1982; De et al., 1983; John et al., 1989a,b; George and Prasad, 1989).

Most estimates of legume N contribution in crop rotation are made by fertilizer N replacement value, i.e., the amount of inorganic N required to produce an equivalent yield. This method assumes that increased yield of succeeding nonlegume crop is solely the result of legume N contribution and that fertilizer N and residue N are equally available. Although some researchers have attributed yield enhancement to N alone (Baldock and Musgrave, 1980), data by other workers suggest that there are effects other than those of residual N (Higgs et al., 1976; Voss and Shrader, 1984; Kurtz et al., 1984). Yield-enhancing effects not directly associated with N contribution are referred to as rotation effects, and as much as 25% of the corn yield increase has been attributed to rotation effects (Baldock et al., 1981). The rotation effect may include reduced disease and insect problems, residual treatment effects, and unexplained effects. Use of ^{15}N fertilizer in recent experiments has permitted some degree of separation of residual and legume N effects.

The net amount of symbiotically fixed N in legume residue returned to the cropping system depends on the amount of symbiotic activity, the amount and type of residue left on or in the soil, and the availability of soil N to the legume (Heichel, 1985). For example, soybean seed production may actually deplete rather than increase soil N because more N may be removed in the seed than is biologically fixed (Ebelhar and Welch, 1981; Heichel, 1985). Estimates of legume residue N availability to the first succeeding crop range from 10% to 35% (Yaacob and Blair, 1980; Ladd et al., 1981, 1983; Ebelhar et al., 1984; Hesterman et al., 1987).

Ebelhar et al. (1984) conducted field experiments from 1977 through 1981 on a Maury soil (near Lexington, Kentucky) to determine the amount of N provided by winter cover crops of hairy vetch (*Vicia villosa* Roth), big flower vetch (*Vicia grandiflora* Koch), crimson clover (*Trifolium incarnatum* L.), and rye (*Secale cereale* L.) and corn residue from the preceding crop to NT corn. Hairy vetch added 209, crimson clover 60, and rye 36 kg N ha^{-1}. The greatest corn yield (9 Mg ha^{-1}) was obtained with hairy vetch plus 100 kg ha^{-1} of fertilizer N. Corresponding yields after big flower vetch, crimson clover, rye, and corn residue were 6.6, 7.4, 7.6, and 6.8 Mg ha^{-1}, respectively. With corn residue, yields declined steadily over time without N fertilizer; remained constant with time with 50 kg fertilizer N ha^{-1}, and increased slightly with 100 kg fertilizer N ha^{-1}.

Hesterman et al. (1987), using ^{15}N in field studies on a Hubbard loamy sand, a mixed Udorthentic Haploboroll, and Waukegan silt loam, a mixed, mesic Typic Hapludoll, at Becker and Rosemount, Minnesota, found that uptake of legume N by corn was greater following alfalfa than following soybean. Use efficiency of alfalfa N varied from 27% to 40% depending on cultivar, location, and cutting treatment. Use efficiency of N from nodulated soybean residue was 18% to 19%, whereas that of nonnodulated soybean was 39% to 50%. Higher efficiency for nonnodulated soybean resulted from greater fertilizer N uptake for those residues. These legume N efficiencies are greater than those reported by Ladd et al. (1981, 1983) for wheat after alfalfa or for corn after *Sesbania aculeate* Pers as reported by Azam et al. (1985). The portion of N derived from different sources by corn at Becker are given in Figure 7. The legume N contribution from alfalfa was nearly three times that from soybean. Application of fertilizer N largely influenced the contribution of legume N. These data suggest considerable priming effect of legume as well as fertilizer N and indicate that N credits commonly attributed to legumes in crop rotation may be inflated. A common credit is 112 to 156 kg N ha^{-1} for a good stand of alfalfa (Bundy, 1985; Warncke et al., 1985; Voss and Shrader, 1984; Jokela et al., 1981; Rasmussen and Smiley, 1989). However, these conclusions are based only on the first-year residual effect of legume N. Legume N residual effects probably also contribute to the second, third, and succeeding crops of corn.

Power et al. (1986), working on a Crete–Butler silty loam (near Lincoln, Nebraska), found that little of the N in corn stover left on the soil surface at harvest of the 1980 crop was utilized by the 1981 crop. However, up to 95% of the N in the 1980 soybean [*Glycine max* (L.) Merr.] residues, when added in large amounts, was taken up by the 1981 soybean crop (Table 8).

A number of studies have been made of NT seeding of cotton into legume mulches (Harris and Smith, 1980; Roach, 1981; Touchton et al., 1982a,b; Touchton and Johnson, 1982). Touchton et al. (1984) studied NT

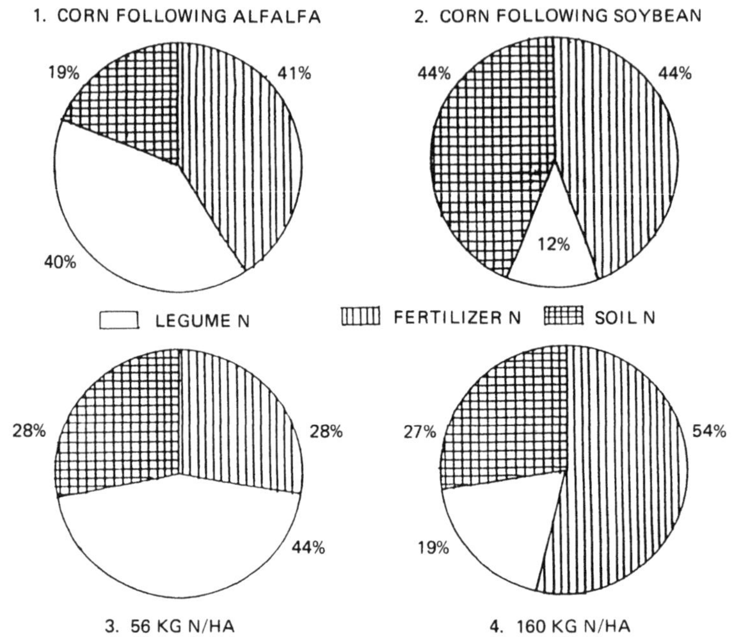

Figure 7. Percentage of N in whole corn plans derived from legume residue, fertilizer N, and soil N as affected by legume (1,2) and rate of N applied (3,4). (From Hesterman et al., 1987)

Table 8. Grain yield and N uptake by corn as affected by corn and soybean residues

		Source of N (kg ha^{-1})				
Crop residue rate (%)†	Grain yield (Mg ha^{-1})	Crop residues	Residual fertilizer	Current fertilizer	Native soil N	Total
Corn						
0	2.64	0	5	4	73	82
50	3.34	0	6	7	97	110
100	3.67	2	6	7	114	129
Soybean						
0	1.66	0	2	14	84	100
50	2.04	1	2	21	124	148
100	2.23	38	7	16	116	177
150	2.36	63	6	20	106	195

*Source: Power et al. (1986).

†Percentage of residues produced by previous crop.

planting of cotton into crimson clover, common vetch (*Vicia sativa* L.), and fallow on a Norfolk fine sandy loam, a siliceous, thermic Typic Paleudult. Without fertilizer N, cotton no-tilled planted in clover or vetch gave yields similar to those obtained with 34 kg N ha^{-1} in fallowed soil. Application of N in clover plots delayed maturity because of excess N and reduced cotton yield.

Few reports are available from tropical and subtropical regions using ^{15}N. Goswami et al. (1988) reported from New Delhi that grain and straw yield of rice (*Oryza sativa* L.) with 60 Kg N ha^{-1} after green manuring with *Sesbania* were slightly higher than with 120 kg N ha^{-1} after summer fallow. Thus, summer green manuring led to a savings of 60 kg N ha^{-1} for rice. Labeled nitrogen fertilizer (^{15}N) data showed that green manuring contributed 4.7% to 5.6% of the total N taken up by rice and 1.9% to 4.8% of that taken up by the succeeding wheat, depending on the level of fertilizer N applied to rice. In a study at the International Rice Research Institute, Los Banos, John et al. (1989a) found that green manuring with cowpea (*Vigna unguiculata* L) substituted for 35 kg urea N ha^{-1}, whereas the cowpea residue (grain harvested and straw residue incorporated in soil) substituted for 44 to 50 kg N ha^{-1} in rice. In Texas, Lemon et al (1990) found that legume green manures provided the equivalent of 60 kg fertilizer N per ha.

The contribution of legume N to succeeding crops in rotation is summarized as follows:

1. Legume N can contribute substantially to N needs of succeeding crops in rotation, but amounts depend on the time during which legumes are grown, amount of biomass incorporated, and species of succeeding crop grown.
2. Increased yields of crops following legumes may result from increased N availability and/or from the "rotation effect."
3. Use of ^{15}N can greatly help in determining the amount of N derived from different sources, i.e., native soil N, fertilizer N, and legume N.
4. Factors other than N affecting the growth and yield of crops succeeding legumes in rotation (rotation effect) need to be identified, and techniques for measuring their effects need to be developed.

B. Phosphorus and Potassium

Triplett (1975), reviewing early literature on P nutrition of NT corn, concluded that surface broadcast application of P is a satisfactory practice. Hargrove (1985) reported more P uptake for NT than CT for corn: 35 kg P ha^{-1} for NT and 31 kg P ha^{-1} for CT. However, Eckert and Johnson (1985) found that subsurface placement (5 cm below and 5 cm to the side of the seed) of P produced greater corn yields and gave better fertilizer efficiency. Differences in P content of corn leaves at silking were noted only in plots having low soil test value and receiving 29 kg P ha^{-1}. Kunishi et al. (1986) also found that there were no differences in P content in above-ground corn residues between NT and CT plots.

Johnson and Eckert (1987) recommended 15% more K for corn and soybean under NT than under CT in Ohio. Potash placement also had a positive impact in yields with NT. Strip banding of K increased NT soybean yields by more than 0.27 Mg ha^{-1} over broadcast K.

VII. Environment

Leaving crop residues in the field and associated NT crop production have been advocated as a cost-effective means of reducing pollution caused by runoff, nutrient loss, and erosion from agricultural lands (Christensen and Norris, 1983). However, Baker and Laflen (1983) and Logan and Adams (1981) were of the opinion that the stratified distribution of P in soil profiles with NT may lead to increased losses of soluble P even though sediment-bound P losses were reduced. Eckert and Johnson (1985) reported that as compared to broadcast application, subsurface banding significantly reduced surface P accumulation. These results suggest advantages for P placement in reducing the pollution of ground water.

In a field study at Sidney, Nebraska (Follett and Peterson, 1988), NT as compared to CT had significantly more extractable Pb and Ni in the 0- to 5- and 5- to 10-cm depth but decreased Ni in the 10- to 20-cm depth. Extractable Pb and Cd were similarly greater with NT at the 0- to 5-cm depth, as was Pb in the 5- to 10-cm depth. These effects were attributed to organic matter accumulation in the 0- to 5-cm depth under NT. A strong correlation of extractable Pb and Cd levels with organic matter content was reported by Miller and McFee (1983). However, in the Sidney study extractable Pb, Cd, and Ni in NT plots were not significantly different from those in native grass sod. Therefore, as compared to native sod, long-term NT cultivation did not result in accumulation of extractable Pb, Cd, and Ni in the surface soil, but CT reduced their concentrations.

Atmospheric pollution by smoke originating from straw burning is of concern in the U.K. Of special concern are problems caused by smoke and ash deposition and by damage to property, trees, and hedgrows (Graham et al., 1986).

From the environmental viewpoint, by far the greatest effect of reduced- and no-tillage systems is to reduce runoff and soil erosion and subsequent problems of surface water quality, sedimentation, and loss of soil productivity. In the United States, present government price support programs require that producers on erodible soils develop and follow a conservation program designed to significantly reduce erosion. The maintenance of crop residues on the soil surface is a key practice in these conservation production plans.

VIII. Summary and Conclusions

In developed and affluent countries, utilization of crop residues is a major concern for environmental protection and sustained productivity. In drier regions of the United States, such as the Midwest and Great Plains areas, leaving cereal residues in situ has been utilized to control soil erosion and conserve water. No-tillage (NT) planting machinery has been developed and is widely used for this purpose. When adequate weed control is achieved on well-drained soils, crop yields for NT are usually similar to or better than those obtained with conventional tillage (CT), especially in drier years. In wetter and cooler regions of the world or on poorly drained soils, crop yields may be lower when crop residues are left in situ, and NT planting is utilized. When excessive residues are present, the most economical and simplest way of disposing of this excess is to burn crop residues on the farm. However, this approach creates serious environmental problems and fire hazards as well as eventually resulting in reduced soil organic matter content. Management of crop residue is less difficult in developing countries because crop residues are routinely removed for use as a source of fodder for farm animals. As a matter of fact, in arid regions of developing countries, where subsistence farming is practiced and a good harvest may be possible only once in 3 to 5 years, coarse cereals such as sorghum and millet are grown mainly for obtaining fodder. However, removal of crop residues from fields in these developing countries, for whatever reason, reduces soil organic matter content and often leaves soils susceptible to accelerated soil erosion. This is a major problem of resource development and use in developing nations that has major consequences, especially in semiarid regions where the quantity of residues produced is limited.

A. Achievements

During the 1980s data were published from a number of long-term studies concerning the effects of crop residue management practices on chemical and physical properties of soils. In general, leaving crop residues on the soil surface, as in no-tillage (NT) farming, has resulted in more organic C, total N, sodium-bicarbonate- or dilute-acid-extractable P, exchangeable K, and other plant nutrients in the 0- to 5-cm soil layer; soil pH, on the other hand, was lowered. Burning may initially have a similar effect, but continued burning eventually results in soil deterioration. Leaving crop residues on the surface as in NT farming results in higher storage and retention of soil water and lower temperatures; these may be of advantage for germination and early seedling growth of crops under drier conditions and harmful under cooler and wetter climates. Effects of crop residue management on soil bulk density, cone index, porosity, and hydraulic conductivity show considerable variation, making it difficult to generalize.

The ^{15}N studies on fertilizer needs of crops as affected by crop residue management showed that only small amounts (about 5%) of N contained in cereal residues left on the soil surface with NT becomes available to the succeeding crop. At low N rates, recovery of applied N is often greater in conventionally tilled (CT) plots than in NT or MT (minimum tillage) plots, whereas at higher N rates, these differences may disappear, or recovery may even be greater for NT. In general, higher N rates are needed to obtain maximum yield on NT than on CT plots, but yields are also often greater for NT. About 30% of the applied N usually cannot be accounted for in crop and soil, and presumably is lost into the environment. Legume residues can substantially contribute to N needs of succeeding crops.

B. Research Needs

Research is needed on the effects of crop residue management practices on secondary and micronutrient composition of soil and their availability to crops grown. Soil biological properties also need to be carefully studied; detailed information on soil macro- and microfauna under different agroecological conditions is needed. More research is needed on the effects of residue management on soil physical properties, especially soil water movement and availability. The effects of tillage practices on the soil microenvironment need to be better quantified to enable us to predict changes in soil microbial activity resulting from tillage.

Other than contributing to N needs of the succeeding crops, the effects of legume residues on succeeding crops are poorly defined. Since many legumes have deep root systems, it would be interesting to study their contribution to the accumulation of nutrients other than N in the upper soil layers.

Because field research is time consuming and expensive, it would be desirable to develop suitable computer simulations to predict the effects of a number of factors on soil properties and finally on crop growth and yield. Knowledge gained at one research station could then be put to use on a large area and different locations having similar agroecological situations. At present, this goal is not attainable, but continued development of the research data bases required plus advances in computer modeling should result in development of such models in the future.

Acknowledgments

The authors are grateful to the Government of India, Department of Science and Technology, Indian Council of Agricultural Research and Indian Agricultural Research Institute and to the Agricultural Research Service in the U.S. Department of Agriculture, U.S. Agency for International Development, and the University of Nebraska, Lincoln, for providing funds and facilities under the Indo-U.S. Senior Scientists Panel (STI) Program that permitted our visits and exchange of ideas, which resulted in this review.

References

Aase, J.K., and F.H. Siddoway. 1980. Stubble height effects on seasonal microclimate, water balance, and plant development of no-till winter wheat. *Agric. Meteorol.* 21:1–20.

Abrol, I.P., and S.P. Palaniappan. 1988. Green manuring in irrigated and rainfed lowland rice systems in South Asia. pp. 71–82. In: *Green manure in rice farming*. International Rice Research Institute, Los Banos, Philippines.

Abshahi, A., F.J. Mills, and F.E. Broadbent. 1984. Nitrogen utilization by wheat from residual sugarbeet fertilizer and soil incorporated sugar crops. *Agron. J.* 76:954–958.

Addala, M.S.A. 1982. Effect of application methods on the performance of soil applied herbicides. Ph.D. Thesis, Univ. Reading, U.K.

Al-Darby, A.M., and B. Lowery. 1986. Evaluation of corn (*Zea mays* L.) growth and productivity with three conservation tillage systems. *Agron. J.* 78:901–907.

Al-Darby, A.M., and B. Lowery. 1987. Seed zone soil temperature and early corn growth with three conservation tillage systems. *Soil Sci. Soc. Am. J.* 51:768–774.

Al-Darby, A.M., B. Lowery, and T.C. Daniel. 1987. Corn leaf water potential and water use efficiency under three conservation tillage systems. *Soil Tillage Res.* 9:241–254.

Alessi, J., and J.F. Power. 1971. Corn emergence in relation to soil temperature and seeding depth. *Agron. J.* 63:717–719.

American Society of Agronomy. 1978. Crop residue management system. *Am. Soc. Agron.* Spec. Publ. 31.

American Society of Agronomy. 1983. *Dryland agriculture*. Monograph No. 23. American Society of Agronomy, Madison, WI.

Armbrust, D.V., J.D. Dickerson, E.L. Skidmore, and O.G. Rus. 1982. Dry soil aggregation as influenced by crop and tillage. *Soil Sci. Soc. Am. J.* 46:390–393.

Aston, A.R., and R.A. Fischer. 1986. The effect of conventional cultivation, direct drilling and crop residues on soil temperature during the early growth of wheat at Murrumbateman, New South Wales. *Aust. J. Soil Res.* 24:49–60.

Aulakh, M.S., D.A. Rennie, and E.A. Paul. 1984. The influence of plant residues on denitrification and rates in conventional and zero-tilled soils. *Soil Sci. Soc. Am. J.* 48:790–794.

Australian Society of Agronomy. 1986. *Tillage–the Australian perspective.* CSIRO, Canberra, Australia.

Azam, F., K.A. Malik, and M.I. Sajjad. 1985. Transformations in soil and availability to plants of ^{15}N applied as inorganic fertilizer and legume residue. *Plant Soil* 86:3–13.

Baker, J.L., and J.M. Laflen. 1983. Water quality consequences of conservation tillage. *J. Soil Water Conserv.* 38:186–193.

Baldock, J.O., R.L. Higgs, W.H. Paulson, J.A. Jackobs, and W.D. Shrader. 1981.

Legume and mineral N effects on crop yields in several crop sequences in the upper Mississippi Valley. *Agron. J.* 73:885–890.

Baldock, J.O., and R.B. Musgrave. 1980. Manure and mineral effects in continuous and rotational crop sequences in central New York. *Agron. J.* 72:511–518.

Barley, K.P. 1961. The abundance of earthworms in agricultural land and their possible significance in agriculture. *Adv. Agron.* 13:249–268.

Barnes, B.T., and F.B. Ellis. 1979. Effects of different methods of cultivation and direct drilling and disposal of straw residues on population of earthworms. *J. Soil Sci.* 30:669–679.

Bauder, J.W., G.W. Randall, and J.B. Swann. 1981. Effect of four continuous tillage systems on mechanical impedance of a clay loam soil. *Soil Sci. Soc. Am. J.* 45:802–806.

Bauer, A., and A.L. Black. 1981. Soil carbon nitrogen and bulk density comparisons in two cropland tillage systems after 25 years and in virgin grassland. *Soil Sci. Soc. Am. J.* 45:1160–1170.

Baumhardt, R.L., R.E. Zartman, and P.W. Unger. 1985. Grain sorghum response to tillage method used during fallow and to limited irrigation. *Agron. J.* 77:643–646.

Biederbeck, V., C.A. Campbell, K.E. Bowren, M. Schnitzer, and R.N. McIver. 1980. Effect of burning cereal straw on soil properties and grain yields in Saskatchewan. *Soil Sci. Soc. Am. J.* 44:103–111.

Blacklow, W.M. 1972. Influence of germination and temperature on elongation of the radical and shoot of corn. *Crop Sci.* 12:647–650.

Blevins, R.L., G.W. Thomas, and P.L. Cornelius. 1977. Influence of no-tillage and nitrogen fertilization on certain soil properties after five years of continuous corn. *Agron. J.* 69:383–386.

Blevins, R.L., G.W. Thomas, M.S. Smith, W.W. Frye, and P.L. Cornelius. 1983. Changes in soil properties after 10 years of non-tilled and conventionally tilled corn. *Soil Tillage Res.* 3:135–145.

Blumberg, A.Y., and D.A. Crossley, Jr. 1983. Comparison of soil surface arthropod populations in conventional tillage, no tillage and old field systems. *Agro. Ecosyst.* 8:247–253.

Bond, J.J., and W.O. Willis. 1969. Soil water evaporation in surface-residue rate and placement effects. *Soil Sci. Soc. Am. Proc.* 33:445–448.

Bouldin, D.R. 1988. Effect of green manure on soil organic matter content and nitrogen availability. pp. 151–163. In: *Green manure in rice farming*. International Rice Research Institute, Los Banos, Philippines.

Bundy, L.G. 1985. Corn Fertilization. *Univ. Wisconsin Coop. Ext. Serv. Bull.* A3340.

Cannell, R.Q., F.B. Ellis, D.G. Christian, J.P. Graham, and J.J. Douglas. 1980. The growth and yield of winter cereals after direct drilling shallow cultivation and ploughing on non-calcareous clay soils (1974–1978). *J. Agric. Sci.* 94:345–359.

Capriel, P., T. Beck, H. Borchert, and P. Härter. 1990. Relationship between soil aliphatic fraction extracted with supercritical hexane, soil microbial biomass, and soil aggregrate stability. *Soil Sci. Soc. Am. J.* 54:415–420.

Carter, M.R., and D.A. Rennie. 1982. Changes in soil quality under zero tillage farming systems: distribution of microbial biomass and mineralizable C and N potentials. *Can. J. Soil Sci.* 62:587–597.

Carter, M.R., and D.A. Rennie. 1984a. Dynamics of soil microbial biomass N under zero and shallow tillage for spring wheat using ^{15}N urea. *Plant Soil* 76:157–164.

Carter, M.R., and D.A. Rennie. 1984b. Nitrogen transformations under zero and shallow tillage. *Soil Sci. Soc. Am. J.* 48:1077–1081.

Carter, M.R., and D.A. Rennie. 1987. Effects of tillage on deposition and utilization of ^{15}N residual fertilizer. *Soil Tillage Res.* 9:33–43.

Christian, D.G., and D.P. Miller. 1986. Straw incorporation by different tillage systems and the effect on growth and yield of winter oats. *Soil Tillage Res.* 8:239–252.

Christensen, L.A., and P.E. Norris. 1983. A comparison of tillage systems for reducing soil erosion and water pollution. *USDA Economic Res. Serv., Agric. Econ. Rept. 499.* U.S. Govt. Printing Office, Washington, D.C.

Clark, R.L., and J.T. Reid. 1984. *Topographical mapping of soil properties. ASAE Tech. Paper No. 84-1092.* Am. Soc. Agric. Engr., St. Joseph, MO.

Costamagna, C.A., R.K. Stivers, H.M. Galloway, and S.A. Barber. 1982. Three tillage systems affect selected properties of a tiled, naturally poorly drained soil. *Agron. J.* 74:442–444.

Crossley, D.A. Jr. 1977. The roles of terrestrial saprophagus arthropods in forest soils: Current status of concepts. pp. 49–56. In: W.J. Mattson (ed.), *The role of arthropods in forest ecosystems.* Springer-Verlag, New York.

Crossley, D.A. Jr., G.J. House, R.M. Snider, R.J. Snider, and B.R. Stinner. 1984. The positive interactions in agroecosystems. pp. 73–81. In: R.R. Lawrence, B.R. Stinner, and G.H. House (eds.), *Agricultural ecosystems—Unifying concepts.* John Wiley & Sons, New York.

Cruse, R.M., G.A. Yakle, T.C. Colvin, D.R. Timmons, and A.L. Musseleman. 1983. Tillage effects on corn and soybean production in farmer-managed, university-monitored field plots. *J. Soil Water Conserv.* 38:512–514.

Culley, J.L.B., W.E. Larson, and G.W. Randall. 1987a. Physical properties of a typic haplaquoll under conventional and no-tillage. *Soil Sci. Soc. Am. J.* 51:1587–1592.

Culley, J.L.B., W.E. Larson, R.R. Allmaras, and M.J. Shaffer. 1987b. Soil-water regimes of a typic haplaquoll under conventional and no-tillage. *Soil Sci. Soc. Am. J.* 51:1604–1610.

Cummins, D.G., and W.L. Parks. 1961. The germination of corn and wheat as affected by various fertilizer salts at different soil temperature. *Soil Sci. Soc. Am. Proc.* 25:47–49.

Cussans, G.W., S.R. Moss, R.J. Hance, S.J. Embling, D. Caverley, T.G. Marks, and J.J. Palmes. 1987. The effect of tillage method and soil factors on the performance of chlorotoluron and isoproturon. *Proc. Br. Crop Protec. Conf. Weeds* 153–160.

De, R., Y. Yogeswara Rao, and W. Ali. 1983. Grain and fodder legumes as preceding crops affecting the yield and N economy of rice. *J. Agric. Sci.* 101:463–666.

Dick, W.A. 1983. Organic carbon, nitrogen and phosphorus concentration and pH in soil profiles as affected by tillage intensities. *Soil Sci. Soc. Am. J.* 47:102–107.

Doran, J.W. 1980. Soil microbial and biochemical changes associated with reduced tillage. *Soil Sci. Soc. Am. J.* 44:765–771.

Doran, J.W., and M.S. Smith. 1987. Organic matter management and utilization of soil and fertilizer nutrients. J.J. Mortvedt, D.R. Buxton and S.H. Mickelson (eds.) *Soil Fertility and Organic Matter as Critical Components of Production System*. Soil Sci. Soc. Am. Spec. Publ. 19:53–72.

Dowdell, R.J., and R. Crees. 1980. The uptake of ^{15}N-labelled fertiliser by winter wheat and its immobolisation in clay soil after direct drilling or ploughing. *J. Sci. Food. Agric.* 31:992–996.

Dowdell, R.J., R. Crees, and R.Q. Cannell. 1983. A field study on effects of contrasting methods of cultivation on soil nitrate content during autumn, winter and spring. *J. Soil Sci.* 34:367–379.

Drew, M.C., and L.R. Saker. 1978. Effects of direct drilling and ploughing on root distribution in spring barley and on the concentration of extractable phosphate and potassium in the upper horizons of a clay soil. *J. Sci. Food Agric.* 29:201–206.

Drew, M.C., and L.R. Saker. 1980. Direct drilling and ploughing: Their effects on the distribution of extractable phosphorus and potassium and of roots in the upper horizons of two clay soils under winter wheat and spring barley. *J. Agric. Sci.* 99:411–423.

Ebelhar, M.W., and L.F. Welch. 1981. Variation in grain and non-grain nitrogen from selected soybean cultivars. *Agron. Abstr.* 1981:174.

Ebelhar, S.A., W.W. Frye, and R.L. Blevins. 1984. Nitrogen from legume cover crops for no-tillage corn. *Agron. J.* 76:51–55.

Eckert, D.J., and J.W. Johnson. 1985. Phosphorus fertilization in no tillage corn production. *Agron J.* 77:789–792.

Edwards, C.A. 1975. Effects of direct drilling on the soil fauna. *Outlook Agric.* 8:243–244.

Edwards, C.A., and J.R. Lofty. 1977. *Biology of earthworms*. Chapman & Hall, London.

Edwards, C.A., and J.R. Lofty. 1982. Nitrogenous fertilizers and earthworm populations in agricultural soils. *Soil Biol. Biochem.* 14:515–521.

El-Haris, M.K., V.L. Cochran, L.P. Elliot, and D.F. Bezdicek. 1983. Effect of tillage, cropping, and fertilizer management on soil nitrogen mineralization potential. *Soil Sci. Soc. Am. J.* 47:1157–1161.

Elliott, L.F., and J.M. Lynch. 1984. The effect of available carbon and nitrogen in straw on soil and ash aggregation and acetic acid production. *Plant Soil* 78:335–343.

Elliott, L.P., and R.I. Papendick, 1986. Crop residue management for improved soil productivity. *Biol. Agric. Hort.* 3:131–142.

Ellis, F.B., and J.M. Lynch. 1977. Why burn straw? *ARC Res. Rev.* 2:29–33.

Embling, S.J., E.G. Cotterill, and R.J. Hance. 1983. Effect of heat treating soil and straw in the subsequent adsorption of chlortoluron and atrazine. *Weed Res.* 23:357–363.

FAO. 1987. *World crop and livestock statistics, 1948–85*. pp. 7–123. FAO, Rome.

Fischer, R.A. 1982. Direct drilling and crop performance in Southern New South Wales. *Biannual report Division of Plant Industries*. pp. 26–32. CSIRO, Canberra, Australia.

Follett, R.F., and G.A. Peterson. 1988. Surface soil nutrient distribution as affected by wheat–fallow tillage systems. *Soil Sci. Soc. Am. J.* 52:141–147.

Fox, R.H., J.M. Kern, and W.P. Piekielek. 1986. Nitrogen fertilizer source and method and time of application effects on no-till corn yields and nitrogen uptakes. *Agron J.* 78:741–746.

Fox, R.H., and W.P. Piekielek. 1983. Response of corn to nitrogen fertilizer and prediction of soil nitrogen availability with chemical tests in Pennsylvania. *Penn. State. Univ. Agri. Exp. Stn. Bull.* 843:1–32.

Frederickson, J.K., F.E. Koehler, and H.H. Cheng. 1982. Availability of $_{15}N$ labeled nitrogen in fertilizer and in wheat straw to wheat in tilled and no-till soil. *Soil Sci. Soc. Am. J.* 46:1218–1222.

Freebairn, D.M., L.D. Ward, A.L. Clarke, and G.D. Smith. 1986. Research and development of reduced tillage systems for vertisols in Queensland, Australia. *Soil Tillage Res.* 8:211–229.

Gallaher, R.N. 1984. Soybean root as affected by tillage in old tillage studies. pp. 102–105. In: J.T. Touchton and R.E. Stevenson (eds.), *Proc. Southeast No. Tillage Systems 7th Conf.* Auburn University, Auburn, AL.

Gantzer, C.J., and G.R. Blake. 1978. Physical characteristics of Le Sueur clay loam following no-till and conventional tillage. *Agron. J.* 70:853–857.

George, M., and R. Prasad. 1989. Studies on the effect of legumes on fertilizer N utilization by rice using ^{15}N technique in rice-based multiple cropping systems. *Res. Dev. Agric.* 6:115–119.

Gerik, T.J., and J.E. Morrison, Jr. 1984. No tillage of grain sorghum on a shrinking clay soil. *Agron. J.* 76:71–76.

Germann, P., and K. Beven. 1981. Water-flow in soil macropores. I. An experimental approach. *J. Soil Sci.* 32:1–13.

Goswami, N.N., R. Prasad, M.C. Sarkar, and S. Singh. 1988. Studies on the effect of green manuring in nitrogen economy in a rice-wheat rotation using ^{15}N. *J. Agric. Sci.* 111:413–417.

Graham, J.P., F.B. Ellis, D.G. Christian, and R.O. Cannel. 1986. Effect of straw residues on the establishment, growth and yield of autumn sown cereals. *J. Agric. Eng. Res.* 33:39–49.

Great Plains Agricultural Council. 1976. *Conservation tillage. Great Plains Council Publ. No. 77.* Colo. Agr. Expt. Sta., Fort Collins, CO.

Great Plains Agricultural Council. 1984. *Conservation tillage. Great Plains Council Publ. No. 110.* Univ. of Nebraska, Lincoln, NE.

Greenland, D.J. 1977. Soil damage by intensive arable cultivation: Temporary or permanent. *Phil. Trans. R. Soc. Lond.* 281:193–208.

Greenland, D.J. 1985. Nitrogen and food production in the tropics: Contribution from fertilizer nitrogen and biological nitrogen fertilizers. pp. 9–38. In: *Nitrogen management in farming systems in humid and sub-humid tropics.* International Institute for Tropical Agriculture, Ibadan, Nigeria.

Grevers, M.C., J.A. Kirkland, E. De Jong, and D.A. Rennie. 1986. Soil water conservation under zero- and conventional tillage systems on the Canadian prairies. *Soil Tillage Res.* 8:265–276.

Griffith, D.R., J.V. Mannering, and W.C. Moldenhauer. 1977. Conservation tillage in the eastern corn belt. *J. Soil Water Conserv.* 32:20–28.

Gupta, S.C. 1985. Predicting corn planting dates for moldboard and no-till tillage systems in the corn belt. *Agron. J.* 77:446–455.

Gupta, V.V.S.R., and J.J. Germida. 1988. Distribution of microbial biomass and its activity in different soil aggregrate size classes as affected by cultivation. Soil Biol. Biochem. 20:777–786.

Hargrove, W.L. 1985. Influence of tillage on nutrient uptake and yield of corn. *Agron. J.* 77:763–768.

Hargrove, W.L., and W.S. Hardcastle. 1984. Conservation tillage practices for winter wheat production in the Appalachian Piedmont. *J. Soil Water Conserv.* 39:324–326.

Hargrove, W.L., J.T. Reid, J.T. Touchton, and A.N. Gallahar. 1982. Influence of tillage practices on the fertility status of an acid soil double-cropped to wheat and soybeans. *Agron. J.* 74:684–687.

Harper, S.H.T., and J.M. Lynch. 1981. The kinetics of straw decomposition in relation to its potential to produce the phototoxin acetic acid. *J. Soil Sci.* 32:627–637.

Harris, C.H., and C.W. Smith. 1980. Cotton production affected by row profile and nitrogen rates. *Agron. J.* 72:919–922.

Havlin, J.L., D.E. Kissel, L.D. Maddox, M.M. Claasen, and J.H. Long. 1990. Crop rotation and tillage effects on soil organic carbon and nitrogen. Soil Sci. Soc. Am. J. 54:448–452.

Hay, R.K.M. 1977. Effect of tillage and direct drilling on soil temperature in winter. *J. Soil Sci.* 28:403–409.

Hazarika, B., and A.R. Baruah. 1981. Multiple cropping with rice under rainfed conditions. *J. Res. Assam. Agric. Univ.* 2:234–239.

Heichel, G.H. 1985. Symbiosis: Nodule bacteria and leguminous plants. pp. 64–71. In: M.E. Heath, R.C. Barnes, and D.S. Metcalf (eds.), *Forages: The science of grassland agriculture*. Iowa State University, Ames, IA.

Hesterman, O.B., M.P. Russelle, C.C. Sheaffer, and G.H. Heichel. 1987. Nitrogen utilization from fertilizer and legume residues in legume–corn rotations *Agron. J.* 79:726–731.

Higgs, R.L., W.H. Paulson, J.W. Pendleton, A.E. Peterson, J.A. Jacobs, and W.D. Shrader. 1976. Crop rotation and nitrogen: Crop sequence comparisons on soils of the driftless area of southwestern Wisconsin. *Univ. Wisc. Coll. Agric. Life Sci. Res. Bull.* 2761.

Hill, R.L., and R.M. Cruse. 1985. Tillage effects on bulk density and soil strength of two mollisols. *Soil Sci. Soc. Am. J.* 49:1270–1273.

Hill, R.L., R. Horton, and R.M. Cruse. 1985. Tillage effects on soil water retention and pore size distribution of two mollisols. *Soil Sci. Soc. Am. J.* 49:1264–1270.

Hooker, M.L., G.M. Herron, and P. Penas. 1982. Effects of residue burning, removal, and incorporation on irrigated cereal crop yields and soil chemical properties. *Soil Sci. Soc. Am. J.* 46:122–126.

House, G.J., and J.N. All. 1981. Carbid beetles in soybean agroecosystems. *Environ. Entomol.* 10:194–196.

House, G.J., and R.W. Parmele. 1985. Comparison of soil anthropods and earthworm from conventional and non-tillage agro-ecosystems. *Soil Tillage Res.* 5:351–360.

House, G.J., and B.R. Stinner. 1983. Arthropods in no-tillage soybean agroecosystems: Community composition and ecosystem interactions. *Environ. Manage. J.* 23–28.

International Soil Tillage Research Organization. 1986. Abstracts of Papers: ISTRO Conference - July 85. University of Guelph, Ontario, Agric. College, Guelph, Ontario, Canada. *Soil Tillage Res.* 8:321–376.

Jakhro, A.A., and G. Faroque. 1982. Number of tillers, nitrogen concentration, and grain yield of kharif rice crops as influenced by dubari crops and nitrogen level. *Int. Rice Res. Newslett.* 7(6):22–23.

John, P.S., R.K. Pandey, R.J. Buresh, and R. Prasad. 1989a. Lowland rice response to urea following three cowpea cropping systems. *Agron. J.* 81:853–857.

John, P.S., R. Prasad, S. Singh, M. George, and S.N. Sharma. 1989b. Effect of lentil residues on the yield and response of succeeding rice to nitrogen. *J. Agron. Crop Sci.* 163:177–180.

Johnson, J.W., and D.J. Eckert. 1987. Potassium requirement for no-till corn and soybeans. *Agriviews* 6:4.

Johnson, M.D., and B. Lowery. 1985. Effect of three conservation tillage practices on soil temperature and thermal properties. *Soil Sci. Soc. Am. J.* 49:1547–1552.

Jokela, W.E., W.E., Fenster, C.J. Overdahl, C.A. Simkens, and J. Grava. 1981. Guide to computer programmed soil test recommendations for field crops in Minnesota. *Univ. Minn. Ext. Bull.* 416.

Kitur, B.K., M.S. Smith, R.L. Blevins, and W.W. Frye. 1984. Fate of ^{15}N-depleted ammonium nitrate applied to no-tillage and conventional tillage corn. *Agron. J.* 76:240–243.

Klute, A., and C. Dirksen. 1986. Hydraulic conductivity of saturated soils: Laboratory methods. In A. Klute (ed.), *Methods of soil analysis*. Pt. I. 2nd ed. *Agronomy* 9:687–734.

Knapp, E.B., L.F. Elliott, and G.S. Campbell. 1983a. Microbial respiration and growth during the decomposition of wheat straw. *Soil Biol. Biochem.* 15:319–323.

Knapp, E.B., L.F. Elliott, and G.S. Campbell. 1983b. Carbon, nitrogen and microbial biomass interrelationships during the decomposition of wheat straw: A mechanistic simulation model. *Soil Biol. Biochem.* 15:455–461.

Knipscheer, H.C., K.M. Menz, and L. Verinumbe. 1983. The evaluation of preliminary farming systems technologies: Zero tillage systems in West Africa. *Agric. Syst.* 11:95–103.

Kraft, J.M., and R.R. Allmaras. 1985. Pea root pathogen populations in relation to soil structure, compaction, and water content. pp. 203–205. In: C.A. Parker, A.D. Rovita, K.J. Moore, and P.T.W. Wong (eds.), *Ecology and management of soil-borne plant pathogens*. American Pathological Society, St. Paul, MN.

Kunishi, H.M., V.A. Bandel, and F.R. Mulford. 1982. Measurement of available soil phosphorus under conventional and no-till management. *Commun. Soil Sci. Plant Anal.* 13:607–618.

Kunishi, H.M., V.A. Bandel, and F.R. Mulford. 1986. Seasonal P uptake by corn under no-till and conventional till management. *Commun. Soil Sci. Plant Anal.* 17:591–600.

Kurtz, L.T., L.V. Boone, T.R. Peck, and R.G. Hoeft. 1984. Crop rotation for efficient nitrogen use. pp. 295–300. In: R.D. Hauck (ed.), *Nitrogen in crop production*. American Society of Agronony, Madison, WI.

Ladd, J.N., M. Amato, R.B. Jackson, and J.H.A. Butler. 1983. Utilization by wheat crops of nitrogen from legume residues decomposing in soils in the field. *Soil Biol. Biochem.* 15:231–238.

Ladd, J.N., J.M. Oades, and M. Amato. 1981. Distribution and recovery of nitrogen from legume residues decomposing in soils sown to wheat in the field. *Soil Biol. Biochem.* 13:251–256.

Lamb, J.A., G.A. Peterson, and C.R. Fenster. 1985. Wheat-fallow tillage systems' effect on a newly cultivated grassland soil's nitrogen budget. *Soil Sci. Soc. Am. J.* 49:352–356.

Langdale, G.W., W.L. Hargrove, and J. Giddens. 1984. Residue management in double-crop conservation tillage systems. *Agron. J.* 76:689–694.

Leitch, M.H., and L.V. Vaidyanathan. 1983. N use by winter wheat established in cultivated and direct drilled soils. *J. Agric. Sci.* 100:461–471.

Lemon, R.G., F.M. Hons, and V.A. Saladino. 1990. Tillage and clover cover crop effects on grain sorghum yield and nitrogen uptake. J. Soil Water Conserv. 45:125–127.

Locke, M.A., and F.M. Hons. 1988. Fertilizer placement effects on seasonal nitrogen accumulation and yield of no-tillage and conventional tillage sorghum. *Agron. J.* 80:180–185.

Logan, T.J., and J.R. Adams. 1981. *Lake Erie waste-water management study. Tech. Rept. Series.* U.S. Army Engineer District, Buffalo, NY.

Lynch, J.M. 1977. Phototoxicity of acetic acid produced in the anaerobic decomposition of wheat straw. *J. Appl. Bacteriol.* 42:81–87.

Lynch, J.M. 1981. Promotion and inhibition of soil aggregate stabilization by micro-organisms. *J. Gen. Microbiol.* 126:371–375.

Lynch, J.M. 1984. Interactions between biological processes, cultivation, and soil structure. *Plant Soil* 76:307–318.

Lynch, J.M., F.B. Ellis, S.H.T. Harper, and D.G. Christian. 1981. The effect of straw on the establishment and growth of winter cereals. *Agric. Environ.* 5:321–328.

Lynch, J.M., K.B. Gunn, and L.M. Panting. 1980. On the concentration of acetic acid in straw and soil. *Plant Soil* 56:93–98.

Martin, J.P., and D.E. Stott. 1981. Characteristics of organic matter decomposition and retention in arid soils. pp. 107–114. In: J.J. Skujins (ed.), *Proc. 6th Int. Symp. Environ. Biochem.* UNEP, Paris.

Maurya, P.R. 1986. Effect of tillage and residue management on maize and wheat yield and on physical properties of an irrigated sandy loam soil in northern Nigeria. *Soil Tillage Res.* 8:161–170.

McCalla, T.M. 1958. Microbial and related studies of stubble mulching. *J. Soil Water Conserv.* 13:255–258.

McCalla, T.M., and T.J. Army. 1961. Stubble mulch farming. *Adv. Agron.* 13:125–196.

McMahon, M.A., and G.W. Thomas. 1976. Anion leaching in two Kentucky soils under conventional tillage and killed-sod mulch. *Agron. J.* 68:437–442.

Meisinger, J.J. 1984. Evaluating plant available nitrogen in soil-crop systems. pp. 391–416. In: R.D. Hauck (ed.), *Nitrogen in crop production.* American Society of Agronomy, Madison, WI.

Meisinger, J.J., V.A. Bandel, G. Stanford, and J.O. Legg. 1985. Nitrogen utilization of corn under minimal tillage and moldboard plow tillage. *Agron. J.* 77:602–611.

Mengel, D.B., D.W. Nelson, and D.C. Martena. 1982. Placement of nitrogen fertilizers for no-till and conventional till corn. *Agron. J.* 74:515–518.

Mielke, L.N., W.W. Wilhelm, K.A. Richards, and C.R. Fenster. 1984. Soil physical characteristics of reduced tillage in wheat-fallow system. *Trans ASAE.* 27:1724–1728.

Miller, W.P., and W.W. McFee. 1983. Distribution of cadmium, zinc, copper, and lead in soils of industrial northwestern Indiana. *J. Environ. Qual.* 12:29–33.

Mock, J.J., and D.C. Erbach. 1977. Influence of conservation tillage environments on growth and productivity of corn. *Agron. J.* 69:337–340.

Moss, S.R. 1979. The influence of tillage and method of straw disposal on the survival and growth of black grass, *Alopecurus myosuroides* and its control by chlorotoluron and isoproturon. *Ann. Appl. Biol.* 91:91–100.

Moss, S.R., and E.G. Cotterill. 1985. The influence of straw ash on some soil properties that can affect herbicide performance. *Soil Tillage Res.* 5:361–370.

Musick, J.T., A.F. Wiese, and P.R. Allen. 1977. Management of bed-furrow irrigated soil with limited and no-tillage systems. *Trans ASAE* 20:666–672.

Myers, R.K.F. 1979. Nitrogen and phosphorus nutrition of dryland grain sorghum at Katherine, Northern Territory 4: ^{15}N studies on nitrogen carrier and method of application. *Aust. J. Exp. Agric. Anim. Husb.* 18:481–487.

Nagarajah, S. 1988. Transformation of green manure nitrogen in lowland rice soils. pp. 193–208. In: *Green manure in rice farming.* International Rice Research Institute, Los Banos, Philippines.

National Agriculture Centre. 1984. *Straw incorporation in heavy soils.* NAC Arable Unit, Stoneleigh, Kenilworth, Works, England.

Negi, S.C., G.S.V. Raghavan, and F. Taylor. 1981. Hydraulic characteristics of conventionally and zero tilled field plots. *Soil Tillage Res.* 2:281–292.

Nelson, L.R., R.N. Gallaher, R.R. Bruce, and M.R. Holmes. 1977. Production of corn and sorghum grain in double-cropping systems. *Agron J.* 69:41–45.

Nuttall, W.F., K.E. Bowren, and C.A. Campbell. 1986. Crop residue management practices and nitrogen and phosphorus fertilizer effects on crop response and some physical and chemical properties of a black chernozem over 25 years in a continuous wheat rotation. *Can. J. Siol Sci.* 66:159–171.

Ohio State University. 1972. *Proceedings: No-tillage Systems Symposium.* Ohio State University, Columbus, OH.

Oliphant, J.M. 1982. The effect of stubble on the yield of winter wheat after cultivation or direct drilling. *Exp. Husb.* 38:60–68.

Olsen, S.R., and L.A. Dean. 1965. Phosphorus. In: C.A. Black (eds.), *Methods of soil analysis,* Pt. 2. *Agronomy* 9:1035–1049.

Osuji, G.E. 1984. Water storage, water use and maize yield for tillage systems on a tropical alfisol in Nigeria. *Soil Tillage Res.* 4:339–348.

Papendick, R.I., and D.E. Miller. 1977. Conservation tillage in the Pacific Northwest. *J. Soil Water Conserv.* 32:49–56.

Parr, J.F., and R.I. Papendick. 1978. Factors affecting the decomposition of crop residues by microorganisms. pp. 109–124. In: W.R. Oschwald (ed.), *Crop residues management systems.* American Society of Agronomy, Madison, WI.

Peterson, A.E., and G.E. Bubenzer. 1986. Intake rate: Sprinker Infiltrometer. In: A. Klute (ed.), *Methods of soil analysis,* Pt. I, 2nd ed. *Agronomy* 9:845–870.

Peterson, H., and M. Luxton. 1982. A comparative analysis of soil fauna populations and their role in decomposition processes. *Oikos* 39:287–288.

Pidgeon, J.D., and B.D. Soane. 1977. Effects of tillage and direct drilling on soil properties during the growing season in a long term barley mono-culture system. *J. Agri. Sci.* 88:431–442.

Pikul, J.L., Jr., and R.R. Allmaras. 1986. Physical and chemical properties of a haploxeroll after 50 years of residue management. *Soil Sci. Soc. Am. J.* 50:214–219.

Plant Protection Limited. 1973. Minimum tillage issue. *Outlook Agric.* 7:142–200.

Plant Protection Limited. 1975. Reduced cultivation and Direct Drilling Issue. *Outlook Agric.* 8:211–260.

Power, J.F., W.W. Wilhelm, and J.W. Doran. 1986. Crop residue effects on soil environment and dryland maize and soya bean production. *Soil Tillage Res.* 8:101–111.

Powlson, D.S., and D.S. Jenkinson. 1981. A comparison of the organic matter biomass, adenosine triphosphate, and mineralizable nitrogen contents of ploughed and direct-drilled soils. *J. Agric. Sci.* 97:713–721.

Radcliffe, D.E., E.W. Tollner, W.L. Hargrove, R.L. Clark, and M.H. Golabi. 1988. Effect of tillage practices on infiltration and soil strength of a typic hapludult soil after ten years. *Soil Sci. Soc. Am. J.* 52:798–803.

Rasmussen, P.E., R.R. Allmaras, C.R. Rohde, and N.C.J. Roager. 1980. Crop residue influence on soil carbon and nitrogen on wheat–fallow system. *Soil Sci. Soc. Am. J.* 44:496–500.

Rasmussen, P.E., and R.W. Smiley. 1989. Long-term management effects on soil productivity and crop yield in semi-arid regions of eastern Oregon. Station Bull. 675. Columbia Basin Agricultural Research Center, Pendleton, Oregon. 57 pp.

Rees, R.G., and G.J. Platz. 1986. *Yellow spot infectivity of residues. Biennial Report 1982–84.* Queensland Wheat Research Institute Toowoomba, Australia.

Reinertsen, S.A., L.F. Elliott, V.L. Cochran, and G.S. Campbell. 1984. Role of available carbon and nitrogen in determining the rate of wheat straw decomposition. *Soil Biol. Biochem.* 16:459–464.

Reynolds, W.D., D.E. Elrick, and G.C. Topp. 1983. A re-examination of the constant head well permeameter method for measuring saturated hydraulic conductivity above the water table. *Soil Sci.* 136:250–268.

Rice, C.W., and M.S. Smith. 1982. Denitrification in non-till and plowed soils. *Soil Sci. Soc. Am. J.* 46:1168–1173.

Rickerl, D.H., W.B. Gordon, E.A. Curl, and J.T. Touchton. 1988. Winter legume and tillage effects in cotton growth and soil ecology. *Soil Tillage Res.* 11:63–71.

Roach, S.H. 1981. Reduced vs. conventional tillage practices in cotton and tobacco:—a comparison of insect populations and yields in Northeastern South Carolina, 1977-1979. *J. Econ. Entomol.* 74:688–695.

Rosegrant, M.W., and J.A. Roumasset. 1988. Economic feasibility of green manure in rice-based cropping systems. pp. 11–27. In: *Sustainable agriculture–green manure in rice farming.* International Rice Research Institute, Los Banos, Philippines.

Schneider, E.C., and S.C. Gupta. 1985. Corn emergence as influenced by soil temperature, matric potential, and aggregate size distribution. *Soil Sci. Soc. Am. J.* 49:415–422.

Seastedt, T.R. 1984. The role of microarthropods in decomposition and mineralization processes. *Annu. Rev. Entomol.* 29:25–46.

Seastedt, T.R., and D.A. Crossley, Jr. 1980. Effects of microarthropods on the seasonal dynamics of nutrients in forest litter. *Soil Biol. Biochem.* 12:337–342.

Sharpe, R.R., L.A. Harper, J.E. Giddens, and G.W. Langdale. 1988. Nitrogen use efficiency and nitrogen budget for conservation tilled wheat. Soil Sci. Soc. Am. J. 52:1394–1398.

Shipley, J.L., and C. Regier. 1977. Water response in the production of irrigated grain sorghum—High plains of Texas. *Texas Agric. Exp. Stn. Rep.* MP-1348 C.

Singh, R., A.C. Vig, and N.S. Singh. 1982. Nitrogen substitution with green manure in maize–wheat rotation. *Indian J. Agron.* 27:371–376.

Skidmore, E.L., J.B. Layton, D.V. Armbrust, and M.L. Hooker. 1986. Soil physical properties as influenced by cropping and residue management. *Soil Sci. Soc. Am. J.* 50:415–419.

Smith, K.A., and R.S. Howard. 1980. Field studies of nitrogen uptake using ^{15}N tracer methods. *J. Sci. Food Agric.* 31:639–840.

Soane, B.D., J.W. Dickson, and D.J. Campbell. 1982. Compaction by agricultural vehicles: A review. III. Incidence and control of compaction in crop production. *Soil Tillage Res.* 2:3–36.

Soil Conservation Society of America. 1973. *Proc. Natl. Conf. Conservation Tillage Symp.* Soil Conservation Society of America, Ankeny, IA.

Soil Conservation Society of America. 1977. *Conservation tillage—problems and potentials. Spec. Publ. No. 20.* Soil Conservation Society of America, Ankeny, IA.

Soil Conservation Society of America. 1979. *Effect of tillage and crop residue removed on erosion, run-off, and plant nutrients. Spec. Publ. No. 25.* Soil Conservation Society of America, Ankeny, IA.

Stanford, G. 1973. Rationale for optimum nitrogen fertilization in corn production. *J. Environ. Qual.* 2:159–166.

Staniforth, A.R. 1979. *Cereal straw.* Clarendon Press, Oxford.

Stevenson, F.J. 1982. *Humus chemistry, genesis, composition, reactions.* John Wiley & Sons, New York.

Tanaka, D.L. 1985. Chemical and stubble-mulch fallow influences on seasonal water content. *Soil Sci. Soc. Am. J.* 49:728–733.

Taylor, H.M., and E. Burnett. 1964. Influence of soil strength on the root growth habits of plants. *Soil Sci.* 98:174–180.

Taylor, H.M., and H.R. Gardner. 1963. Penetration of cotton seedling taproots as influenced by bulk density, moisture content, and strength of soil. *Soil Sci.* 96:153–156.

Thomas, G.W., R.L. Blevins, P.E. Phillips, and M.A. McMahon. 1973. Effect of killed soil mulch on nitrate movement and corn yield. *Agron. J.* 65:736–739.

Thompson, J.P., J. Mackenzie, J. McCulloach, and T.G. Clewett. 1982. *Integrated control of root lesion nematods. Biennual Report 1982–84.* Queensland Wheat Research Institute, Toowoomba, Australia.

Thurlow, D.L., C.B. Elkins, and A.E. Hilthold. 1984. Effect of in-row chisel at planting on yield and growth of full season soybean. pp. 124–126. In: J.T. Touchton and R.E. Stevenson (eds.), *Proceedings Southeast No-tillage 7th Conf.* Auburn University, Auburn, AL.

Toth, J., P.J. Milhamn, and J.M. Raison. 1981. Ash from rice stubble inactivates thiobencarb and molinate. *Weed Res.* 21:113–117.

Touchton, J.T., F. Adams, and G.H. Burmaster. 1982a. Nitrogen fertilizer rates and cotton petiole analyses in Alabama field experiments. *Alabama Agric. Exp. Sta. Bull.* 528:1–24.

Touchton, J.T., W.A. Gardener, W.L. Hargrove, and R.R. Duncan. 1982b. Reseeding crimson clover as a N source for no tillage grain sorghum production. *Agron. J.* 74:283–287.

Touchton, J.T., and J.W. Johnson. 1982. Soybean tillage and planting method effects on yield of double cropped wheat and soybeans. *Agron. J.* 74:57–59.

Touchton, J.T., D.H. Rickerl, R.H. Walker, and C.E. Snipes. 1984. Winter legumes as a nitrogen source for no tillage cotton. *Soil Tillage Res.* 4:391–401.

Triplett, G.B., Jr. 1975. Fertilizer use for no-tillage systems. *Proceedings TVA Fertility Conference, July, 29–31, Louisville, KY* Tennessee Valley Authority, Muscle Shoals. AL.

Tyler, D.D., and G.W. Thomas. 1977. Lysimeter measurements of nitrate and chloride losses from soil under conventional and no-tillage corn. *J. Environ. Qual.* 6:63–66.

Unger, P.W. 1978. Straw mulch effects on soil temperatures and sorghum germination and growth. *Agron J.* 70:858–864.

Unger, P.W. 1984. Tillage and residue effects on wheat, sorghum and sunflower grown in rotation. *Soil Sci. Soc. Am. J.* 48:885–891.

Unger, P.W. 1986. Wheat residue management effects on soil water storage and corn production. *Soil Sci. Soc. Am. J.* 50:764–770.

Unger, P.W. 1988. Grain and forage sorghum production with no tillage on dryland. *Agron. J.* 80:193–197.

Unger, P.W., and T.M. McCalla. 1980. Conservation tillage systems. *Adv. Agron.* 33:2–53.

Unger, P.W., and B.A. Stewart. 1976. Land preparation and seedling establishment in multiple cropping systems. pp. 255–273. In: R.I. Papendick, P.A. Sanchez, and G.B. Triplett (eds.), *Multiple cropping. Spec. Publ. No. 27.* American Society of Agronomy, Madison, WI.

Unger, P.W., and A.F. Wiese. 1979. Managing irrigated winter wheat residues for water storage and subsequent dryland grain sorghum production. *Soil Sci. Soc. Am. J.* 43:582–588.

United States Department of Agriculture. 1977. *Research progress and needs—Conservation tillage. ARS-NC Publ. 57.* U.S. Government Printing Office, Washington, DC.

Van der Heide, J. 1988. The role of green manure in the low input farming in humid tropics. pp. 185–191. In: *Green manure in rice farming.* International Rice Research Institute, Los Banos, Philippines.

Van Doren, D.M., Jr., and R.R. Allmaras. 1978. Effect of residue management practices on the soil physical environment, microclimate, and plant growth. pp. 49–83. In: W.R. Oschwald (ed.), *Crop residue management systems. Spec Publ. No. 31.* American Society of Agronomy, Madison, WI.

Videnovic, Z., G. Vasik, and F. Kolcar. 1986. Effect of fertilizers and soil tillage on corn yield under dry farming and irrigated conditions. *Soil Tillage Res.* 8:113–118.

Voss, R.D., and W.D. Shrader. 1979. Rotation effects and legume sources of nitrogen for corn. pp. 61–68. In: D.A. Bezdicek and J.F. Power. (eds.), *Organic farming—Current technology and its role in sustainable agriculture Spec. Publ. No. 46* American Society of Agronomy, Madison, WI.

Wallwork, J.A. 1983. Oribatids in forest ecosystems. *Annu. Rev. Entomol.* 28:109–130.

Ward, L.D., and R.C. Rees, 1985. The effect of burning wheat residues on wheat grain yields in a commercial crop in 1984. pp. 118–120. In: R.J. Martin and

W.L. Felton (eds.), *No-tillage crop production in Northern N.S. Wales*. Tamworth, N.S. Wales, Australia.
Warncke, D.D., D.R. Christianson, and M.L. Vitosh. 1985. Fertilizer recommendations: Vegetable and Field Crops in Michigan. *Michigan State Univ. Coop. Ext. Serv. Bull. E.* 550.
Watanabe, I. 1984. Use of green manures in northeast Asia. pp. 224–234. In: *Organic matter and rice*. International Rice Research Institute, Los Banos, Philippines.
Watts, W.R. 1973. Soil temperature and leaf expansion in *Zea mays*. *Exp. Agric.* 9:1–18.
Webster, C.P., R.J. Dowdell, and R.O. Cannel. 1985. Uptake of labelled nitrate by roots of winter barley on a direct-drilled or ploughed silt loam soil. *Soil Tillage Res.* 5:381–389.
Weil, R.R., P.W. Bendetto, L.J., Sikora, and V.A. Bandel. 1988. Influence of tillage practices on phosphorus distribution and forms in three ultisols. *Agron. J.* 80:503–509.
Westcott, M.P., and D.S. Mikkelsen. 1988. Effect of green manure on rice soil fertility in the United States. pp. 257–274. In: —*Green manure in rice farming*. International Rice Research Institute, Los Banos, Philippines.
Wiggins, E.A., and E.A. Curl. 1979. Interactions of Collembola and microflora of cotton rhizosphere. *Phytopathology* 69:224–249.
Wildermuth, G.M., D.M. Freebairn, and R.B. Namara. 1986. Influence of tillage on crown rot of wheat. Biennial Report. 1982–84. Queensland Wheat Research Institute, Toowoomba, Australia.
Wilhelm, W.W., H. Bouzerzour, and J.F. Power. 1989. Soil disturbance–residue management effect on winter wheat growth and yield. *Agron. J.* 81:581–588.
Wilhelm, W.W., J.W. Doran, and J.F. Power. 1986. Corn and soybean yield response to crop residue management under no-tillage production systems. *Agron. J.* 78:184–189.
Willis, W.W., W.E. Larson, and D. Kirkham. 1957. Corn growth as affected by soil temperature and mulch. *Agron. J.* 49:323–327.
Xiao, W.Q. 1984. Utilization of organic materials in rice production in China. pp. 45–56. In: *Organic matter and rice*. International Rice Research Institute, Los Banos, Philippines.
Yaacob, O., and G.J. Blair. 1980. The growth and nitrogen uptake of Rhodes grass grown on soils with varying histories of legume cropping. *Plant Soil* 57:249–255.
Zuzel, J.F., J.L. Pikul, and P.E. Rasmussen. 1990. Tillage and fertilizer effects on water infiltration. Soil Sci. Soc. Am. J. 54:205–208.

Index

MIX
Papier aus verantwortungsvollen Quellen
Paper from responsible sources
FSC® C105338

If you have any concerns about our products,
you can contact us on
ProductSafety@springernature.com

In case Publisher is established outside the EU,
the EU authorized representative is:
Springer Nature Customer Service Center GmbH
Europaplatz 3, 69115 Heidelberg, Germany

Printed by Libri Plureos GmbH
in Hamburg, Germany